Cell Assembly with 3D Bioprinting

Cell Assembly with 3D Bioprinting

Yong He
Qing Gao
Yifei Jin

WILEY-VCH

Authors

Prof. Yong He
School of Mechanical Engineering
Room 123, Teaching Building 1,
Yuquan Campus, Zhejiang University
No. 38, Zheda Road
310027 Hangzhou
China

Dr. Qing Gao
School of Mechanical Engineering
Room 123, Teaching Building 1,
Yuquan Campus, Zhejiang University
No. 38, Zheda Road
310027 Hangzhou
China

Prof. Yifei Jin
University of Nevada, Reno
Department of Mechanical Engineering
Room 230
1664 N. Virginia Street
NV
United States

Cover Image: Yong He

Library of Congress Card No.: applied for

British Library Cataloguing-in-Publication Data
A catalogue record for this book is available from the British Library.

Bibliographic information published by the Deutsche Nationalbibliothek
The Deutsche Nationalbibliothek lists this publication in the Deutsche Nationalbibliografie; detailed bibliographic data are available on the Internet at <http://dnb.d-nb.de>.

Print ISBN: 978-3-527-34796-4
ePDF ISBN: 978-3-527-82857-9
ePub ISBN: 978-3-527-82858-6
oBook ISBN: 978-3-527-82859-3

Cover Design Adam-Design, Weinheim, Germany
Typesetting Straive, Chennai, India
Printing and Binding CPI Group (UK) Ltd, Croydon, CR0 4YY

Printed on acid-free paper

C107397_021121

Contents

Preface

3D bioprinting has opened up a frontier in biomedical research, aiming at additive manufacturing or assembling living structures with cells, which provides the possibility to generate significant breakthroughs, yielding new treatments, and change the foundation of regenerative medicine. It is truly an interdisciplinary field, cross-fertilized ranging from mechanical engineering, materials science, and computer science to biology, medicine, pharmaceutical science, and so on.

The potential applications for 3D bioprinting include: (i) in vitro 3D tissue/organ models for drug screening, organ development, toxicological and cosmetic research, etc.; (ii) 3D biofabrication of living structures for clinical transplantation or tissue repair.

Compared to conventional additive manufacturing, 3D bioprinting possesses three remarkable characteristics: (i) **bioprinting usually utilizes cell-laden hydrogel** (called bioink) in terms of material use; (ii) **bioprinted structures have to go through hydrogel crosslinking process** (thermal, chemical, or enzymatic) during formation for manufacturing desired tissue structures; (iii) **cross-talking and functionalization of cells to acquire some tissue properties after printing is the goal**. In these regards, bioprinting faces two major challenges: (i) structural controlled manufacturing, which requires a stable printing process to ensure cell-laden hydrogels being accurately assembled. As the bioink is something like a soft tofu, precision manufacturing is difficult; (ii) functional controlled postprocessing, which needs to provide biomimetic microenvironment with physical and chemical stimulation for living constructs realizing functionalization.

Combined with our research experiences in 3D bioprinting over years, this book is outlined in 14 chapters: commonly used 3D bioprinting methods are summarized first; then the design of bioink is put forward; several bioprinting approaches are elaborated afterward including coaxial bioprinting, digital light projection, direct ink writing, and liquid support bath-assisted 3D printing; in the following parts, microgel-based, microfiber-based, and microfluidics-based biofabrication approaches and their applications are meticulously illustrated; and a protocol of 3D bioprinting is well represented in the end to show several examples of complete bioprinting process.

We wish to thank the valuable support from everyone who contributed to this book. This book would never have been published without your effort. Thanks to

Dr. Zeming Gu for help in writing Chapter 1; thanks to Prof. Changxue Xu for help in writing Chapter 2; thanks to Mr. Peng Zhang for help in writing Chapter 3; thanks to Dr. Chaoqi Xie for help in writing Chapter 4; thanks to Dr. Yuan Sun for help in writing Chapter 5; thanks to Mr. Cheng Zhang for help in writing Chapter 6; thanks to Dr. Weijian Hua for help in writing Chapter 7; thanks to Dr. Mingjun Xie for help in writing Chapters 8, 9, and 14; thanks to Prof. Lei Shao for help in writing Chapters 10 and 11; thanks to Dr. Jing Nie for help in writing Chapters 12 and 13. Thanks to Zhengyi Zhang, Danyang Zhao, Lily Raymond, Heqi Xu, Matthew Warner, and Beatriz Godina for their help in completing this book. Thanks to Ms. Katherine Wong for editing the manuscript.

We sincerely hope that our readers will find the book professionally written, richly illustrated, accessible, and most importantly, intriguing. We would be flattered if this book attracts new researchers from different disciplines into the field of 3D bioprinting. Due to limited time and scholars with different backgrounds involved in compilation, there may be some unsatisfactory points in the writing style or content of this book. Therefore, readers are welcome to put forward criticism or suggestions for our further improvement.

4 May 2021

Yong He
Zhejiang University
Hangzhou, China

1

3D Bioprinting, A Powerful Tool for 3D Cells Assembling

1.1 What Is 3D Bioprinting?

3D printing, also known as additive manufacturing, is a layer-by-layer manufacturing approach, and it has been applied in many industrial applications and research fields. It could be thought of as an inverse process of potato cutting, assembling the chips or slices into integrity by certain rules. When 3D printing met biomedical engineering, 3D bioprinting was born. 3D bioprinting is an interdisciplinary science closely related to medicine biology, mechanical engineering, and material science. It can be divided into two concepts. Broadly speaking, 3D bioprinting refers to the use of 3D printing technology to achieve biomedical applications, such as the printing of medical aids, polymers, ceramics, or metal scaffolds [1–3]; in a narrow sense, this concept simply means 3D cells assembling through printing, therefore it can also be identified as cell printing or organ printing [4–6]. Here, this book is mainly focusing on the narrow viewpoint. A cartoon introduction of 3D bioprinting is illustrated in Figure 1.1.

In vitro bio-manufacturing of tissues/organs has always been a great dream pursued by mankind, driven by two needs: organ transplantation and accurate tissue models. First, there is a huge shortage of organs for transplantation. In 2016, there were 160 000 organ transplant recipients, but only 16 000 organ donors in the United States [8]. The complexity of human organs is not only reflected in the mechanism of organ growth that has not been revealed by biology, but also in the reproduction of fine structure manufacturing. The use of 3D bioprinting technology to solve the shortage of organ transplants is far too optimistic at the present stage. Second, traditional methods utilizing 2D cell culture were applied for drug screening and medical mechanism studies. However, microenvironment in vivo is far more complex than the 2D cell culture, and in some cases, 2D models may lead to opposite results. 3D bioprinting technology can realize spatiotemporal directional manipulation of various cells and has become the most ideal method to construct a 3D cell-laden structure in vitro.

In vitro models have undergone a meaningful revolution both in forms and functions: mini-tissue, organ-on-a-chip, and tissue/organ construct. Based on common bioprinting techniques, 3D mini-tissue in forms of spheres, fibers, or other geometric shapes could be fabricated [9, 10]. These models contribute to the simulation

Cell Assembly with 3D Bioprinting, First Edition. Yong He, Qing Gao, and Yifei Jin.

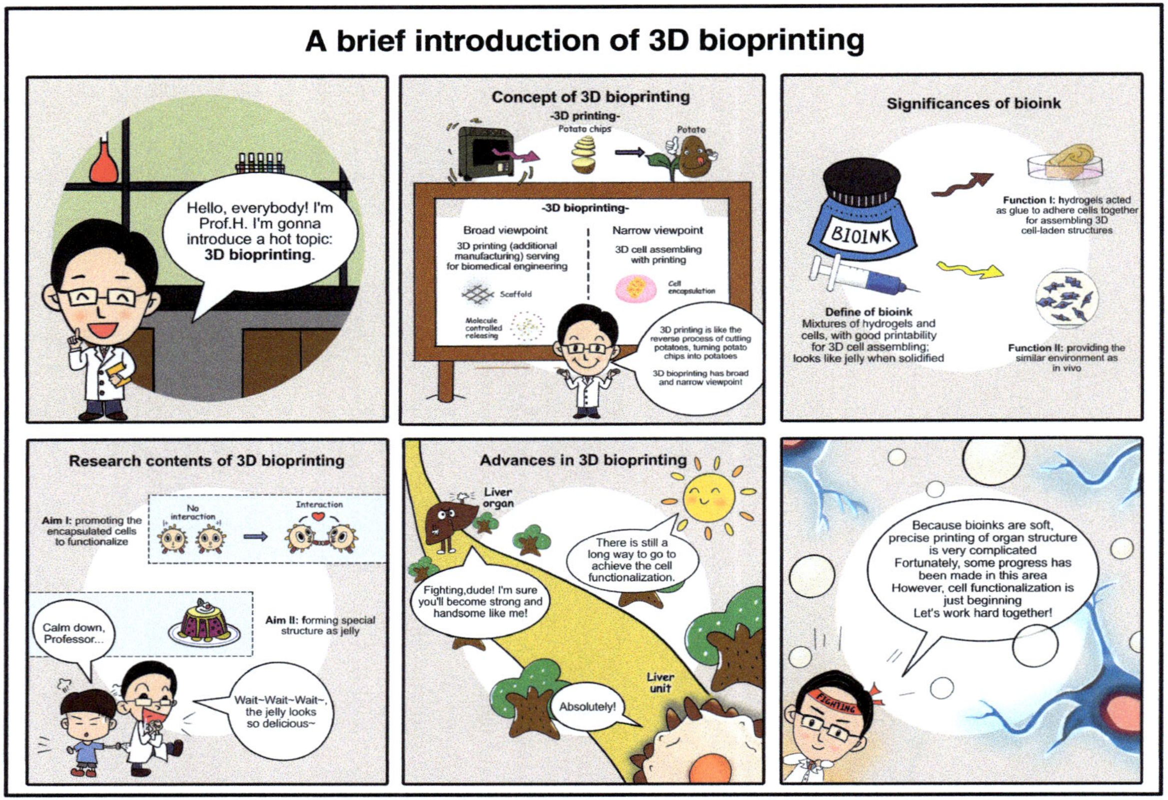

Figure 1.1 Introduction of 3D bioprinting. Source: He et al. [7]. Reproduced with permission of Springer Nature.

of functional units with simple composition and independent operation, which can be applied in high-throughput testing with a low dose. Besides, 3D bioprinting has been gradually involved in the setting up of organ-on-a-chip devices because of its excellent customizability and cell compatibility [11]. Modified microfluidic systems could be constructed with biomaterials through 3D bioprinting, on which specified cells are loaded and routine reactions were carried out. And the interactions and cross-talking between multiple organs can be well simulated by connecting different modules by means of microfluidic methods. Furthermore, 3D bioprinting has been further facilitated in the biofabrication of tissue/organ constructs with an inner channel network. A large number of 3D bioprinting strategies have been adopted in building 3D tissue/organ constructs with a vascular network, including coaxial bioprinting, projection-based bioprinting, as well as the integration of 3D bioprinting and sacrificial templates.

1.2 Evolution of 3D Bioprinting

As mentioned above, it is not practical to realize 3D bioprinting for full-function organ transplantation at present. However, it is an undeniable fact that bioprinting techniques have come a long way. Decades ago, pioneers such as Vladimir Mironov, Gabor Forgacs, and Thomas Boland saw the natural combination of technologies including cell patterning and others, such as commercial inkjet printing, to build living structures that might one day be used for human organ transplantation [6, 12, 13]. A timeline for the evolution of bioprinting technology up to state-of-the-art is illustrated in Table 1.1.

In 1984, Charles Hull invented stereolithography (SLA) for printing 3D objects from digital data, symbolizing the birth of 3D printing. Bioprinting was first demonstrated in 1988 while Klebe using a standard Hewlett–Packard (HP) inkjet printer to deposit cells by cytoscribing technology [14]. In 1996, Forgacs and coworkers drew a conclusion that apparent tissue surface tension was the macroscopic manifestation of molecular adhesion between cells and provided a quantitative measure for tissue cohesion [15]. In 1999, Odde and Renn first utilized laser-assisted bioprinting to deposit living cells for developing analogs with complex anatomy [16]. In 2001, direct printing of a scaffold in the shape of a bladder and seeding of human cells took place [17]. In 2002, the first extrusion-based bioprinting technology was reported by Landers et al., which was later commercialized as "3D-Bioplotter" [18]. Wilson and Boland developed the first inkjet bioprinter in 2003 by modifying an HP standard inkjet printer [19]. Their team implemented cell-loaded bioprinting with a commercial SLA printer a year after [20]. Also in 2004, 3D tissue with only cells (no scaffold) was developed. In 2006, electrohydrodynamic jetting was applied to deposit living cells [21]. Scaffold-free vascular tissue was engineered through bioprinting by Norotte et al. in 2009 [22]. In 2012, in situ bioprinting was attempted by Skardal et al. on mouse models [23]. The following years saw the introduction of many new bioprinting products, such as articular cartilage and artificial liver in 2012, tissue integration with the circulatory system in 2014, and so on [24, 25]. In 2015, coaxial

Table 1.1 Timeline for bioprinting evolution.

Year	Development
1984	Stereo lithography was invented, representing the birth of 3D printing
1988	Bioprinting was first demonstrated by 2D micro-positioning of cells
1996	Cells sticking together during embryonic development was observed
1999	First use of laser technology demonstrating 2D patterning of living cells
2001	3D printed synthetic scaffold for human ladder
2002	First extrusion-based bioprinter was achieved
2003	First inkjet bioprinter was developed
2004	3D tissue with only cells (no scaffold) was presented
2009	Scaffold-free vascular constructs were fabricated
2012	In situ bioprinting was realized on animals
2015	Tubular structure was printed by coaxial technology.
2016	Rapid continuous optical 3D printing based on projection (DLP) was applied
2016	Cartilage model was obtained by ITOP system
2019	Cardioid structure was first bioprinted
2019	Collagen human heart at various scales was built using FRESH technology

technology was adopted by Gao et al. for the fabrication of a tubular structure [4]. In 2016, Pyo et al. applied rapid continuous optical 3D printing based on digital light processing (DLP) [26]. In the same year, a cartilage model was manufactured by Anthony Atala's research group using an integrated tissue-organ printer (ITOP) [27]. In 2019, Noor et al. succeeded in manufacturing a perfusable scale-down heart [28], and a few months later, bioprinting of collagen human hearts at various scales based on the freeform reversible embedding of suspended hydrogels (FRESH) technology was achieved by Lee et al. [29].

1.3 Brief Classification of 3D Bioprinting

Based on different printing principles, cell-laden 3D bioprinting can be divided into three types: extrusion-based, droplet-based, and projection-based bioprinting. Extrusion-based bioprinting generates continuous fibers to set up the structures; droplet-based bioprinting produces droplets as the basic unit for biofabrication and projection-based bioprinting takes advantage of the properties of photosensitive materials by stacking 3D models layer-by-layer. Different approaches possess diverse characteristics aiming at various scenarios and have specific requirements for bioinks.

Extrusion-based bioprinting is the most widely used method, which is suitable for a wide range of biocompatible materials. According to different liquid dispensing modes, pneumatic-driven, piston-driven, and screw-driven extrusion systems are applied to extrude cell-laden bioinks in the form of continuous filaments.

Droplet-based bioprinting which employs discrete droplets stacked into constructs can be roughly divided into inkjet bioprinting [30], electrohydrodynamic jetting (EHDJ) [31], and laser-assisted bioprinting (LAB) [32] based on different droplets forming principles. Thermal and piezoelectric-driven technologies are most commonly used in inkjet bioprinting. EHDJ uses a high voltage motivated electric field to pull droplets out of the nozzle orifice. Changes in voltage certainly affect the size of each droplet, where the higher voltage leads to smaller droplets [33, 34]. LAB is a non-contact, nozzle-free bioprinting strategy used precisely to deposit bioink droplets. LAB technique includes laser-guidance direct writing (LGDW) and laser-induced forward transfer (LIFT). LGDW employs a light trap to guild cells onto a substrate, while LIFT uses a focused pulsed laser to induce partial evaporation of bioink coating to propel the biomaterial toward the receiving layer.

Projection-based bioprinting solidifies light-sensitive biomaterials to form constructs under precisely controlled lighting with high printing precision and fast printing speed. The most common use of projection-based bioprinting is to print cell-free scaffolds, where cells would be seeded post-printing. Currently, however, cell-laden projection-based bioprinting has also been reported using DLP technology.

1.4 Evaluation of Bioinks

Generally speaking, 3D bioprinting has three steps: preparing bioinks, printing the soft live structures with multiple cells, and rebuilding the interaction among cells. And that is why developing appropriate bioinks has always been a significant part, as it affects every step that follows.

The performance of bioinks can be measured by three main factors: printability, biocompatibility, and mechanical property. Printability is to assess the formability of bioinks, where adjustable material viscosity, rapid transition from sol state to gel state, and a broad range of printing parameters are necessary. Biocompatibility is a measure of biomimicry that requires bioink and printed cells to be as similar as possible in the microenvironment in vivo. The mechanical property requires that the cured bioink be strong enough to hold subsequent culture and implantation. Perfusion and degradation might occur during bioprinted constructs culture in vitro, which requires considerable strength to support.

Therefore, the choice of bioink necessitates compromise among printability, biocompatibility, and mechanical property. Considering the requirements of the bioprinting process, cell growth and proliferation, and structural integrity, reasonable bioink design can be carried out according to the actual cell type and printing resolution requirements. But in fact, these three requirements of bioink are inherently contradictory in the mechanism. For example, the higher the viscosity of biological ink, the better the printability, and vice versa, the poorer the biocompatibility.

Hence, bioink selection to meet the specific needs of different applications is a key step in bioprinting.

An ideal bioink would certainly be close to the natural extracellular matrix (ECM), and it would need to be adapted to match different types of cells. Therefore, it could not be better to add specific substances in bioinks that cells possibly need during proliferation and functionalization. For example, when bioprinting chondrocytes, the addition of HA, a common component of cartilage, can significantly promote later culture and functionalization.

Typical bioinks applied in bioprinting may include hydrogels, decellularized matrix components, microcarriers, tissue spheroids and strands, cell pellet, and/or some advanced bioinks such as multi-material, interpenetrating network, nanocomposite, and supramolecular bioink, etc. [35, 36]. Among them, hydrogels are considered to be one of the most important biomaterials in bioinks, because of their outstanding capability of providing a viable microenvironment for cell adhesion, growth, and proliferation. Natural/synthetic hydrogels including alginate, fibrinogen, gelatin, collagen, silk fibroin, chitosan, agarose, pluronics, HA, GelMA, PEG, PEO, etc., have been found in countless applications in bioprinting. They are either ion-sensitive, photosensitive, thermosensitive, enzyme-sensitive, or pH-responsive, so they can be easily gelated to form constructs before, during, and/or after bioprinting [37].

1.5 Outlook and Discussion

3D bioprinting technologies still need further improvement. The complexity of tissues and organs has brought great difficulties to accurate bioprinting. One of the major disadvantages of current bioprinting technologies is the low accuracy of bioprinting compared to natural tissues/organs. Most tissues/organs are more delicate than current bioprinting devices. Another common drawback of bioprinting is the slow speed of bioprinting of complex scale-up structures, especially when it comes to multi-material alternate biofabrication.

Vascularization is the basis of bioprinted structures. Same as the challenge of tissue engineering and regenerative medicine, ensuring adequate vascularization in bio-manufactured structures is a key factor in 3D bioprinting. The effective construction of a multi-scale perfusion vascular network and the promotion of its vascularization by mechanical or chemical stimulation are the basis of the biological fabrication of scale-up constructs.

Functionalization is the primary goal for 3D bioprinting. Most of the current research is still focused on the manufacturing idea-oriented printing process and mechanism, while functionalization is the core factor leading 3D bioprinting from basic research to practical application. In order to be functional, bioink needs to have excellent biocompatibility and mechanical properties to meet the requirements of nutrient perfusion and implantation. In addition, the construction of microenvironments that mimic in vivo scenarios, including mechanical and chemical stimuli such as perfusion culture and growth factors, is also critical for the functionalization of bioprinted structures.

Combined with the outlook of 3D bioprinting, there are several printing methods that are quite promising: DLP, coaxial bioprinting, and embedded bioprinting. Due to its intrinsic principle, DLP has a much higher printing resolution and speed than other bioprinting approaches. As a key application of 3D bioprinting technology, in vitro tissue models need to be standardized not only in sizes, but also in biological and mechanical properties, while DLP owns excellent uniformity and reproducibility compared to other methods. Additionally, coaxial bioprinting has become an increasingly popular extrusion-based bioprinting method since it was introduced into the field of tissue engineering in 2015 [4], especially in the area of blood vessel biofabrication/vascularization. The biggest advantage of coaxial bioprinting is its ability to construct hierarchical tubular structures with tunable biological/mechanical properties. It is well known that hydrogels with good biocompatibility tend to have insufficient mechanical strength. Coaxial bioprinting can partly solve the problem with its core-shell structure: core materials guarantee biocompatibility, while shell materials provide mechanical strength and vice versa. The use of sacrificial materials as the core material would also contribute to the convenient bioprinting of hollow tubular structures. Besides, embedded bioprinting allows anti-gravity writing of 3D freeform constructs within yield stress and gel-based supporting bath, which would be further removed post-printing to retrieve models with desired shapes or channels. Other than traditional bioprinting approaches, it can achieve the fabrication of discrete patterns, which are not mechanically supported [38–40].

In addition to the challenges including bioinks design, bioprinting techniques, vascularization, and functionalization, issues such as cell sources, bioreactor construction, and even ethical problems also require considerable attention. 3D bioprinted fully clinical translation could take a long time until bio-artificial tissues such as cartilage or skin, to be applied in transplantation. We all hope that 3D bioprinting can find its way from structural similarity into functional realization.

This book is organized into 14 chapters. This chapter "3D Bioprinting, A Powerful Tool for 3D Cells Assembling," covers the definition, evolution, and classification of 3D bioprinting. Chapter 2 "Representative 3D Bioprinting Approaches" and Chapter 3 "Bioink Design" demonstrates a variety of commonly used 3D bioprinting methods in detail, and introduces the principle of bioink design. In Chapter 4 "Coaxial 3D Bioprinting," Chapter 5 "Digital Light Projection-Based 3D Bioprinting," Chapter 6 "Direct Ink Writing for 3D Bioprinting Applications," and Chapter 7 "Liquid Support Bath-Assisted 3D Bioprinting," four types of promising 3D bioprinting technologies and their applications are highlighted respectively. Chapter 8 "Bioprinting Approaches of Hydrogel Microgel," and Chapter 9 "Biomedical Applications of Microgels" provides the manufacturing process and medical use of microgels. Chapter 10 "Microfiber-Based Organoids Bioprinting for in vitro Model" and Chapter 11 "Large Scale Tissues Bioprinting" are mainly concerned with biofabricated organoids and scale-up tissues. In Chapter 12 "3D Printing of Vascular Chips" and Chapter 13 "3D Printing of in vitro Models," vascular chips and in vitro models by 3D printing approaches are well presented. Finally, Chapter 14 "Protocol of Typical 3D Bioprinting," comes up with an integrated blueprint for 3D bioprinting.

References

1 He, Y., Xue, G.H., and Fu, J.Z. (2014). Fabrication of low cost soft tissue prostheses with the desktop 3D printer. *Scientific Reports*: 46973.

2 Shao, H., He, Y., Fu, J. et al. (2016). 3D printing magnesium-doped wollastonite/β-tcp bioceramics scaffolds with high strength and adjustable degradation. *Journal of the European Ceramic Society* 36 (6): 1495–1503.

3 Gao, Q., Niu, X., Shao, L. et al. (2019). 3D printing of complex GelMA-based scaffolds with nanoclay. *Biofabrication* 11 (3): 035006.

4 Gao, Q., He, Y., Fu, J.Z. et al. (2015). Coaxial nozzle-assisted 3d bioprinting with built-in microchannels for nutrients delivery. *Biomaterials*: 61203–61215.

5 Zhao, H., Chen, Y., Shao, L. et al. (2018). Airflow-assisted 3D bioprinting of human heterogeneous microspheroidal organoids with microfluidic nozzle. *Small* 14 (39): e1802630.

6 Mironov, V., Boland, T., Trusk, T. et al. (2003). Organ printing: computer-aided jet-based 3D tissue engineering. *Trends in Biotechnology* 21 (4): 157–161.

7 He, Y., Xie, M., Gao, Q., and Fu, J. (2019). Why choose 3D bioprinting? Part I: A brief introduction of 3D bioprinting for the beginners. *Bio-Design and Manufacturing* 2: 221–224.

8 Dong, H., Fang, Y., Wang, D. et al. (2017). Current situation and thinking of organ donation at home and abroad. *Journal of Nursing (China)* 24 (11): 23–26.

9 Xie, M., Gao, Q., Zhao, H. et al. (2019). Electro-assisted bioprinting of low-concentration GelMA microdroplets. *Small* 15 (4): e1804216.

10 Shao, L., Gao, Q., Zhao, H. et al. (2018). Fiber-based mini tissue with morphology-controllable GelMA microfibers. *Small* 14 (44): e1802187.

11 Yi, H.-G., Lee, H., and Cho, D.-W. (2017). 3D printing of organs-on-chips. *Bioengineering* 4 (4).

12 Boland, T., Mironov, V., Gutowska, A. et al. (2003). Cell and organ printing 2: fusion of cell aggregates in three-dimensional gels. *The Anatomical Record Part A* 272 (2): 497–502.

13 Mironov, V. (2003). Printing technology to produce living tissue. *Expert Opinion on Biological Therapy* 3 (5): 701–704.

14 Klebe, R.J. (1988). Cytoscribing: a method for micropositioning cells and the construction of two-and three-dimensional synthetic tissues. *Experimental Cell Research* 179 (2): 362–373.

15 Foty, R.A., Pfleger, C.M., Forgacs, G., and Steinberg, M.S. (1996). Surface tensions of embryonic tissues predict their mutual envelopment behavior. *Development* 122 (5): 1611–1620.

16 Odde, D.J. and Renn, M.J. (1999). Laser-guided direct writing for applications in biotechnology. *Trends in Biotechnology* 17 (10): 385–389.

17 Karzyński, K., Kosowska, K., Ambrożkiewicz, F. et al. (2018). Use of 3D bioprinting in biomedical engineering for clinical application. *Medical Studies* 34 (1): 93–97.

18 Landers, R., Hubner, U., Schmelzeisen, R., and Mülhaupta, R. (2002). Rapid prototyping of scaffolds derived from thermoreversible hydrogels and tailored for applications in tissue engineering. *Biomaterials* 23 (23): 4437–4447.

19 Wilson, W.C. Jr. and Boland, T. (2003). Cell and organ printing 1: protein and cell printers. *The Anatomical Record Part A* 272 (2): 491–496.

20 Dhariwala, B., Hunt, E., and Boland, T. (2004). Rapid prototyping of tissue-engineering constructs, using photopolymerizable hydrogels and stereolithography. *Tissue Engineering* 10 (9–10): 1316–1322.

21 Jayasinghe, S.N., Qureshi, A.N., and Eagles, P.A.M. (2006). Electrohydrodynamic jet processing: an advanced electric-field-driven jetting phenomenon for processing living cells. *Small* 2 (2): 216–219.

22 Norotte, C., Marga, F.S., Niklason, L.E., and Forgacs, G. (2009). Scaffold-free vascular tissue engineering using bioprinting. *Biomaterials* 30 (30): 5910–5917.

23 Skardal, A., Mack, D., Kapetanovic, E. et al. (2012). Bioprinted amniotic fluid-derived stem cells accelerate healing of large skin wounds. *Stem Cells Translational Medicine* 1 (11): 792–802.

24 Duan, B. (2017). State-of-the-art review of 3D bioprinting for cardiovascular tissue engineering. *Annals of Biomedical Engineering* 45 (1): 195–209.

25 Dababneh, A.B. and Ozbolat, I.T. (2014). Bioprinting technology: a current state-of-the-art review. *Journal of Manufacturing Science and Engineering* 136 (6).

26 Pyo, S.H., Wang, P., Hwang, H.H. et al. (2017). Continuous optical 3D printing of green aliphatic polyurethanes. *ACS Applied Materials & Interfaces* 9 (1): 836–844.

27 Kang, H.W., Lee, S.J., Ko, I.K. et al. (2016). A 3D bioprinting system to produce human-scale tissue constructs with structural integrity. *Nature Biotechnology* 34 (3): 312–319.

28 Noor, N., Shapira, A., Edri, R. et al. (2019). 3D printing of personalized thick and perfusable cardiac patches and hearts. *Advancement of Science* 6 (11): 1900344.

29 Lee, A., Hudson, A., Shiwarski, D. et al. (2019). 3D bioprinting of collagen to rebuild components of the human heart. *Science* 365 (6452): 482–487.

30 Iwanaga, S., Arai, K., and Nakamura, M. (eds.) (2015). Inkjet bioprinting. In: *Essentials of 3D Biofabrication and Translation*, 61–79. Elsevier.

31 Huang, Y., Bu, N., Duan, Y. et al. (2013). Electrohydrodynamic direct-writing. *Nanoscale* 5 (24): 12007–12017.

32 Guillotin, B., Ali, M., Ducom, A. et al. (eds.) (2013). Laser-assisted bioprinting for tissue engineering. In: *Biofabrication*, 95–118. Elsevier.

33 Workman, V.L., Tezera, L.B., Elkington, P.T., and Jayasinghe, S.N. (2014). Controlled generation of microspheres incorporating extracellular matrix fibrils for three-dimensional cell culture. *Advanced Functional Materials* 24 (18): 2648–2657.

34 Gasperini, L., Maniglio, D., and Migliaresi, C. (2013). Microencapsulation of cells in alginate through an electrohydrodynamic process. *Journal of Bioactive and Compatible Polymers* 28 (5): 413–425.

35 Hospodiuk, M., Dey, M., Sosnoski, D., and Ozbolat, I.T. (2017). The bioink: a comprehensive review on bioprintable materials. *Biotechnology Advances* 35 (2): 217–239.

36 Zhang, Z., Jin, Y., Yin, J. et al. (2018). Evaluation of bioink printability for bioprinting applications. *Applied Physics Reviews* 5 (4).

37 Heinrich, M.A., Liu, W., Jimenez, A. et al. (2019). 3D bioprinting: from benches to translational applications. *Small* 15 (23): e1805510.

38 Highley, C.B., Rodell, C.B., and Burdick, J.A. (2015). Direct 3D printing of shear-thinning hydrogels into self-healing hydrogels. *Advanced Materials* 27 (34): 5075–5079.

39 Hinton, T.J., Jallerat, Q., Palchesko, R.N. et al. (2015). Three-dimensional printing of complex biological structures by freeform reversible embedding of suspended hydrogels. *Science Advances* 1 (9): e1500758.

40 Bhattacharjee, T., Zehnder, S.M., Rowe, K.G. et al. (2015). Writing in the granular gel medium. *Science Advances* 1 (8): e1500655.

2

Representative 3D Bioprinting Approaches

2.1 Introduction

Three-dimensional (3D) bioprinting, defined as the spatial patterning of living cells, biomaterials, drugs, growth factors, and genes in a layer-by-layer manner, has been rapidly developed in recent years and widely used for fabricating living tissue and organ constructs for various biomedical applications [1–3]. It is envisioned as the first step toward the fabrication of functional replacement human organs in the future by bridging the gap between numerous biologics and integrated 3D living constructs.

In 3D bioprinting, building blocks are the fundamental units to construct living tissue and organ structures. Based on the methodology to form basic building blocks, current 3D bioprinting techniques can be divided into two categories: orifice-based and orifice-free [4]. In orifice-based bioprinting, nozzles with microscale orifices are used as the tools to form either cell-laden spheroids or strands as the building blocks. Inkjet bioprinting and extrusion bioprinting are two representative orifice-based 3D bioprinting techniques. In orifice-free bioprinting, different light sources are used as tools to form cellular layers instead of microscale orifices. Thus, the orifice-free bioprinting is also known as light-based bioprinting, as shown in Figure 2.1.

In orifice-based bioprinting technologies, the typical biofabrication process consists of two main steps: printing and cross-linking. The core function of the printing step is to form building blocks at a liquid state and then deposit them based on the designed trajectories. In inkjet bioprinting, different methods are used to form cell-laden spheroids, namely droplets, as the building blocks, while in extrusion bioprinting, cell-laden filaments are generated with continuous, cylindrical morphology to construct 3D structures. The main purpose of the subsequent cross-linking step is to solidify the deposited droplets and/or filaments rapidly. Thus, the solidified droplets and/or filaments cannot only keep their shapes as designed, but also have the mechanical stiffness to support the following deposited droplets and/or filaments. By repeating the printing and cross-linking steps, 3D structures can be fabricated by orifice-based bioprinting technologies. As a result, classic orifice-based bioprinting is performed in a "solidification-while-printing" fashion.

Orifice-based 3D bioprinting technologies have many advantages. First, the independent cross-linking step makes it possible to print biomaterials with different

Cell Assembly with 3D Bioprinting, First Edition. Yong He, Qing Gao, and Yifei Jin.

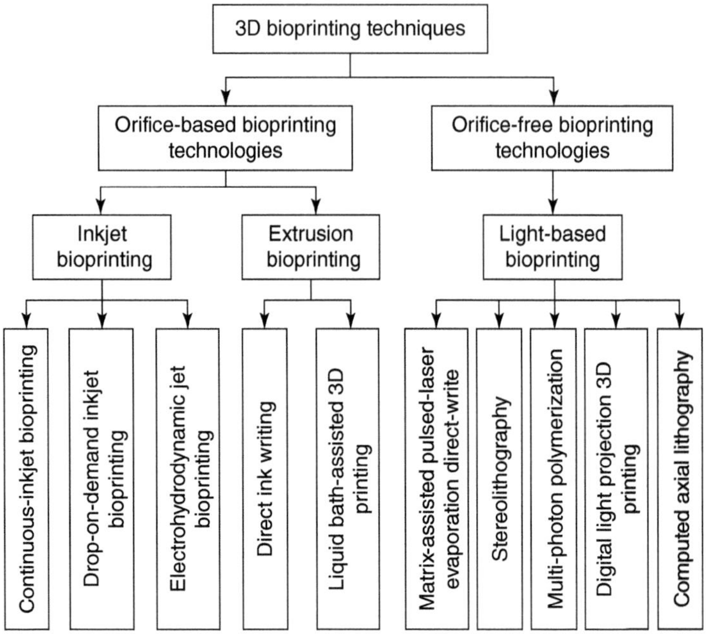

Figure 2.1 Classification of representative 3D bioprinting approaches. Source: Yifei Jin.

cross-linking mechanisms. Thus, orifice-based bioprinting technologies have a wider range of printable materials. Second, it is feasible to improve fabrication efficiency by simultaneously printing via numerous printheads with microscale nozzles. Third, orifice-based bioprinting technologies provide a technical solution to print cellular constructs with different cell types. In both inkjet and extrusion bioprinting, multiple nozzles can be integrated within one printing system to deposit different cells through corresponding nozzles, facilitating the fabrication of spatially heterocellular constructs. Finally, current inkjet bioprinters and extrusion bioprinters can be easily implemented at affordable prices. This is due to the fact that ink-jetting and extrusion are mature technologies and widely used in painting/printing and polymer/metal processing, respectively. However, orifice-based 3D bioprinting technologies have complications. The main challenge during orifice-based bioprinting is the high shear-stress-induced cell damage when cells flow through microscale nozzles. For inkjet bioprinting, the typical cell viability after printing is above 85% [1]. To form droplets with a well-defined shape, cell density in inkjet bioprinting is always controlled at a low level, less than 10^6 cells/ml. For extrusion bioprinting, living cells are propelled through nozzles with microscale diameters to form filaments. During this process, high shear stress may bring damages to cell membranes and kill cells. Thus, typical cell viability in extrusion bioprinting falls in the range of 40–80%, even lower than that in inkjet bioprinting [1].

Orifice-free 3D bioprinting technologies cover several light-based 3D printing approaches. Based on the different functionalities of a light source, they are divided into different subcategories. In matrix-assisted pulsed-laser evaporation (MAPLE) direct-write, the laser beam is illuminated on biomaterial-coated quartz to repel the localized materials away from the quartz in the morphology of droplets. This approach is also known as laser-assisted 3D bioprinting. Since MAPLE uses droplets as the basic building blocks to construct 3D structures, the fabrication process is also composed of printing and cross-linking steps, similar to that of inkjet bioprinting. However, different from inkjet bioprinting, living cells do not experience orifice-induced high shear stress during droplet formation. As a result, cell viability in MAPLE is above 95%, much higher than those in both inkjet bioprinting and extrusion bioprinting. Despite the high cell viability, MAPLE is also constrained by some complications. On the one hand, preparation of coated quartz is time-consuming, which makes the fabrication efficiency relatively low. On the other hand, the high price of laser system increases the cost of MAPLE significantly, which is another concern for clinical tissue engineering applications.

Aside from MAPLE, other light-based 3D printing approaches combine the printing step with the cross-linking step. Thus, cellular biomaterials are loaded in a container before printing and then cross-linked in situ via different light sources. In particular, during stereolithography (SLA) and multi-photon polymerization (MPP), one or multiple light spots are used to selectively cross-link cellular biomaterials in a pixel-by-pixel or voxel-by-voxel manner, while during digital light projection (DLP) and computed axial lithography (CAL), each cross-section of the designed 3D structure is projected via light pattern to cause the solidification of biomaterials in a layer-by-layer manner. Since no orifice is used during printing, the cell viability of these light-based 3D printing approaches can be very high. In addition, the resolution of the printed 3D structures mainly depends on the size of the light spot and the accuracy of the light pattern, which can be easily controlled and improved. As a result, these 3D printing approaches are promising when creating 3D structures with high resolution. The disadvantage of the aforementioned light-based 3D printing approaches is the limited material selection. To facilitate the printing process, biomaterials must be photocurable, which severely constrains the applications of these 3D printing approaches to fabricate living tissue and organ constructs from various non-photocurable materials.

In the following sections, the mechanism, printable biomaterials, and representative biomedical applications of each 3D bioprinting strategy will be introduced in detail. In particular, Section 2.2 will focus on inkjet bioprinting technology. Section 2.3 will cover basic information on extrusion bioprinting, which will further be discussed in detail in Chapters 6 and 7. Section 2.4 will introduce some well-developed light-based 3D printing techniques such as MAPLE, SLA, MPP, and several newly proposed printing strategies, such as DLP and CAL.

2.2 Inkjet Bioprinting

Inkjet bioprinting is a droplet-based 3D bioprinting technique in which cell-laden droplets are used as the basic building blocks to construct complex 3D structures. Since this technique makes it feasible to accurately control the volumes and locations of the deposited bioink, it has been widely used in tissue engineering to fabricate spatially heterocellular constructs from various cellular inks. In this section, we will introduce this 3D bioprinting technique from three aspects: droplet formation mechanisms, materials, and representative biomedical applications.

2.2.1 Mechanisms of Droplet Formation

In inkjet bioprinting, a cellular bioink is squeezed through a microscale nozzle to generate cell-laden droplets. By leveraging multiple factors including gravity, surface tension, and the fluid mechanics of the bioinks, the droplets are ejected from the nozzle and land on a receiving substrate. Based on different mechanisms to form droplets, inkjet bioprinting can be classified into three subcategories: (i) continuous-inkjet (CIJ), (ii) drop-on-demand (DOD), and (iii) electrohydrodynamic (EHD) jet, as shown in Figure 2.2.

2.2.1.1 Continuous-Inkjet Bioprinting

In CIJ bioprinting, cellular bioink is pushed through a nozzle under constant pressure to form a continuous liquid jet. Due to Rayleigh–Plateau instability, this liquid jet rapidly breaks up into a stream of droplets before landing on a substrate, as shown in Figure 2.2a. During the droplet formation, Rayleigh–Plateau instability plays an important role in which the liquid jet is perturbed by several factors including potential energy (gravity), surface energy (surface tension), and kinetic energy [5, 6]. When the wavelength of the perturbed jet is larger than the initial radius by a

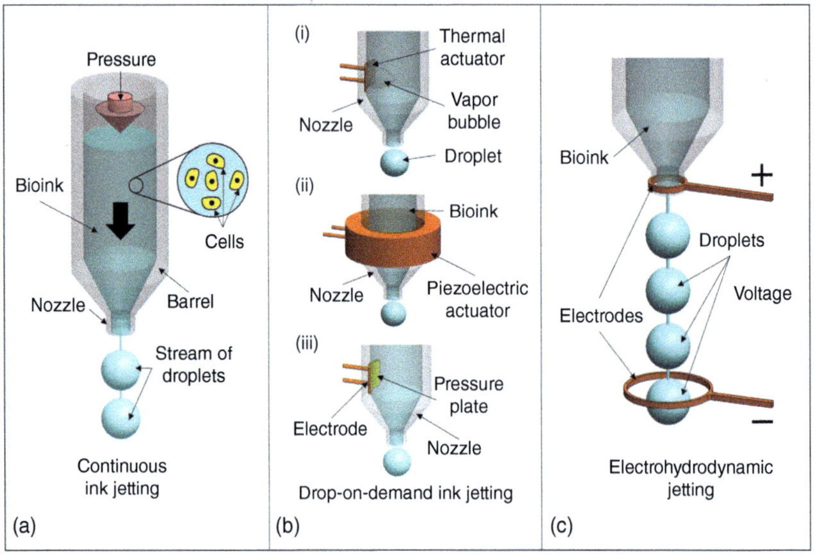

Figure 2.2 Mechanisms of droplet formation in inkjet bioprinting. (a) Continuous ink-jetting based on Rayleigh–Plateau instability. (b) Drop-on-demand ink-jetting including: (i) thermal ink-jetting, (ii) piezoelectric ink-jetting, and (iii) electrostatic ink-jetting. (c) Electrohydrodynamic jetting based on an electric field. Source: Yifei Jin.

certain limit, the perturbation grows rapidly, breaking the jet into a series of droplets to minimize its potential energy. The condition to form droplets in CIJ can be simply modeled as a function of the number of waves per unit length (k) on the perturbed jet and the initial jet radius (R_0). When kR_0 is less than 1, the jet can distort itself into continuous droplets. The detailed information regarding Rayleigh–Plateau instability can be found in other published reports [7].

2.2.1.2 Drop-on-Demand Inkjet Bioprinting

Different from CIJ, DOD inkjet bioprinting can generate droplets based on requirements. Thus, it is much easier to accurately control the locations of the deposited droplets and pattern cells and other biologics accordingly. As a result, DOD is the most popular inkjet bioprinting technique for cell bioprinting purposes. In DOD inkjet bioprinting, bioinks are transferred from fluid chambers to nozzles and squeezed out of the nozzles using different mechanisms. Since multiple nozzles can be used to print simultaneously, DOD inkjet bioprinting has the capacity of printing 3D cellular constructs with high efficiency. Currently, there are three main mechanisms to generate droplets during DOD ink-jetting including: (a) thermal inkjet, (b) piezoelectric inkjet, and (c) electrostatic inkjet.

A. *Thermal inkjet bioprinting*. In thermal inkjet bioprinting, a thermal actuator is integrated with a nozzle and heats bioink under a voltage pulse. This localized heating leads to the formation of a vapor bubble as shown in Figure 2.2b(i), which rapidly expands in the nozzle and generates pressure to the localized bioink, pushing it to the exit of the nozzle. Eventually, the bioink overcomes surface tension at the nozzle's exit and forms a droplet. Thermal inkjet bioprinting has been widely used to print living tissue constructs from various liquid bioinks such as polyethylene glycol dimethacrylate (PEGDMA), alginate, and cell suspensions. For example, Xu et al. [8] printed complex and heterogeneous 3D tissue constructs using a modified thermal inkjet printer. In their study, they mixed human amniotic fluid-derived stem cells (hAFSCs), canine smooth muscle cells (cSMCs), and bovine aortic endothelial cells (bECs) separately with calcium chloride ($CaCl_2$) as the cell-laden inks and then printed them layer-by-layer to the predetermined locations in a sodium alginate-collagen composite, as shown in Figure 2.3a. The ionic cross-linking between alginate and $CaCl_2$ led to the formation of a solid composite gel with encapsulated heterogeneous cells. The printed cellular constructs were able to survive and mature into functional tissues after incubation and generated adequate vascularization *in situ*, which validates the feasibility of inkjet bioprinting heterogeneous living tissue constructs with multiple cell types.
B. *Piezoelectric inkjet bioprinting*. In piezoelectric inkjet bioprinting, a piezoelectric actuator is used to generate a localized pressure pulse in the nozzle. The piezoelectric actuator can provide a radial deformation when a voltage pulse is applied, leading to the change of the bioink volume in the nozzle. Thus, a given volume of bioink is pushed toward the exit of the nozzle. After overcoming the surface tension at the exit, the bioink is ejected in the form of droplets, as

shown in Figure 2.2b(ii). By depositing the resulting droplets layer by layer, 3D constructs can be fabricated. Different from thermal ink-etting in which the bioink is heated periodically, the pressure pulse generated in piezoelectric ink-jetting is caused by the deflection of the actuator. Thus, the entire droplet formation is considered as an isothermal process and the effect of temperature change on the cell viability can be ignored. As a result, piezoelectric inkjet bioprinting is popularly used in tissue engineering to fabricate 3D tissue constructs. For example, Xu et al. [9] used a piezo-based inkjet bioprinting system to successfully print overhang tubular constructs as shown in Figure 2.3b. In their study, they used sodium alginate with a low concentration (2% [w/v]) as the matrix and added fibroblasts (3T3 cells) to prepare the cell-laden bioink. Thus, droplets containing 3T3 cells were deposited based on the designed trajectory in a $CaCl_2$ bath to form solid cellular constructs. The post-printing cell viability was above 82%, proving the cytocompatibility of the piezoelectric inkjet bioprinting approach.

C. *Electrostatic bioprinting*. Similar to piezoelectric inkjet bioprinting, in electrostatic bioprinting, droplets are generated by the pressure pulse resulting from the deformation of a pressure plate, without heating the bioink as in thermal inkjet bioprinting. When a voltage pulse is applied between the pressure plate and an electrode, the pressure plate deflects to increase the volume of bioink in the nozzle. After removing the voltage pulse, the pressure plate rapidly regains its original shape and subsequently ejects droplets, as shown in Figure 2.2b(iii). This inkjet 3D bioprinting approach has also been used to fabricate 3D cellular and/or acellular constructs. For example, Nishiyama et al. [10] built up a homemade 3D printer based on an EPSON SEA-Jet™ inkjet nozzle head (static electricity-actuated inkjet). They used alginate as the main component of the bioink and printed different acellular and cellular constructs including gel lines, gel sheets, gel laminations, and gel tubes with HeLa cells (as shown in Figure 2.3c) to prove the feasibility of this bioprinting strategy.

2.2.1.3 Electrohydrodynamic Jet Bioprinting

In both CIJ and DOD bioprinting, bioink is propelled under different mechanisms through a nozzle to form droplets. However, when the nozzle diameter is extremely small, a high-level pressure must be required and the resulting high shear stress in the nozzle is fatal to living cells in the bioink. In particular, when the concentration of the bioink and/or cell concentration is high, the extremely high shear stress can tear cell membranes in the nozzle, leading to cell death during printing. As a result, these two bioprinting approaches cannot achieve microscale droplets with high cell viability. To overcome this challenge, a new inkjet bioprinting approach EHD jet has been developed. In EHD bioprinting, an electric field is used to pull the bioink droplets from the microscopic nozzle without the utilization of high pressure. The working mechanism of EHD bioprinting is illustrated in Figure 2.2c in which a back pressure is applied to the bioink in a metallic nozzle and forms a spherical meniscus at the exit of the nozzle. Then, an electric field with a high voltage range of 0.5–20 kV

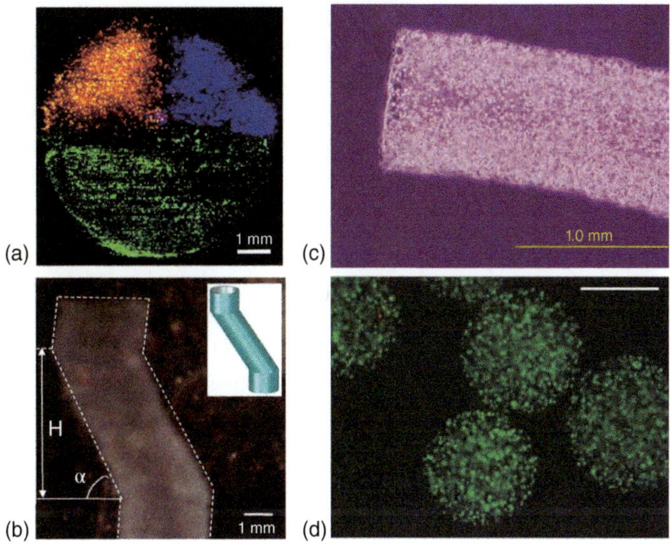

Figure 2.3 Representative inkjet bioprinting results based on different droplet formation mechanisms. (a) 3D multi-cell "pie" construct printed by thermal inkjet bioprinting. Source: Xu et al. [8]. Reproduced with permission of Elsevier. (b) Cell-laden zigzag tubes printed by piezoelectric inkjet bioprinting. Source: Xu et al. [9]. Reproduced with permission of John Wiley and Sons. (c) Gel tube including HeLa cells printed by electrostatic bioprinting. Source: Nishiyama et al. [10]. Reproduced with permission of Elsevier. (d) Cell encapsulations printed by electrohydrodynamic jet bioprinting. Source: Workman et al. [11]. Reproduced with permission of PMC.

is applied between the metallic nozzle tip and the receiving substrate, which accumulates mobile ions in the bioink near the surface of the bioink meniscus. Due to the electrostatic repulsions between ions, the bioink meniscus deforms into a conical shape and eventually forms a droplet when the electrostatic stress exceeds the surface tension at the orifice. By increasing the applied voltage, the jetting mode varies from dripping to cone-jet, and the droplet size decreases. EHD bioprinting is significant in the biomedical applications that require a nozzle diameter of less than 100 μm and bioink concentration higher than 20% (w/v). One of EHD bioprinting is illustrated in Figure 2.3d. In this study, bio-electrospraying was used to produce microspheres containing THP-1 cells, collagen, and alginate as a model cell line. It was found that bio-electrospraying technology did not affect cell viability and cell proliferation ability [11].

2.2.2 Hydrogel-Based Bioinks for Inkjet Bioprinting

Among various biocompatible materials, hydrogels have unique properties including tunable printability, excellent biocompatibility, suitable biodegradability, and enhanced cell adhesive properties. Thus, hydrogels have been widely used as the extracellular matrix (ECM) materials for preparing cellular and/or acellular bioinks. In this chapter, the main properties of hydrogel-based bioinks for inkjet bioprinting will be introduced and some commonly used hydrogels will be summarized.

2.2.2.1 Material Properties for Inkjet Bioprinting Applications

From the 3D bioprinting aspect, there are two main properties to evaluate the functionality of hydrogels for inkjet bioprinting applications: rheological and cross-linking properties. Rheological properties are used to determine printability, which is the ability that hydrogels can generate droplets. Cross-linking properties are adapted to solidify the deposited droplets as designed to form solid 3D constructs.

2.2.2.1.1 Rheological Property

Viscosity is one of the primary rheological properties of hydrogel-based bioinks during inkjet bioprinting, which is predominantly determined by the polymer concentration and molecular weight. Since surface tension is the main driven force to generate droplets in most inkjet bioprinting approaches, the viscosity of hydrogel-based bioink is required to be relatively low to enable a fluid jet to break up into droplet(s). The optimal viscosities of hydrogels for inkjet bioprinting fall in the range of 3.5–12 mPa·s [1].

Shear thinning is another important rheological property of hydrogel-based bioinks, which indicates the phenomenon of decreasing viscosity with increasing shear rate [12–14]. It has resulted from the reorganization of polymer chains to a stretched conformation under shear stress. Thus, the entanglement of polymer chains reduces from microscopic analysis and further leads to the decrease of viscosity from macroscopic behavior. After removing shear stress, polymer chains regain the original entangled status, and the viscosity recovers to the original value. This shear-thinning property effectively facilitates the squeezing of bioinks out of the nozzle or orifice to form droplet(s).

2.2.2.1.2 Cross-linking Property

Cross-linking property is defined as the capacity of gelation/curing/solidification of a bioink after printing, which is necessary to preserve the shape of a printed 3D construct [15]. This property is particularly important in inkjet bioprinting in which bioinks usually have low viscosities and cannot prevent the shape change of printed structures without cross-linking. Physical cross-linking and chemical cross-linking are two main categories of cross-linking mechanisms during 3D bioprinting.

Physical cross-linking depends on nonchemical interactions between molecular polymer chains such as entanglements, ionic interactions, and hydrogen bridges. These interactions are always reversible. Ionic and thermal cross-linking are two primary methods to design cross-linkable bioinks. In ionic cross-linking, di/trivalent cations are used as cross-linkers to bond different polymer chains together. Alginate is an exemplary ionic cross-linked hydrogel, which consists of mannuronic and glucuronic acid residues. After adding Ca^{2+} or Al^{3+}, these acid residues can bond with each other rapidly via cations, leading to the gelation of alginate [16]. In thermal cross-linking, hydrogels are composed of thermosensitive polymer chains, which can form stable 3D networks at one temperature, but collapse into independent polymer chains at another temperature. Gelatin is an example of the thermal cross-linkable hydrogel. At higher temperatures such as physiological temperature, the protein chains in gelatin are short and lack the ability to self-assemble into well-organized fibers, leading to a liquid status of gelatin. However, upon cooling, some segments in gelatin return to the triple-helical conformation and form junctions between gelatin molecular chains. Thus, gelatin transitions from its liquid to solid states [17].

Chemical cross-linking mainly relies on the formation of covalent bonds between hydrogel polymer chains, which can provide cross-linked hydrogels with excellent mechanical strengths. Chemical cross-linking is always achieved by mixing monomers and initiators. Monomers provide hydrogel polymer chains, while initiators start the cross-linking reaction. Thus, when the gel point is reached, hydrogel polymer chains can rapidly develop into 3D networks, resulting in the increase of viscosity and transition of hydrogel from liquid to solid [18].

2.2.2.2 Commonly Used Hydrogels in Inkjet Bioprinting

Due to the severe requirement of viscosity, only a handful of hydrogels have been used in inkjet bioprinting including alginate, collagen, fibrin, gelatin-methacryloyl (GelMA), and polyethylene glycol (PEG).

Alginate, also called alginic acid, is naturally derived from the cell walls of brown algae. It consists of a family of unbranched binary copolymers of 1,4 linked β-D-mannuronic acid (M units) and α-L-guluronic acid (G units). It undergoes physical cross-linking when it interacts with divalent ions (such as Ca^{2+}) and/or trivalent ions (such as Al^{3+}). Gelation occurs as such cations form interchain ionic bonds between G units, forming a stable 3D network of calcium and/or aluminum alginate. Due to its tunable viscosities and mild cross-linking conditions, alginate has been used in inkjet bioprinting to make cellular tissue constructs. For example, Xu et al. [9] printed zigzag cell-laden tubular constructs using alginate-based bioink composed of 2.0% (w/v) sodium alginate solution and 3T3 fibroblast cells, as shown in Figure 2.3b.

Collagen is a triple-helical protein and also a primary component of ECM materials in the human body. Collagen is a thermal cross-linkable biomaterial, which is a liquid state at low temperature (~4 °C), but cross-links at room temperature or physiological temperature [19]. It has been extensively utilized for the fabrication of living tissue constructs due to its excellent biocompatibility and cell adhesive properties [20]. For example, Skardal et al. [21] mixed amniotic fluid-derived stem (AFS) cells and bone marrow-derived mesenchymal stem cells (MSCs) with fibrin-collagen gel and inkjet-printed over the wound site on a mouse model. It was found that cellular construct-driven wound closure and re-epithelialization were significantly better than closure and re-epithelialization in wounds treated by acellular fibrin-collagen gel only.

Fibrin is a hydrogel formed by the reaction of protease thrombin on fibrinogen. Since it can support cell growth and proliferation [21, 22], fibrin has been widely used to fabricate skin grafts in wound healing. For example, Cui et al. [22] printed human microvascular endothelial cells (HMVEC) and fibrin into micron-sized fibrin channels to investigate microvasculature construction. It was found that the printed cells aligned themselves inside the channels and proliferated to form confluent linings, which promoted HMVEC proliferation and microvasculature formation.

GelMA is a semi-synthetic hydrogel consisting of gelatin and methacrylamide and methacrylate functional groups. The derivation with these functional groups switches the cross-linking mechanisms from physical to chemical, significantly improving the mechanical strength of gelled hydrogel while maintaining its excellent biocompatibility. As a result, GelMA has been increasingly used for engineering various tissue constructs. For example, Gurkan et al. [23] successfully printed an anisotropic biomimetic fibrocartilage microenvironment using GelMA, MSCs, bone morphogenetic protein 2, and transforming growth factor β1.

PEG is a synthetical hydrogel with water solubility and tunable mechanical properties. After functionalized with some functional groups such as diacrylate (DA) or methacrylate (MA) [24], some PEG-based hydrogels are prepared, which can be cross-linked by ultraviolet (UV) radiation. Due to the better mechanical strength than natural hydrogels (e.g. alginate, collagen, and fibrin), PEG-based hydrogels such as poly(ethylene glycol) diacrylate (PEGDA) and poly(ethylene) glycol dimethacrylate (PEGDMA) are employed for cartilage tissue engineering. For example, Cui et al. [25] successfully printed human articular chondrocytes with PEGDMA at various densities. It was found that the bioprinted samples treated with growth factors had the best chondrogenic properties due to synergistic stimulation of cell proliferation and chondrogenic phenotype.

The detailed information of the aforementioned hydrogels in inkjet bioprinting is summarized in Table 2.1 as follows:

2.2.3 Representative Cell Printing Applications

Inkjet bioprinting has been used to fabricate engineered tissues and organs from various biocompatible materials and cells. Some representative applications include

Table 2.1 Hydrogels applied for inkjet bioprinting of 3D constructs.

Hydrogels	Fabrication technique	Polymer concentration (w/v)	Cross-linking mechanism	Cell viability	References
Alginate	Thermal	2.3%	Ionic	75–90%	[26–28]
	Piezo	1.0%	Ionic	82% 90%	[9] [29]
	Electrostatic	1.0%	Ionic	70%	[10]
	Electro-hydrodynamic jetting	2.0%	Ionic	>90%	[11]
Collage/alginate	Thermal	1.0%/0.3%	Ionic	90%	[20]
Collagen	Electro-hydrodynamic jetting	3% (v/v)	Thermal	N/A	[30]
Fibrin	Thermal	200 unit/ml	Enzymatic	82%	[22, 31]
GelMA	Electro-hydrodynamic jetting	5%	Photo	>90%	[32]
PEG-based	Thermal	10–20%	Photo	~85%	[25, 33–35]

artificial bone and cartilage tissues, organoids, such as heart and liver tissues, skin tissues, and vascular networks.

2.2.3.1 Bone and Cartilage Tissues

Gao et al. [35] prepared the cell-laden bioink composed of hMSCs, PEGDMA, and hyaluronic acid (HA) nanoparticles. They fabricated the bone-like tissues using thermal inkjet bioprinting, as shown in Figure 2.4a. It was found that the HA nanoparticles were able to mimic the native bone microenvironment and facilitate the differentiation of stem cells on the printed bone-like tissues.

Cui et al. [34] used the thermal inkjet bioprinting technique to fabricate cartilage tissues, which possessed characteristics comparable to native cartilages. In their study, they printed double-layered patches with human articular chondrocytes and PEGDMA to repair circular deflects in osteochondral explants. The repaired explants

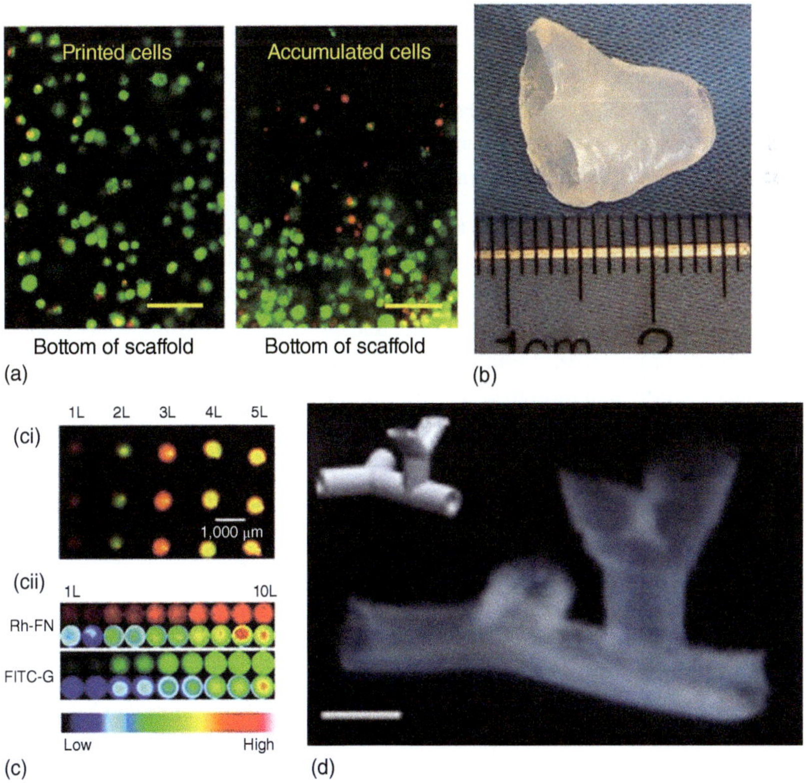

Figure 2.4 Representative biomedical applications of inkjet bioprinting. (a) 3D printed PEG-HA scaffolds with hMSCs. Source: Gao et al. [35]. Reproduced with permission of John Wiley and Sons. (b) 3D cardiac constructs ("half heart") from alginate/gelatin. Source: Xu et al. [28]. Reproduced with permission of IOP Publishing. (c) Multi-layered 3D liver structures consisting of HepG2 cells and HUVECs. Source: Matsusaki et al. [36]. Reproduced with permission of Elsevier. (d) Inkjet-printed cellular structures with both horizontal and vertical bifurcations. Source: Christensen et al. [29]. Reproduced with permission of John Wiley and Sons.

were incubated in vitro for four weeks and more chondrocytes were observed in the printed cartilage tissues, which accelerated chondrogenesis and ECM production during cartilage regeneration.

2.2.3.2 Organoids

Heart tissue was successfully printed by Xu et al. [28], as shown in Figure 2.4b. In this study, HL1 cardiac muscle cells were mixed with alginate to prepare a cell-laden alginate suspension bath. Calcium chloride was 3D printed via thermal inkjet printing into a predefined pattern on a substrate. After that, the substrate was first submerged into the cell-laden alginate suspension bath to cross-link alginate on the substrate into the cell-laden pattern and then moved upward to receive the next layer of calcium chloride. By repeating these steps, 3D heart tissue was fabricated. The beating response of the heart tissue proved that HL1 cardiac muscle cells were attached to and/or entrapped in the cross-linked alginate, which can effectively mimic the native microenvironment in heart tissues.

Liver tissue was engineered by Akashi's group [36]. In their study, they used piezoelectric inkjet bioprinting to deposit hepatocytes, endothelial cells, fibronectin, and gelatin into multi-layered liver tissue models, as shown in Figure 2.4c, which were used to investigate drug metabolism of anti-diabetic drug troglitazone.

2.2.3.3 Skin Tissues

One example of 3D printing skin tissues is the study from Boland's group in which they fabricated three-layered skin grafts with the help of thermal inkjet bioprinting [37]. The skin grafts were composed of a layer of neonatal human dermal fibroblasts-laden collagen with pipetted fibrinogen solution, a 3D printed layer of human umblical vein endothelial cells (HUVECs)-laden thrombin, and a pipetted layer of neonatal human epidermal keratinocytes (NHEKs)-laden collagen. After incubation for 24 hours, the grafts were transplanted onto the surgical wounds of nude mice. It was found that the wounds with skin grafts were completely healed in 14–16 days.

2.2.3.4 Vascular Networks

Vascular tissues or vascular networks have been 3D printed by many groups. Christensen et al. [29] used piezoelectric inkjet bioprinting to successfully fabricate complex vascular networks with both horizontal and vertical bifurcations, as shown in Figure 2.4d. In their study, alginate solution was mixed with 3T3 fibroblasts as the cell-laden bioink and calcium chloride bath was utilized to cross-link the printed constructs as well as provides *in situ* buoyant support during printing. The cell viability was around 92% immediately after printing and remained above 90% after 24-hour incubation.

2.2.4 Summary

Inkjet bioprinting has been widely used to fabricate cell-laden living tissue constructs due to its unique advantages such as simplicity, agility, high printing efficiency, and excellent controllability of on-demand depositing living cells. Although different inkjet bioprinting techniques enable the fabrication of heterocellular tissue constructs in a reproducible fashion, there are still some limitations including weak mechanical properties of printed tissues and/organs due to the use of low-viscosity bioink as well as limited biomaterial options, inhibiting the translation of inkjet bioprinting into clinics. Despite these constraints, novel breakthroughs, such as *in situ* bioprinting, make the eventual clinical translation of inkjet bioprinting technologies ineluctable.

2.3 Extrusion Bioprinting

Extrusion bioprinting is the most commonly used 3D bioprinting strategy. Around 30 000 extrusion-based 3D printers are sold all over the world every year and universities and academic institutions are increasingly applying various extrusion 3D bioprinting techniques in tissue engineering. The detailed information of different extrusion-based bioprinting strategies will be introduced in Chapters 6 and 7. Herein, only some basic concepts during extrusion bioprinting will be discussed including: (i) mechanisms of extruding biocompatible materials, (ii) primary extrusion bioprinting strategies, and (iii) main categories of extrudable biomaterials.

2.3.1 Mechanisms of Extruding Biocompatible Materials

A typical extrusion-based 3D bioprinting system has many components. The temperature control module controls the temperatures of bioinks and/or the surrounding environment during and/or after printing. The motion control module (stage(s)) moves the printhead(s) and/or substrate along the *x*-, *y*-, and *z*-axes based on the designed trajectory. The dispensing system contains one or multiple printheads to deposit biomaterials as required. The accessory modules, UV radiation, and imaging system facilitate and monitor the entire 3D printing process, respectively. Among all of the aforementioned modules, the dispensing system is of great significance. The dispensing system is the module in which bioinks are extruded through dispensing nozzles under different driving mechanisms to form continuous, cylindrical filaments. These filaments are the basic units to construct complex 3D structures in a layer-by-layer fashion.

The most common driving mechanisms to extrude biomaterials are pneumatic [38–41] and mechanical [42–44]. In a pneumatic dispensing system, high-pressure air is used to propel bioinks flowing through dispensing nozzles and filament diameter can be directly controlled by increasing or decreasing dispensing pressure, as shown in Figure 2.5a. However, due to the delay of the compressed air volume in the system, under-deposition and over-deposition phenomena can be observed at the beginning and the end of the material extrusion, respectively, which may affect the resolution of the printed 3D structures. In addition, due to the limitation of achievable pressure, some highly viscous bioinks are not easily extruded by pneumatic dispensing system. The development of mechanical dispensing system

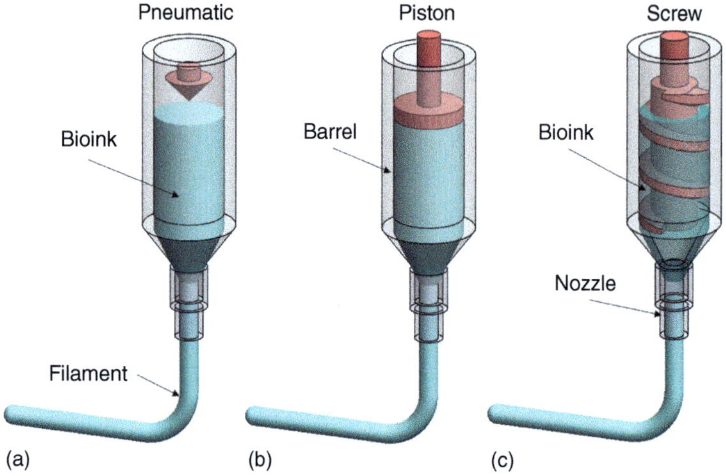

Figure 2.5 Different mechanisms for biomaterials deposition. (a) High pressure-driven, (b) piston-driven, and (c) screw-driven extrusion bioprinting. Source: Yifei Jin.

overcomes the main challenges in pneumatic dispensing system. In a mechanical dispensing system, either a piston or screw is applied to provide repelling force for material extrusion. Generally, piston-driven deposition can directly control the flow rate or the volume of the bioinks being extruded (Figure 2.5b), while screw-driven deposition may provide more spatial control by adjusting the screw rotation speed and more propelling force for high-viscosity material deposition (Figure 2.5c). For bioprinting applications, the high shear stress in the screw-driven deposition system can potentially be harmful to living cells suspended in the bioinks. Thus, the geometries of the screw must be specifically designed to accommodate bioprinting purposes.

2.3.2 Primary Extrusion Bioprinting Strategies

Currently, two main extrusion bioprinting strategies are developed and widely used to fabricate complex 3D constructs: direct ink writing (DIW) and liquid bath-assisted printing. The core of DIW is how to match the printing speed with cross-linking speed and make the extruded filament rapidly transit from a liquid state to a solid or solid-like state. Thus, the deposited layer has the mechanical stiffness to maintain its shape after printing and provides support to the subsequently deposited layers, as shown in Figure 2.6a. Two methodologies have been developed to address this technical issue. First, different approaches are investigated to rapidly cross-link bioinks during or after printing such as the utilization of pre-cross-linked bioinks [45–47] and co-extrusion of bioinks with cross-linkers [33, 48]. Second, some biomaterials with a self-supporting property are investigated and used as bioinks for 3D bioprinting purposes, which usually possess yield-stress property and can switch their status between liquid and solid-like upon stressed and unstressed conditions. The details of DIW will be comprehensively discussed in Chapter 6.

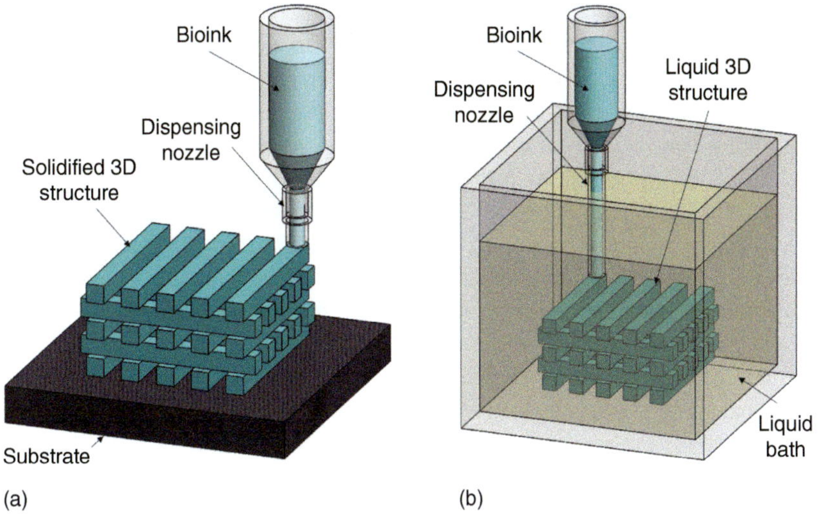

Figure 2.6 Two main extrusion bioprinting strategies: (a) direct ink writing and (b) liquid bath-assisted 3D printing. Source: Yifei Jin.

Different from DIW, liquid bath-assisted 3D bioprinting undergoes in a liquid bath, which provides *in situ* support during printing [16, 49]. Thus, the 3D structure can remain at a liquid state in the liquid bath and be cross-linked after the entire structure is finished, as shown in Figure 2.6b. Since it does not require the rapid solidification property of bioinks, liquid bath-assisted 3D bioprinting has the potential to print living tissue constructs from various biocompatible materials, especially the materials with a long cross-linking period such as collagen. Furthermore, both of the commonly used bioinks and liquid baths are aqueous solvent-based. Thus, the interfacial tension or surface tension during liquid bath-assisted 3D bioprinting can be ignored, which prevents the surface tension-induced droplet formation as illustrated in inkjet bioprinting. As a result, biomaterials with low viscosity can be easily printed into complex 3D structures in a liquid bath. More information regarding different liquid bath-assisted 3D bioprinting approaches will be discussed in detail in Chapter 7.

2.3.3 Main Categories of Extrudable Biomaterials

Extrusion bioprinting is very versatile in printing various bioink types with a wide range of viscosities (from 30 to over 6×10^7 mPa·s) including hydrogels [15, 50], microcarriers [51], cell aggregates [52–54], and decellularized matrix (dECM) components [55].

2.3.3.1 Hydrogels

A wide variety of hydrogels have been printed using extrusion bioprinting. These hydrogels include but are not limited to alginate, chitosan, gelatin and its derivatives (e.g. GelMA), HA, PEG, and its derivatives (e.g. PEGDA), agarose, collagen, pluronic, matrigel, and fibrin.

Chitosan is a nontoxic, biodegradable, and antibacterial/antifungal material produced by the deacetylation of chitin. Chitin has been widely used in bone, cartilage, and skin tissue engineering [56–58]. It can be dissolved in acid solutions to prepare bioink and cross-linked by ionic and covalent agents. Due to its long cross-linking period (~10 minutes), only highly viscous chitosan samples can be printed by DIW, which can hold the deposited shape by itself [59]. Adding other hydrogels to adjust the gelation rate is also an alternative plan for chitosan printing [60, 61]. Since the liquid bath can help to hold the shape of deposited structures before or during cross-linking for a long time, liquid bath-assisted 3D bioprinting will be the best option for chitosan printing in the future.

HA is a natural nonsulfated glycosaminoglycan in connective tissues [62], which plays an important role in the regulation of cell behaviors. However, due to the poor mechanical properties and rapid degradation [63], pure HA has been rarely used in extrusion bioprinting, and various HA-based bioinks such as UV curable HA-MA [64] and HA-PEG were developed for printing purpose.

Agarose is a thermosensitive hydrogel extracted from seaweed. Depending on the hydroxyethylation and concentration, agarose has different cross-linking temperatures. For extrusion bioprinting purposes, the most suitable agarose should have a low melting temperature to be easily extruded out as well as a low cross-linking temperature in the range of 26–30 °C to be rapidly solidified.

Pluronic is a tri-block copolymer (PEO-PPO-PEO) with thermosensitivity, which has been used as a drug carrier and injectable gel for wound healing applications [39, 65]. Usually, Pluronic presents liquid state at room temperature or lower temperatures and switches to solid-like state at physiological temperature or higher temperatures. Thus, for Pluronic printing, an extrusion-based 3D printing system must have an accurate temperature control module. One representative result is illustrated in Figure 2.7a. Currently, Pluronic has been used as a fugitive ink [67] and/or support bath material [68] to fabricate vascular networks.

Matrigel is also a thermosensitive hydrogel. But different from other thermosensitive hydrogels such as gelatin, agarose, and Pluronic, the cross-linking of Matrigel is irreversible. The gelation temperature of Matrigel is 24–37 °C. After gelation, the decreasing temperature cannot let Matrigel switch back to a liquid state. Thus, similar to Pluronic printing, an accurate temperature control module is also required for Matrigel printing.

2.3.3.2 Micro-Carriers

Micro-carriers usually possess porous microstructures and can carry cells inside. During incubation, cells can rapidly proliferate in the micro-carriers. Micro-carriers are commonly made of hard materials such as dextran [69], plastic [70], and glass [71] with spheroid shapes. For printing purposes, micro-carriers are mixed with hydrogels which function as both delivery matrix to facilitate the printing process as well as binder to bond numerous micro-carriers together after cross-linking. Thus, the resulted micro-carrier-hydrogel suspensions are highly viscous and extrusion bioprinting is the best way to print 3D structures from these suspensions as shown in Figure 2.7b.

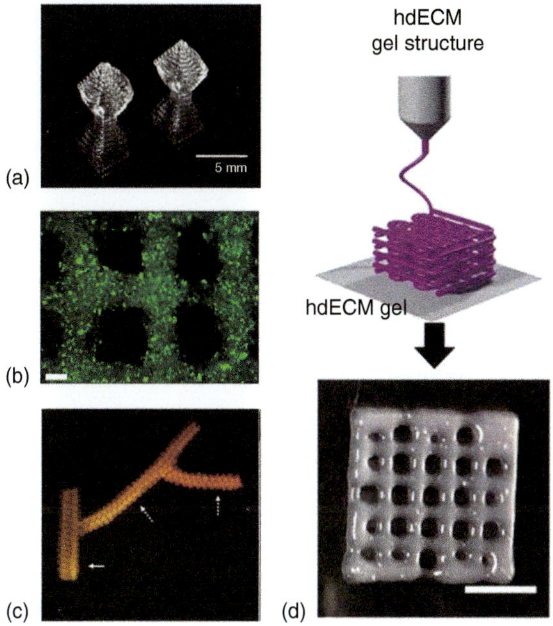

Figure 2.7 Extrusion bioprinting of complex 3D structures from different types of biomaterials. (a) 3D structures printed using F127 hydrogel. Source: Chang et al. [39]. Reproduced with permission of John Wiley and Sons. (b) Scaffolds printed using microcarrier/cell-laden GelMA-based bioink.Source: Levato et al. [51]. Reproduced with permission of IOP Publishing. (c) Scaffold-free bioprinting of a branched vascular network using 300 μm human skin fibroblast (HSF) spheroids. Source: Norotte et al. [66]. Reproduced with permission of Elsevier. (d) 3D printing scaffold from dECM gel. Source: Pati et al. [55]. Reproduced with permission of Nature Communications.

2.3.3.3 Cell Aggregates

Cell aggregate-based bioinks are emerging recently and are promising for 3D bioprinting in the future. Cell aggregates are the aggregations of homogenous and/or heterogeneous cells in the form of spheroid or strand. Since there are already extremely high cell numbers in the aggregates, it is unnecessary to expect cells to proliferate in the commonly used cell-laden hydrogel inks. In addition, such numerous cells are easy to deposit ECM. Thus, engineered tissues can be fabricated in a relatively short period. It is much easier to investigate homocellular and heterocellular interactions using cell aggregate-based bioinks than the conventional hydrogel-based bioinks. As seen from Figure 2.7c, a branched vascular network has been successfully printed using human skin fibroblast spheroids [66]. Despite these advantages, cell aggregates also have some limitations. First, it is extremely time-consuming to prepare a large number of cells and cell aggregate-based bioinks. Second, cell aggregates usually possess poor mechanical properties resulting from insufficient ECM components secreted by parenchymal cells in highly metabolic organs. Finally, due to the diffusion limitations of nutrients and oxygen, the dimensions of cell aggregations are severely constrained. For example, when cell spheroids with the diameter over 400 μm, hypoxia will occur, making highly metabolic cells inside the spheroids difficult to survive [72].

2.3.3.4 Decellularized Matrix Components

dECM has been considered as a new component of bioinks, which is achieved by organ decellularization. These dECM components are chopped into small fragments and mixed with cells as bioinks. Usually, extrusion bioprinting is used to deposit such highly viscous suspension bioinks into complex 3D structures. Decellularized matrix-based bioink is in the initial stage. The feasibility of these new extrudable bioinks has been proved by Dong-Woo and his coworkers [55], as shown in Figure 2.7d. In their study, they used extrusion bioprinting technique to print PCL frames along with dECM-cell suspensions and they observed the natural differentiation of three different cells in the native dECM.

2.3.4 Summary

Due to easy implementation, wide range of printable materials, and high printing efficiency, extrusion bioprinting is the mainstream cell printing methodology and has been widely used to fabricate complex 3D cellular constructs for various tissue engineering applications. The future investigations of extrusion bioprinting may focus on three directions: (i) innovation of extrusion-based bioprinting techniques, (ii) development of new extrudable bioinks, and (iii) 3D bioprinting full-scale organs or tissues.

2.4 Light-Based Bioprinting

In light-based 3D bioprinting techniques, various lights are used as the tool to either eject cell-laden biomaterials to form droplets or in situ cross-link photo-curable cellular bath to form 3D structures. The representative light-based bioprinting approaches include MAPLE direct-write, also known as laser-assisted bioprinting, SLA, MPP, DLP, and CAL. In this chapter, the detailed information of these 3D printing approaches will be introduced including mechanisms, suitable biomaterials, and biomedical applications.

2.4.1 Laser-Assisted Bioprinting

2.4.1.1 Mechanism

The fundamental mechanism of laser-assisted bioprinting is based on laser-induced forward transfer (LIFT) in which a laser pulse is used to directly eject metals from metal coating glass slides. However, for cell printing applications, since it is difficult to coat pure cells on glass slides without cell medium and the laser pulse may kill living cells in the coated cell layer, a new laser-induced transfer mechanism, called MAPLE, has been developed. In this new mechanism, a metal coating glass slide is replaced by a "ribbon," which is composed of a donor transport support, usually made from quartz, an optional energy-absorbing layer (EAL), and a layer of hydrogel-based cell suspension, as shown in Figure 2.8a. During printing, a pulsed and focused laser beam is exposed on the EAL to generate a high-pressure bubble which propels the localized hydrogel-based cell suspension on the adjacent layer to form a jet toward the substrate. The jet subsequently breaks up into droplets in air, which land on the substrate finally. Then, the laser beam moves to another location and repeats this droplet formation process. Simultaneously, a motion control system is used to move the substrate along x-, y-, and z-directions based on the design and let the deposited droplets construct 3D structures layer by layer [74–76]. The utilization of "ribbon" can effectively overcome the challenges in traditional

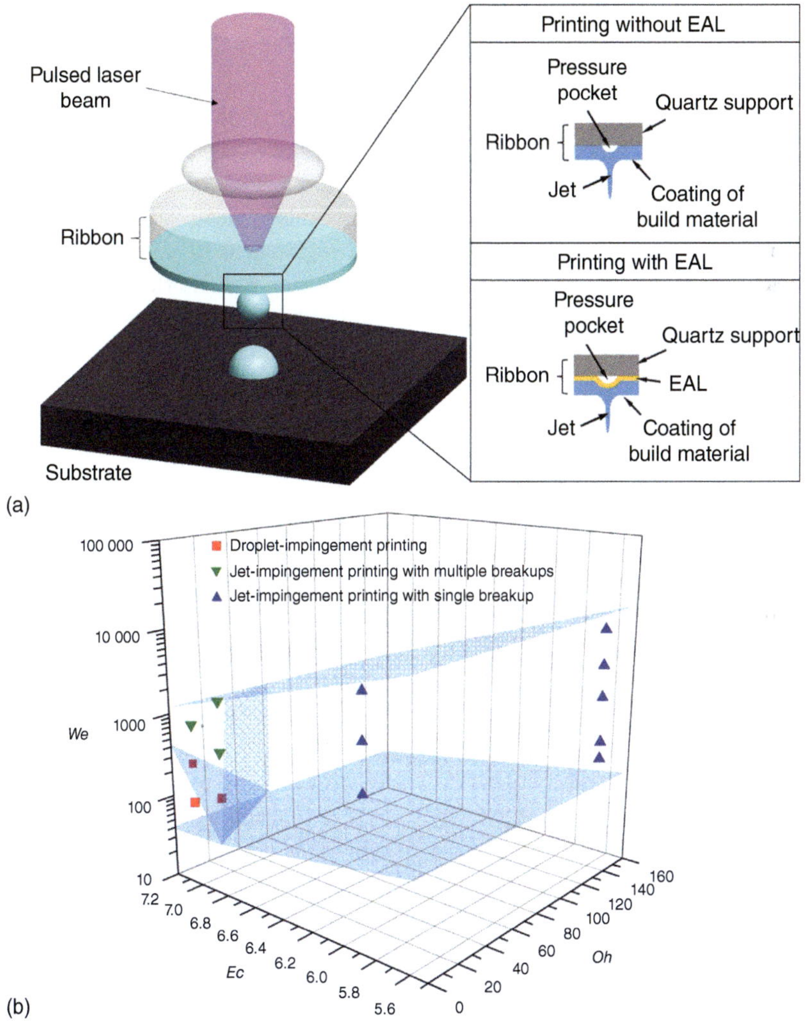

Figure 2.8 (a) Mechanism of laser-assisted bioprinting without and with EAL. Source: Yifei Jin. (b) Phase diagram of printing types during laser-assisted bioprinting. Source: Zhang et al. [73]. Reproduced with permission of American Chemical Society.

LIFT, providing a relatively mild working environment for cell printing. As a result, MAPLE direct-write has become one of the representative 3D bioprinting approaches and is increasingly used for organ/tissue printing applications [74, 75].

The resolution of the 3D cellular structures fabricated by MAPLE direct-write is mainly affected by the process parameters such as laser fluence, air gap between ribbon and substrate, path speed, as well as the physical and rheological properties of hydrogels which are used as the matrices of cell suspension. For example, the surface tension and viscosity of the matrix hydrogels can affect the droplet formation process. Only a suitable combination of these factors can generate well-defined droplets to print cellular constructs. Zhang et al. have already built up a phase diagram to unveil the relationship between these printing factors and droplet/jet types, as shown in Figure 2.8b [73].

2.4.1.2 Materials

Materials used in laser-assisted 3D bioprinting mainly include matrix hydrogels and corresponding cross-linking agents as well as EAL materials. Since laser-assisted bioprinting is nozzle-free, the problem of nozzle clogging with cells in inkjet bioprinting and extrusion bioprinting can be effectively avoided. Thus, cell density in laser-assisted 3D bioprinting can be relatively high (up to 10^8 cells/ml). For the bioink design, except for excellent cell compatibility and suitable rheological property, bioink for laser-assisted printing also needs to have fast gelation property [77]. Thus, deposited droplets/layers can rapidly transit the states from liquid to solid to support subsequently deposited droplets/layers.

Alginate is the most commonly used matrix hydrogel to prepare the cellular layer due to its excellent cell compatibility, tunable rheological properties, and fast gelation rate. Zhang et al. [73] investigated the effects of alginate concentration on the formation of jets/droplets and they found that the most suitable alginate concentration for laser-assisted bioprinting falls in the range of 6–8% (w/v). During alginate printing, a substrate is usually submerged in a calcium chloride bath to rapidly cross-link alginate droplets and facilitate the printing process [74, 75, 78]. Besides alginate, collagen [79], DNA [80], peptide fibrils [81] have also been reported for cell printing.

The investigation on EAL materials is in the beginning stage and only a handful of candidate materials are available. For example, 10% (w/v) gelatin has been used as EAL in cell printing experiments. It was found to be comparable with the traditional laser-assisted printing approach without EAL, the cell viability with gelatin EAL can be effectively improved (77.1% with EAL versus 68.1% without EAL). This is because EAL can reduce mechanical stress generated during laser printing. In addition, the printed cellular constructs have thinner walls, which enhance the exchange of nutrients and oxygen from the culture medium [75].

2.4.1.3 Biomedical Applications

Laser-assisted 3D bioprinting has been used for biofabrication of engineered tissues. For example, Gruene et al. laser printed scaffold-free autologous grafts [82] using 2%

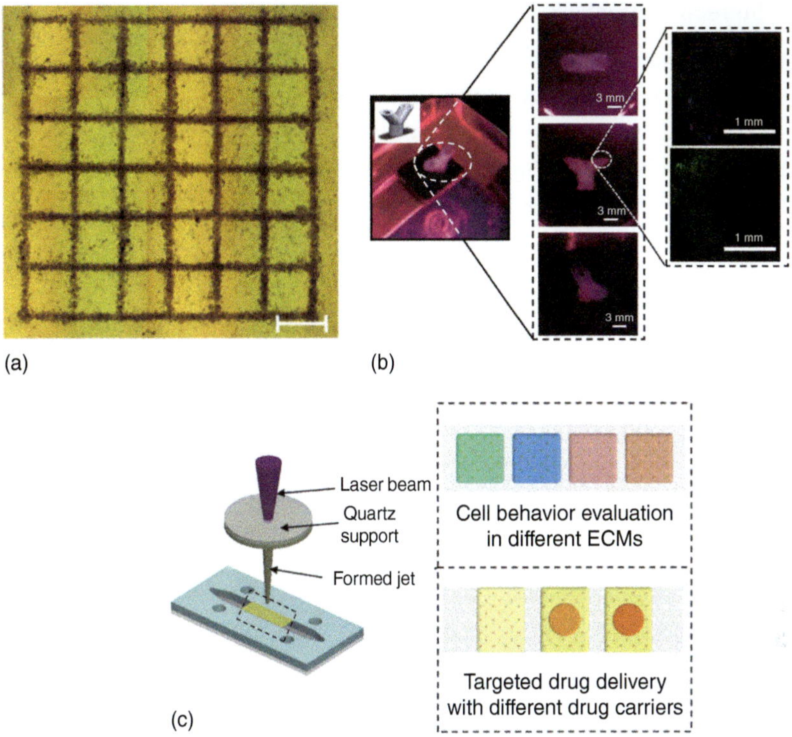

Figure 2.9 Representative biomedical applications of laser-assisted bioprinting. (a) Microscopic image of the MSC pattern directly after laser printing. Source: Gruene et al. [82]. Reproduced with permission of Mary Ann Liebert, Inc. (b) Y-shaped cellular tube printed with gelatin EAL. Source: Xiong et al. [75]. Reproduced with permission of IOP Publishing. (c) Laser bioprinting of different cellular patterns in lab-on-a-chip devices. Source: Xiong et al. [76]. Reproduced with permission of Royal Society of Chemistry.

(w/v) alginate and MSCs, as shown in Figure 2.9a. It was found that laser printing did not cause much cell damage and most pre-differentiated MSCs survived from printing with good functionality. Furthermore, Xiong et al. used laser-assisted 3D bioprinting to successfully fabricate a Y-shaped vascular network with well-defined geometries, as shown in Figure 2.9b. In their study, they used alginate as the matrix hydrogel and mixed it with 3T3 fibroblasts to construct a *Y*-shaped vascular network [75]. The cell viability immediately after printing was about 77% and after a 24-hour incubator, the cell viability increased slightly to 80% [75].

Except for tissue/organ printing, laser-assisted printing has also been used to fabricate biomedical devices. For example, Xiong et al. used laser-assisted bioprinting to create heterogeneous cellular patterns in a lab-on-a-chip device, as shown in Figure 2.9c [76]. Two applications including parallel evaluation of cellular behavior and targeted drug delivery to cancer cells have been implemented using the proposed devices, which have the potential to be used for drug screening and disease modeling in the future.

2.4.2 Stereolithography

2.4.2.1 Mechanism

Stereolithography is a traditional additive manufacturing technique, which has been widely used to fabricate complex 3D structures from various photosensitive (photocurable) polymers, elastomers, and composites. An SLA-based 3D printing system is generally composed of four parts: a laser source/generator, a scanner system (x- and y-movements), a photosensitive liquid bath, and a build platform. During printing, the laser source generates a laser beam, usually UV light or other visible light, and illuminates the beam to the scanner system. In the scanner, the laser beam is transferred into a light spot with a suitable size as the tool to cross-link the photosensitive liquid in the bath. In addition, the scanner also controls the movement of the light spot horizontally based on the designed path. Thus, one layer of polymerized photosensitive material can be printed into the required pattern on the build platform. The main function of the build platform is to control the movement of the printed structure along the vertical direction. Based on the moving direction, there are two strategies of SLA-based 3D printing: bottom-up and top-down. In the bottom-up strategy, the laser spot is illuminated to the photosensitive liquid bath atop the platform. As a result, the bottom layer of the printed 3D structure is cross-linked on the platform. Then the platform moves downward to receive the next layer of cross-linked material, as shown in Figure 2.10a. In the top-down strategy, the laser beam is illuminated to the bath below the platform. Thus, the top layer of the printed structure is created firstly on the platform. Then, the platform is controlled to move upward with a given step distance to receive the next layer of cross-linked material. By moving the platform along the vertical direction and simultaneously controlling the light spot path horizontally by the scanner, 3D structures can be printed in the photosensitive liquid bath in a layer-by-layer fashion as shown in Figure 2.10a.

Since SLA is a nozzle-free printing technique, the negative effects of shear pressure on cells can be effectively eliminated. Thus, SLA is one of the most cell-friendly 3D bioprinting techniques and has been increasingly used for cell printing applications

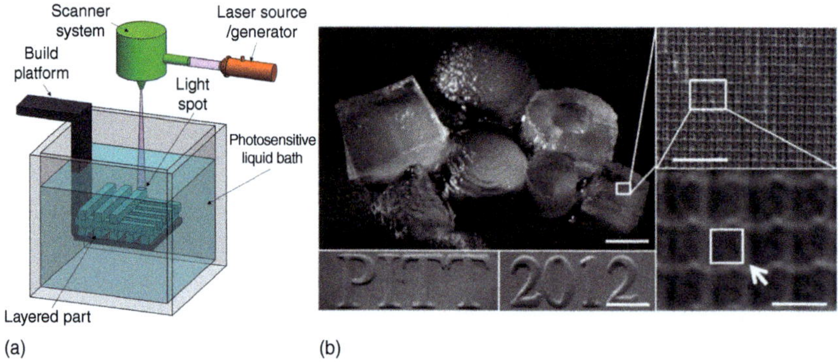

Figure 2.10 (a) Schematic of stereolithography bioprinting. Source: Yifei Jin. (b) Representative high-resolution scaffolds from PEG-based hydrogels. Source: Lin et al. [83]. Reproduced with permission of Springer Nature.

to fabricate 3D constructs with high cell viability. Generally, the printing speed of SLA is fast, and the resolution of printed structures is relatively high with a range of 5–300 μm [84, 85].

2.4.2.2 Materials

For 3D bioprinting applications, the photosensitive liquid bath is usually filled by UV curable hydrogel precursors and corresponding photoinitiators. The commonly used hydrogel precursors include PEG-based hydrogels and GelMA as aforementioned. Photoinitiators are chemical molecules, which can create reactive agents and initiate the photocuring process when exposed to light energy. By reacting with monomers in photocurable hydrogel precursors, polymer chains can be connected to form 3D networked microstructures, which transits the status of hydrogel from liquid to solid. Photoinitiators are sensitive to different ranges of wavelength. In 3D bioprinting cases, UV light is one of the most commonly used light sources. Thus, Irgacure 2959 [86] becomes the most popular photoinitiator for UV curable material printing applications, which has the least cytotoxicity. For example, Liew and Zhang [87] printed cell-laden fibers with diameters of 25–75 μm. In their study, they mixed HUVECs with GelMA, PEGMA, and Irgacure 2959 and used a UV beam to cross-link the hydrogel precursors and encapsulate cells. The cell viability after 3-day incubation was around 95%, which proved the cytocompatibility of the photoinitiator as well as the fabrication approach.

Except Irgacure 2959, Lithium phenyl-2,4,6-trimethylbenzoylphosphinate (LAP) [83, 88] is another commonly used UV-sensitive photoinitiator, which can create reactive agents under near-UV blue light [86]. For example, Lin et al. [83] prepared the liquid bath using human adipose-derived stem cells (hADSCs)-PEGDA/LAP suspension and 3D printed scaffold structure via SLA as shown in Figure 2.10b. They found that although near-UV blue light with the wavelength of 400–490 nm brought damages to mammalian cells, the encapsulated hADSCs presented high metabolic activity after printing and the metabolic activity even increased by 75% and 50% after 5-day and 7-day incubation, respectively [89].

When visible light is used for cross-linking hydrogel precursors, different photoinitiators should be selected. Among the available visible light-sensitive photoinitiators, eosin Y is the favorite which has cytotoxicity even lower than Irgacure 2959 [86]. This photoinitiator can generate reactive agents at the wavelength of 514 nm. Wang et al. [90] mixed PEG with eosin Y and GelMA to investigate the effects of GelMA on cell viability. They found that the samples without GelMA had the lowest cell viability, while the addition of 5% and 7.5% (w/v) GelMA can effectively increase the cell viability by improving the cell adhesion of the hydrogel matrix.

2.4.2.3 Biomedical Applications

Currently, SLA has been widely used for various biomedical applications. For example, Wang et al. successfully printed scaffolds from GelMA via visible light-based SLA [91]. In their study, they mixed NIH-3T3 fibroblasts with eosin Y as a photoinitiator and the scaffold was cross-linked at the wavelength of 522 nm. It was found that after 5-day incubation, most of the cells adhered to the printed scaffold.

2.4.3 Multi-Photon Polymerization

2.4.3.1 Mechanism

MPP is a type of laser-based 3D printing technology developed from SLA. But different from SLA, MPP operates by projecting laser light into a 3D voxel, instead of a 2D surface, within the polymerizable material bath. During printing, high-intensity pulsed lasers with femtosecond pulse duration are used to provide two or more photons, as shown in Figure 2.11a. Only the material at the laser's focal point undergoes polymerization by nonlinearly absorbing two or more photons. The surrounding regions are transparent to the wavelength of the used laser and cannot absorb enough energy to catalyze polymerization. By tracking the laser's focal point based on the path design, 3D structures can be free-form fabricated in the polymerizable material bath.

Since MPP is based on the multi-photon absorption phenomenon in which the probability of multi-photon absorption by a molecule is associated with the square of light intensity [94], MPP can fabricate 3D structures with single-cell resolution or even an extremely high resolution of 100 nm [95], making it promising for

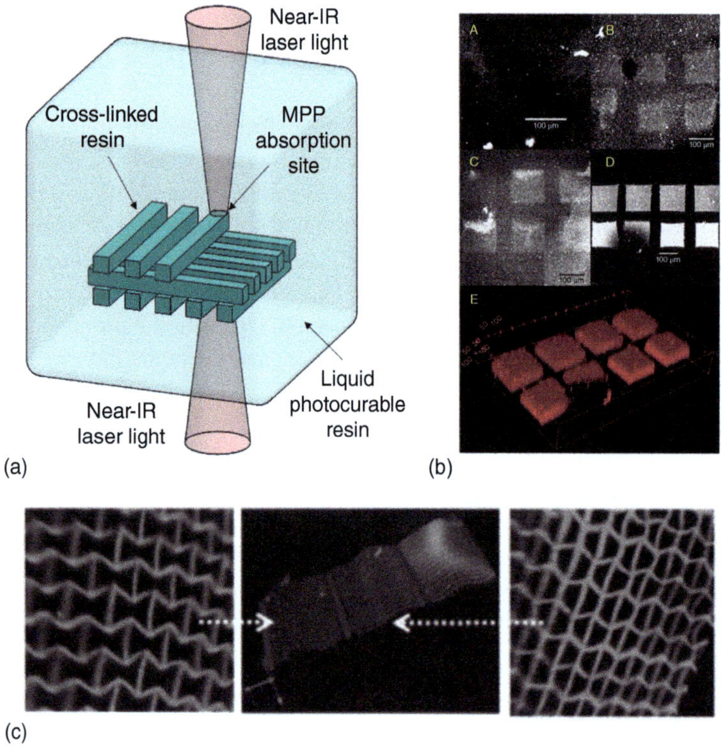

Figure 2.11 (a) Mechanism of MPP bioprinting. Source: Yifei Jin. (b) Collagen complex 3D structures fabricated via MPP bioprinting. Source: Bell et al. [92]. Reproduced with permission of Elsevier. (c) Suspended web structures from PEGDA with positive and negative Poisson's ratios. Source: Zhang et al. [93]. Reproduced with permission of Elsevier.

nanoscale and microscale printing applications. However, such a high resolution is traded off by limited printable sizes and printing speed. The typical printing speed of MPP-based printers falls in the range of 200–1600 mm/s, which is comparable to inkjet bioprinting and faster than extrusion bioprinting.

2.4.3.2 Materials

The material selections in MMP are severely constrained by its printing mechanism. Thus, only the biocompatible polymers with free-radical polymerization property can be used in MMP. Such polymers include type I collagen [92], bovine serum albumin [96], laminin [97], streptavidin [98], and PEG-based hydrogels [93]. For example, Bell et al. [92] printed complex 3D structures with a resolution of 1 μm from type I collagen with flavin mononucleotide photosensitizer, as shown in Figure 2.11b. Zhang et al. [93] used two-photon polymerization to fabricate suspended web structures from PEGDA with positive and negative Poisson's ratios, as shown in Figure 2.11c, which demonstrated biaxial expansion/compression behavior. The fabrication methodology can be used to tune the Poisson's ratio of several photocurable biomaterials and could have potential implications in the field of mechanobiology.

2.4.3.3 Biomedical Applications

The main biomedical application of MPP is to fabricate high-resolution scaffolds from synthetic and natural hydrogels for orienting cell proliferation. For example, Ma et al. [99] created a human diseased cardiac tissue model using MPP. In their study, they printed the filament array scaffold with filament diameters of 5 and 10 μm seeded iPSC-CMs on the scaffold. It was found that cells aligned along the vertical filaments and both healthy and diseased cardiomyocytes were observed. In addition, the dose responses to various drugs were recorded. Gao et al. printed GelMA scaffolds using MPP, which had the filament width of 15 μm and the filament length of 100 μm. After printing, iPSC-CMs, endothelial cells, and smooth muscle cells were seeded on the polymerized scaffold and the calcium transients and conduction velocities across the scaffold were analyzed. Finally, the scaffold samples were implanted on a mouse model. It was found that the samples can effectively improve ejection fraction and fractional shortening in four weeks [100], which indicated that 3D printed scaffolds by MPP have the potential to recover cardiac function.

2.4.4 Digital Light Projection 3D Printing

Digital light projection 3D printing (DLP-3DP) is a derived additive manufacturing technique from SLA, which is proposed in the recent five years. Different from SLA in which a single laser beam is raster-scanned across the surface of photopolymerizable hydrogel precursors, in DLP-3DP, a spatial light modulating (SLM) element is used to project a 2D plane of specifically patterned light as the basic building unit. Since Chapter 5 mainly focuses on DLP-3DP, only the basic concepts will be introduced briefly in this chapter.

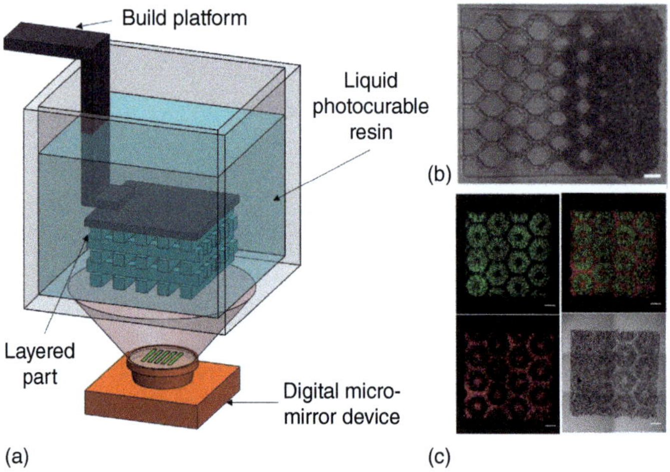

Figure 2.12 (a) Mechanism of digital light projection 3D printing. Source: Yifei Jin. (b) Bioprinted cellular construct with HUVECs and 10T1/2 (50 : 1) encapsulated in the intended channels. Source: Zhang et al. [93]. Reproduced with permission of Elsevier. (c) 3D bioprinted hepatic construct from GelMA. Source: Ma et al. [101]. Reproduced with permission of PNAS.

2.4.4.1 Mechanism

DLP-3DP system is commonly composed of a digital micro-mirror device (DMD), a polymerizable biomaterial bath, and a motorized translational stage, as shown in Figure 2.12a. The DMD chip consists of two million micro-mirrors. By controlling the on/off status of each micro-mirror, precise light projection patterns can be created. Thus, the illumination of these patterns by UV and/or other light sources onto the hydrogel precursors can cause the localized polymerization of the biomaterials in the bath to form a solid pattern on a receiving substrate. After that, the motorized translational stage moves the substrate along the vertical direction with a given step distance and the DMD chip illuminates the adjacent projection pattern based on design to solidify the adjacent layer of photopolymerizable biomaterial in the bath. By repeating these steps, a complex 3D structure can be fabricated in a layer-by-layer manner.

Generally, the resolution of 3D parts fabricated by DLP-3DP is at the microscale level, which can be further improved by decreasing the focal size of the light beam from each micro-mirror. In addition, the mechanical properties of the printed parts are better than those of the parts printed by inkjet bioprinting and extrusion bioprinting. During DLP-3DP, the entire layer is illuminated and polymerized at the same time and there is no interface in each layer. Thus, the inner-layer integration is much better than that in either inkjet bioprinting or extrusion bioprinting in which interfaces are formed between adjacent droplets or filaments in one layer. Also, since projection is performed layer by layer, the fabrication efficiency of DLP-3DP is higher than the conventional 3D bioprinting strategies such as inkjet bioprinting, extrusion bioprinting, and other light-based 3D bioprinting approaches (including MAPLE direct-write, SLA, and MPP).

2.4.4.2 Materials

Almost all of the commonly used photopolymerizable biocompatible hydrogels can be printed by DLP-3DP including GelMA [102, 103], PEGDA, and glycidyl methacrylate hyaluronic acid (GMHA). For example, Hribar et al. [104] used 20% (w/v) PEGDA as the photopolymerizable hydrogel to print concave structures for breast cancer cell spheroids and Ma et al. [101] printed hepatic models from GelMA and GelMA/GMHA blend, respectively.

2.4.4.3 Biomedical Applications

Due to its high fabrication efficiency and resolution, DLP-3DP is promising to 3D bioprint complex tissue constructs with microscale resolution. Currently, some representative 3D structures have been constructed such as vascular networks [105] (as shown in Figure 2.12b), cardiac scaffolds, and liver micro-models [101], as shown in Figure 2.12c.

2.4.5 Computed Axial Lithography

2.4.5.1 Mechanism

CAL is a derived DLP-based 3D printing strategy, which was proposed in 2019 [106]. The mechanism of polymerization in CAL is similar to that in DLP in which a DLP projector is used to deliver light energy to the material volume as a set of 2D images. But different from DLP, there is no receiving substrate used in CAL. Instead of translationally moving substrate along the vertical direction, the photopolymerizable material container rotates in CAL and each image projection propagates through the material from a different angle. Thus, the exposures from different angles can provide sufficient energy dose to cross-link the material into the desired geometry, as shown in Figure 2.13a. After a 360° rotation of the container, complex 3D structures are successfully printed *in situ*.

Since there is no translational movement in CAL, 3D parts are no longer printed in the conventional layer-by-layer fashion. Thus, no interface is generated between adjacent layers, making the printed structures have excellent mechanical properties. In addition, using spatial rotation to replace the translational movements along *x*-, *y*-, and *z*-directions can significantly improve the printing efficiency: large 3D structures with lateral sizes up to 55 mm can be printed in only 30–300 seconds based on the light intensities selected. As a result, this new 3D printing strategy is extremely promising for full-scale organ printing in the future.

2.4.5.2 Materials and Biomedical Applications

Currently, photocurable resin has been reported to print some proof-of-concept structures via CAL such as dental model, lattice geometry, and smooth sphere without layering artifacts as shown in Figure 2.13b, to name a few. But based on the cross-linking mechanism, all photocurable biocompatible hydrogels such as PEG-based hydrogels, GelMA, and MAHA can be printed into cell-laden living tissue constructs using CAL in the near future.

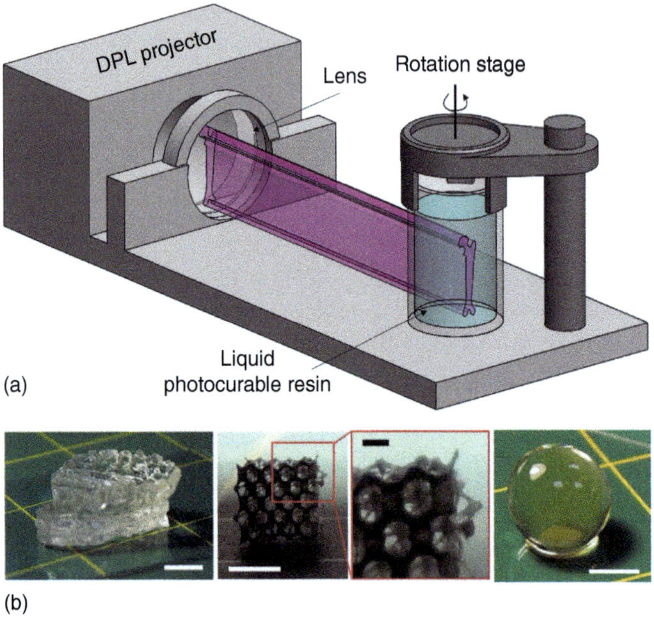

Figure 2.13 (a) Mechanism of computed axial lithography-based printing. Source: Yifei Jin. (b) Representative 3D structures including dental model, lattice, and smooth sphere. Source: Kelly et al. [106]. Reproduced with permission of The American Association for the Advancement of Science.

2.4.6 Summary

Light-based 3D bioprinting is becoming a promising cell printing methodology due to its high resolution, excellent cell viability, and high fabrication efficiency. In particular, DLP and CAL will be used as the mainstream printing approaches in the future since these two methods have the potential to rapidly and effectively print full-scale human organs/tissues with excellent mechanical properties. However, limited material selection is the main challenge for light-based 3D bioprinting. Thus, future investigations may focus on developing more photocurable biomaterials and biocompatible photoinitiators.

References

1 Murphy, S.V. and Atala, A. (2014). 3D bioprinting of tissues and organs. *Nature Biotechnology* 32 (8): 773–785.

2 Cui, H., Nowicki, M., Fisher, J.P., and Zhang, L.G. (2017). 3D bioprinting for organ regeneration. *Advanced Healthcare Materials* 6 (1): 1601118.

3 Mobaraki, M., Ghaffari, M., Yazdanpanah, A. et al. (2020). Bioinks and bioprinting: a focused review. *Bioprinting* 18: e00080.

4 Ma, X., Liu, J., Zhu, W. et al. (2018). 3D bioprinting of functional tissue models for personalized drug screening and in vitro disease modeling. *Advanced Drug Delivery Reviews* 132: 235–251.

5 Derby, B. (2008). Bioprinting: inkjet printing proteins and hybrid cell-containing materials and structures. *Journal of Materials Chemistry* 18 (47): 5717–5721.

6 Gudapati, H., Dey, M., and Ozbolat, I. (2016). A comprehensive review on droplet-based bioprinting: past, present and future. *Biomaterials* 102: 20–42.

7 Rayleigh, L. (1878). On the instability of jets. *Proceedings of the London Mathematical Society* 1 (1): 4–13.

8 Xu, T., Zhao, W., Zhu, J.M. et al. (2013). Complex heterogeneous tissue constructs containing multiple cell types prepared by inkjet printing technology. *Biomaterials* 34 (1): 130–139.

9 Xu, C., Chai, W., Huang, Y., and Markwald, R.R. (2012). Scaffold-free inkjet printing of three-dimensional zigzag cellular tubes. *Biotechnology and Bioengineering* 109 (12): 3152–3160.

10 Nishiyama, Y., Nakamura, M., Henmi, C. et al. (2009). Development of a three-dimensional bioprinter: construction of cell supporting structures using hydrogel and state-of-the-art inkjet technology. *Journal of Biomechanical Engineering* 131 (3).

11 Workman, V.L., Tezera, L.B., Elkington, P.T., and Jayasinghe, S.N. (2014). Controlled generation of microspheres incorporating extracellular matrix fibrils for three-dimensional cell culture. *Advanced Functional Materials* 24 (18): 2648–2657.

12 Guvendiren, M., Lu, H.D., and Burdick, J.A. (2012). Shear-thinning hydrogels for biomedical applications. *Soft Matter* 8 (2): 260–272.

13 Highley, C.B., Rodell, C.B., and Burdick, J.A. (2015). Direct 3D printing of shear-thinning hydrogels into self-healing hydrogels. *Advanced Materials* 27 (34): 5075–5079.

14 Liu, W., Heinrich, M.A., Zhou, Y. et al. (2017). Extrusion bioprinting of shear-thinning gelatin methacryloyl bioinks. *Advanced Healthcare Materials* 6 (12): 1601451.

15 Malda, J., Visser, J., Melchels, F.P. et al. (2013). 25th anniversary article: engineering hydrogels for biofabrication. *Advanced Materials* 25 (36): 5011–5028.

16 Jin, Y., Compaan, A., Bhattacharjee, T., and Huang, Y. (2016). Granular gel support-enabled extrusion of three-dimensional alginate and cellular structures. *Biofabrication* 8 (2): 025016.

17 Shi, W., Sun, M., Hu, X. et al. (2017). Structurally and functionally optimized silk-fibroin–gelatin scaffold using 3D printing to repair cartilage injury in vitro and in vivo. *Advanced Materials* 29 (29): 1701089.

18 Wong, S.S. and Wong, L.J.C. (1992). Chemical crosslinking and the stabilization of proteins and enzymes. *Enzyme and Microbial Technology* 14 (11): 866–874.

19 Shoulders, M.D. and Raines, R.T. (2009). Collagen structure and stability. *Annual Review of Biochemistry* 78: 929–958.

20 Roth, E.A., Xu, T., Das, M. et al. (2004). Inkjet printing for high-throughput cell patterning. *Biomaterials* 25 (17): 3707–3715.

21 Skardal, A., Mack, D., Kapetanovic, E. et al. (2012). Bioprinted amniotic fluid-derived stem cells accelerate healing of large skin wounds. *Stem Cells Translational Medicine* 1 (11): 792–802.

22 Cui, X. and Boland, T. (2009). Human microvasculature fabrication using thermal inkjet printing technology. *Biomaterials* 30 (31): 6221–6227.

23 Gurkan, U.A., El Assal, R., Yildiz, S.E. et al. (2014). Engineering anisotropic biomimetic fibrocartilage microenvironment by bioprinting mesenchymal stem cells in nanoliter gel droplets. *Molecular Pharmaceutics* 11 (7): 2151–2159.

24 Zalipsky, S. and Harris, J.M. (1997). Introduction to chemistry and biological applications of poly(ethylene glycol). In: *Poly(ethylene glycol) Chemistry and Biological Applications*, Chapter 1, 1–13. ACS Publications.

25 Cui, X., Breitenkamp, K., Lotz, M., and D'Lima, D. (2012). Synergistic action of fibroblast growth factor-2 and transforming growth factor-beta1 enhances bioprinted human neocartilage formation. *Biotechnology and Bioengineering* 109 (9): 2357–2368.

26 Boland, T., Tao, X., Damon, B.J. et al. (2007). Drop-on-demand printing of cells and materials for designer tissue constructs. *Materials Science and Engineering: C* 27 (3): 372–376.

27 Xu, T., Kincaid, H., Atala, A., and Yoo, J.J. (2008). High-throughput production of single-cell microparticles using an inkjet printing technology. *Journal of Manufacturing Science and Engineering* 130 (2).

28 Xu, T., Baicu, C., Aho, M. et al. (2009). Fabrication and characterization of bio-engineered cardiac pseudo tissues. *Biofabrication* 1 (3): 035001.

29 Christensen, K., Xu, C., Chai, W. et al. (2015). Freeform inkjet printing of cellular structures with bifurcations. *Biotechnology and Bioengineering* 112 (5): 1047–1055.

30 Kim, H.S., Lee, D.Y., Park, J.H. et al. (2007). Optimization of electrohydrodynamic writing technique to print collagen. *Experimental Techniques* 31 (4): 15–19.

31 Xu, T., Gregory, C.A., Molnar, P. et al. (2006). Viability and electrophysiology of neural cell structures generated by the inkjet printing method. *Biomaterials* 27 (19): 3580–3588.

32 Xie, M., Gao, Q., Zhao, H. et al. (2019). Electro-assisted bioprinting of low-concentration GelMA microdroplets. *Small* 15 (4): 1804216.

33 Gao, G., Yonezawa, T., Hubbell, K. et al. (2015). Inkjet-bioprinted acrylated peptides and PEG hydrogel with human mesenchymal stem cells promote robust bone and cartilage formation with minimal printhead clogging. *Biotechnology Journal* 10 (10): 1568–1577.

34 Cui, X., Breitenkamp, K., Finn, M.G. et al. (2012). Direct human cartilage repair using three-dimensional bioprinting technology. *Tissue Engineering Part A* 18 (11–12): 1304–1312.

35 Gao, G., Schilling, A.F., Yonezawa, T. et al. (2014). Bioactive nanoparticles stimulate bone tissue formation in bioprinted three-dimensional scaffold and human mesenchymal stem cells. *Biotechnology Journal* 9 (10): 1304–1311.

36 Matsusaki, M., Sakaue, K., Kadowaki, K., and Akashi, M. (2013). Three-dimensional human tissue chips fabricated by rapid and automatic inkjet cell printing. *Advanced Healthcare Materials* 2 (4): 534–539.

37 Yanez, M., Rincon, J., Dones, A. et al. (2015). In vivo assessment of printed microvasculature in a bilayer skin graft to treat full-thickness wounds. *Tissue Engineering Part A* 21 (1–2): 224–233.

38 Khalil, S. and Sun, W. (2007). Biopolymer deposition for freeform fabrication of hydrogel tissue constructs. *Materials Science and Engineering: C* 27 (3): 469–478.

39 Chang, C.C., Boland, E.D., Williams, S.K., and Hoying, J.B. (2011). Direct-write bioprinting three-dimensional biohybrid systems for future regenerative therapies. *Journal of Biomedical Materials Research Part B: Applied Biomaterials* 98 (1): 160–170.

40 Fedorovich, N.E., Swennen, I., Girones, J. et al. (2009). Evaluation of photocrosslinked lutrol hydrogel for tissue printing applications. *Biomacromolecules* 10 (7): 1689–1696.

41 Chang, R., Nam, J., and Sun, W. (2008). Effects of dispensing pressure and nozzle diameter on cell survival from solid freeform fabrication-based direct cell writing. *Tissue Engineering Part A* 14 (1): 41–48.

42 Cohen, D.L., Malone, E., Lipson, H.O.D., and Bonassar, L.J. (2006). Direct freeform fabrication of seeded hydrogels in arbitrary geometries. *Tissue Engineering* 12 (5): 1325–1335.

43 Jakab, K., Damon, B., Neagu, A. et al. (2006). Three-dimensional tissue constructs built by bioprinting. *Biorheology* 43 (3, 4): 509–513.

44 Visser, J., Peters, B., Burger, T.J. et al. (2013). Biofabrication of multi-material anatomically shaped tissue constructs. *Biofabrication* 5 (3): 035007.

45 Ouyang, L., Highley, C.B., Sun, W., and Burdick, J.A. (2017). A generalizable strategy for the 3D bioprinting of hydrogels from nonviscous photo-crosslinkable inks. *Advanced Materials* 29 (8): 1604983.

46 Li, C., Wang, K., Zhou, X. et al. (2019). Controllable fabrication of hydroxybutyl chitosan/oxidized chondroitin sulfate hydrogels by 3D bioprinting technique for cartilage tissue engineering. *Biomedical Materials* 14 (2): 025006.

47 Wilkinson, N.J., Smith, M.A.A., Kay, R.W., and Harris, R.A. (2019). A review of aerosol jet printing—a non-traditional hybrid process for micro-manufacturing. *The International Journal of Advanced Manufacturing Technology* 105 (11): 4599–4619.

48 Gao, Q., He, Y., Fu, J.Z. et al. (2015). Coaxial nozzle-assisted 3D bioprinting with built-in microchannels for nutrients delivery. *Biomaterials* 61: 203–215.

49 Jin, Y., Compaan, A., Chai, W., and Huang, Y. (2017). Functional nanoclay suspension for printing-then-solidification of liquid materials. *ACS Applied Materials & Interfaces* 9 (23): 20057–20066.

50 Khalil, S. and Sun, W. (2009). Bioprinting endothelial cells with alginate for 3D tissue constructs. *Journal of Biomechanical Engineering* 131 (11).

51 Levato, R., Visser, J., Planell, J.A. et al. (2014). Biofabrication of tissue constructs by 3D bioprinting of cell-laden microcarriers. *Biofabrication* 6 (3): 035020.

52 Mironov, V., Visconti, R.P., Kasyanov, V. et al. (2009). Organ printing: tissue spheroids as building blocks. *Biomaterials* 30 (12): 2164–2174.

53 Mehesz, A.N., Brown, J., Hajdu, Z. et al. (2011). Scalable robotic biofabrication of tissue spheroids. *Biofabrication* 3 (2): 025002.

54 Yu, Y. and Ozbolat, I.T. (2014). Tissue strands as "bioink" for scale-up organ printing. In *2014 36th Annual International Conference of the IEEE Engineering in Medicine and Biology Society* (pp. 1428–1431). IEEE.

55 Pati, F., Jang, J., Ha, D.H. et al. (2014). Printing three-dimensional tissue analogues with decellularized extracellular matrix bioink. *Nature Communications* 5 (1): 1–11.

56 Zhang, Y., Venugopal, J.R., El-Turki, A. et al. (2008). Electrospun biomimetic nanocomposite nanofibers of hydroxyapatite/chitosan for bone tissue engineering. *Biomaterials* 29 (32): 4314–4322.

57 Ma, L., Gao, C., Mao, Z. et al. (2003). Collagen/chitosan porous scaffolds with improved biostability for skin tissue engineering. *Biomaterials* 24 (26): 4833–4841.

58 Hong, Y., Song, H., Gong, Y. et al. (2007). Covalently crosslinked chitosan hydrogel: properties of in vitro degradation and chondrocyte encapsulation. *Acta Biomaterialia* 3 (1): 23–31.

59 Hao, T., Wen, N., Cao, J.K. et al. (2010). The support of matrix accumulation and the promotion of sheep articular cartilage defects repair in vivo by chitosan hydrogels. *Osteoarthritis and Cartilage* 18 (2): 257–265.

60 Huang, J., Fu, H., Wang, Z. et al. (2016). BMSCs-laden gelatin/sodium alginate/carboxymethyl chitosan hydrogel for 3D bioprinting. *RSC Advances* 6 (110): 108423–108430.

61 Liu, Q., Li, Q., Xu, S. et al. (2018). Preparation and properties of 3D printed alginate–chitosan Polyion complex hydrogels for tissue engineering. *Polymers* 10 (6): 664.

62 Oxlund, H. and Andreassen, T.T. (1980). The roles of hyaluronic acid, collagen and elastin in the mechanical properties of connective tissues. *Journal of Anatomy* 131 (Pt 4): 611.

63 Jeon, O., Song, S.J., Lee, K.J. et al. (2007). Mechanical properties and degradation behaviors of hyaluronic acid hydrogels cross-linked at various cross-linking densities. *Carbohydrate Polymers* 70 (3): 251–257.

64 Gerecht, S., Burdick, J.A., Ferreira, L.S. et al. (2007). Hyaluronic acid hydrogel for controlled self-renewal and differentiation of human embryonic stem cells. *Proceedings of the National Academy of Sciences* 104 (27): 11298–11303.

65 Xi, L., Wang, T., Zhao, F. et al. (2014). Evaluation of an injectable thermosensitive hydrogel as drug delivery implant for ocular glaucoma surgery. *PLoS One* 9 (6): e100632.

66 Norotte, C., Marga, F.S., Niklason, L.E., and Forgacs, G. (2009). Scaffold-free vascular tissue engineering using bioprinting. *Biomaterials* 30 (30): 5910–5917.

67 Wu, W., DeConinck, A., and Lewis, J.A. (2011). Omnidirectional printing of 3D microvascular networks. *Advanced Materials* 23 (24): H178–H183.

68 Homenick, C.M., de Silveira, G., Sheardown, H., and Adronov, A. (2011). Pluronics as crosslinking agents for collagen: novel amphiphilic hydrogels. *Polymer International* 60 (3): 458–465.

69 Skardal, A., Sarker, S.F., Crabbé, A. et al. (2010). The generation of 3-D tissue models based on hyaluronan hydrogel-coated microcarriers within a rotating wall vessel bioreactor. *Biomaterials* 31 (32): 8426–8435.

70 Curran, S.J., Chen, R., Curran, J.M., and Hunt, J.A. (2005). Expansion of human chondrocytes in an intermittent stirred flow bioreactor, using modified biodegradable microspheres. *Tissue Engineering* 11 (9–10): 1312–1322.

71 Malda, J., Van Blitterswijk, C.A., Grojec, M. et al. (2003). Expansion of bovine chondrocytes on microcarriers enhances redifferentiation. *Tissue Engineering* 9 (5): 939–948.

72 Achilli, T.M., Meyer, J., and Morgan, J.R. (2012). Advances in the formation, use and understanding of multi-cellular spheroids. *Expert Opinion on Biological Therapy* 12 (10): 1347–1360.

73 Zhang, Z., Xiong, R., Corr, D.T., and Huang, Y. (2016). Study of impingement types and printing quality during laser printing of viscoelastic alginate solutions. *Langmuir* 32 (12): 3004–3014.

74 Xiong, R., Zhang, Z., Chai, W. et al. (2015). Freeform drop-on-demand laser printing of 3D alginate and cellular constructs. *Biofabrication* 7 (4): 045011.

75 Xiong, R., Zhang, Z., Chai, W. et al. (2017). Study of gelatin as an effective energy absorbing layer for laser bioprinting. *Biofabrication* 9 (2): 024103.

76 Xiong, R., Chai, W., and Huang, Y. (2019). Laser printing-enabled direct creation of cellular heterogeneity in lab-on-a-chip devices. *Lab on a Chip* 19 (9): 1644–1656.

77 Kattamis, N.T., Purnick, P.E., Weiss, R., and Arnold, C.B. (2007). Thick film laser induced forward transfer for deposition of thermally and mechanically sensitive materials. *Applied Physics Letters* 91 (17): 171120.

78 Kingsley, D.M., Roberge, C.L., Rudkouskaya, A. et al. (2019). Laser-based 3D bioprinting for spatial and size control of tumor spheroids and embryoid bodies. *Acta Biomaterialia* 95: 357–370.

79 Sorkio, A., Koch, L., Koivusalo, L. et al. (2018). Human stem cell based corneal tissue mimicking structures using laser-assisted 3D bioprinting and functional bioinks. *Biomaterials* 171: 57–71.

80 Colina, M., Serra, P., Fernández-Pradas, J.M. et al. (2005). DNA deposition through laser induced forward transfer. *Biosensors and Bioelectronics* 20 (8): 1638–1642.

81 Dinca, V., Kasotakis, E., Catherine, J. et al. (2008). Directed three-dimensional patterning of self-assembled peptide fibrils. *Nano Letters* 8 (2): 538–543.

82 Gruene, M., Deiwick, A., Koch, L. et al. (2011). Laser printing of stem cells for biofabrication of scaffold-free autologous grafts. *Tissue Engineering Part C: Methods* 17 (1): 79–87.

83 Lin, H., Tang, Y., Lozito, T.P. et al. (2017). Projection stereolithographic fabrication of BMP-2 gene-activated matrix for bone tissue engineering. *Scientific Reports* 7 (1): 1–11.

84 Raman, R., Bhaduri, B., Mir, M. et al. (2016). High-resolution projection microstereolithography for patterning of neovasculature. *Advanced healthcare materials* 5 (5): 610–619.

85 Yue, K., Trujillo-de Santiago, G., Alvarez, M.M. et al. (2015). Synthesis, properties, and biomedical applications of gelatin methacryloyl (GelMA) hydrogels. *Biomaterials* 73: 254–271.
86 Mondschein, R.J., Kanitkar, A., Williams, C.B. et al. (2017). Polymer structure-property requirements for stereolithographic 3D printing of soft tissue engineering scaffolds. *Biomaterials* 140: 170–188.
87 Liew, W.L.A. and Zhang, Y. (2018). Laser-based fabrication of 3D hydrogel constructs using bessel beams. *Bioprinting* 9: 44–51.
88 Xu, H., Casillas, J., Krishnamoorthy, S., and Xu, C. (2020). Effects of Irgacure 2959 and lithium phenyl-2,4,6-trimethylbenzoylphosphinate on cell viability, physical properties and microstructure in 3D bioprinting of vascular-like constructs. *Biomedical Materials* 15: 055021.
89 Smith, S., Maclean, M., MacGregor, S.J., et al. (2009). Exposure of 3T3 mouse fibroblasts and collagen to high intensity blue light. In *13th International Conference on Biomedical Engineering* (pp. 1352–1355). Berlin, Heidelberg: Springer.
90 Wang, Z., Abdulla, R., Parker, B. et al. (2015). A simple and high-resolution stereolithography-based 3D bioprinting system using visible light crosslinkable bioinks. *Biofabrication* 7 (4): 045009.
91 Wang, Z., Tian, Z., Jin, X. et al. (2017). Visible light-based stereolithography bioprinting of cell-adhesive gelatin hydrogels. *2017 39th Annual International Conference of the IEEE Engineering in Medicine and Biology Society (EMBC)* (pp. 1599–1602). IEEE.
92 Bell, A., Kofron, M., and Nistor, V. (2015). Multiphoton crosslinking for biocompatible 3D printing of type I collagen. *Biofabrication* 7 (3): 035007.
93 Zhang, W., Soman, P., Meggs, K. et al. (2013). Tuning the Poisson's ratio of biomaterials for investigating cellular response. *Advanced Functional Materials* 23 (25): 3226–3232.
94 Zipfel, W.R., Williams, R.M., and Webb, W.W. (2003). Nonlinear magic: multiphoton microscopy in the biosciences. *Nature Biotechnology* 21 (11): 1369–1377.
95 Truby, R.L. and Lewis, J.A. (2016). Printing soft matter in three dimensions. *Nature* 540 (7633): 371–378.
96 Engelhardt, S., Hoch, E., Borchers, K. et al. (2011). Fabrication of 2D protein microstructures and 3D polymer–protein hybrid microstructures by two-photon polymerization. *Biofabrication* 3 (2): 025003.
97 Su, P.J., Tran, Q.A., Fong, J.J. et al. (2012). Mesenchymal stem cell interactions with 3D ECM modules fabricated via multiphoton excited photochemistry. *Biomacromolecules* 13 (9): 2917–2925.
98 Wylie, R.G., Ahsan, S., Aizawa, Y. et al. (2011). Spatially controlled simultaneous patterning of multiple growth factors in three-dimensional hydrogels. *Nature Materials* 10 (10): 799–806.
99 Ma, Z., Koo, S., Finnegan, M.A. et al. (2014). Three-dimensional filamentous human diseased cardiac tissue model. *Biomaterials* 35 (5): 1367–1377.
100 Gao, L., Kupfer, M.E., Jung, J.P. et al. (2017). Myocardial tissue engineering with cells derived from human-induced pluripotent stem cells and a native-like, high-resolution, 3-dimensionally printed scaffold. *Circulation Research* 120 (8): 1318–1325.

101 Ma, X., Qu, X., Zhu, W. et al. (2016). Deterministically patterned biomimetic human iPSC-derived hepatic model via rapid 3D bioprinting. *Proceedings of the National Academy of Sciences* 113 (8): 2206–2211.

102 Wadnap, S., Krishnamoorthy, S., Zhang, Z., and Xu, C. (2019). Biofabrication of 3D cell-encapsulated tubular constructs using dynamic optical projection stereolithography. *Journal of Materials Science: Materials in Medicine* 30: 36.

103 Krishnamoorthy, S., Wadnap, S., Noorani, B. et al. (2020). Investigation of gelatin methacrylate working curves using dynamic optical projection stereolithography. *European Polymer Journal* 124: 109487.

104 Hribar, K.C., Finlay, D., Ma, X. et al. (2015). Nonlinear 3D projection printing of concave hydrogel microstructures for long-term multicellular spheroid and embryoid body culture. *Lab on a Chip* 15 (11): 2412–2418.

105 Zhu, W., Qu, X., Zhu, J. et al. (2017). Direct 3D bioprinting of prevascularized tissue constructs with complex microarchitecture. *Biomaterials* 124: 106–115.

106 Kelly, B.E., Bhattacharya, I., Heidari, H. et al. (2019). Volumetric additive manufacturing via tomographic reconstruction. *Science* 363 (6431): 1075–1079.

3

Bioink Design: From Shape to Function

3.1 Significance of Bioink Design

Three-dimensional (3D) bioprinting provides an emerging approach for the construction of mimic biological tissues by accurate deposition of cell-laden bioinks. Since the concept of bioprinting was put forward, the research on bioink design, printing strategies, and equipment has emerged one after another. However, most of the printed structures are only morphologically similar to internal tissues, and the cells in the printed structures have a simple function, which is far from the complex physiological function of the real organ. To achieve a functional breakthrough of bioprinting, the key is to choose suitable bioinks. Bioink, which is a biomaterial carrier that wraps cells is the raw material for forming 3D structure in bioprinting and constructs a 3D microenvironment for cell growth. To achieve cell functionalization and tissue formation, bioink must have superior biocompatibility and adjustable physical and chemical properties. At the same time, to form the tissue structure of printing, bioink should have strong printability and fluidity, and can quickly solidify and maintain the final structure after printing [1–7]. Determining the best formulation of bioink is a key step in the success of 3D bioprinting, and the appropriate bioink should be taken into account for both, forming the 3D structure and functional realization of cells [8].

3.2 Categories of Bioink

At present, the biomaterial carriers used as bioink are mainly natural and synthetic biomaterials with specific functions. Natural bioink materials mainly include animal sources and nonanimal sources. Animal-derived natural bioink materials are mainly derived from the animal extracellular matrix (ECM), such as decellularized extracellular matrix (dECM), collagen, fibrin, matrigel, hyaluronic acid, and gelatin. Bioinks made of natural biomaterials usually have good biocompatibility, but their mechanical properties are usually poor, which is not conducive to the implementation of bioprinting. Natural bioink materials from nonanimal sources are mainly polysaccharides, e.g. alginate and agarose. They have excellent mechanical properties and good printability, but their biological properties are not good

Cell Assembly with 3D Bioprinting, First Edition. Yong He, Qing Gao, and Yifei Jin.

enough, which is not conducive for the realization of cell and tissue function. The main synthetic bioink materials are polyethylene glycol (PEG), pluronic F127, and a series of modified photo-cross-linked bioinks mainly used at present, such as polyethylene glycol dipropylene (PEGDA), hyaluronic acid methacryloyl (HAMA), silk fibroin glycidyl methacrylate (Sil-MA), methacryloylated gelatin (GelMA). Modified photo-cross-linked biological hydrogel is a widely used and potential bioink material at present, which is mostly modified by photo-cross-linking of natural biomaterials with excellent biological properties. It not only maintains the original excellent biological properties, but also has adjustable gelation mechanical properties, which is the best choice for bioink.

3.3 Three Evaluation Criteria of Bioink

3.3.1 Printability

Printability refers to the ability of bioink to produce 3D structures with high structural integrity and fidelity. Among them, shape fidelity is the most important index in the evaluation of printability. In order to print structures with high shape fidelity, it is quite essential to study the printability of the bioink. In other words, for a new bioink, in addition to designing printing strategy according to bioink properties, the manufacturing process should also be deeply studied to obtain a suitable formulation of bioink and corresponding printing parameters for successfully printing complex scaffolds. For example, in extrusion-based bioprinting, the manufacturing process can be divided into three key steps, namely, (i) extruding the ink to form filaments, (ii) depositing the filaments to form basic units, and (iii) stacking the filaments to form 3D structures. Therefore, printability analysis was carried out according to the above three steps to determine the optimal printing parameters for fabricating structures with good shape fidelity.

3.3.2 Mechanical Properties

The mechanical properties of bioink are closely related to its application, cytocompatibility, and so on. The bioink with weak mechanical properties is suitable for 3D culture of loaded cells, and the bioink with strong mechanical properties is suitable for tissue engineering. The mechanical properties of bioink are mainly evaluated by the rheological and compression tests. The rheological test mainly includes the viscoelastic behavior and shear thinning properties of bioink (Figure 3.1). The compression test mainly aims at the compressive properties and Young's modulus of bioink.

3.3.3 Biocompatibility

Biocompatibility in bioprinting refers to the positive and controllable influence on the structure of biological and functional components. These effects include interaction with endogenous tissue or immune system, supporting appropriate cellular

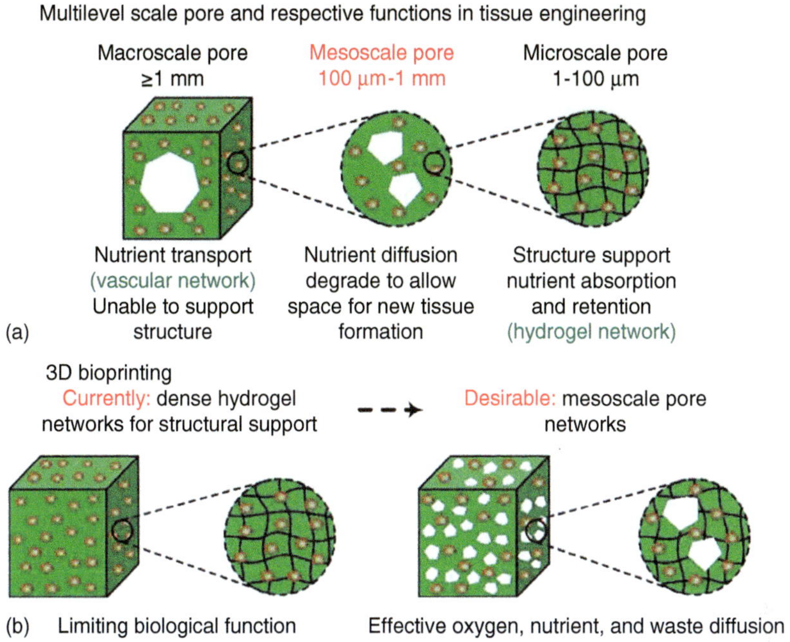

Figure 3.1 (a) Multilevel scale pore and respective functions in tissue engineering. (b) Current situation and expectation of pore networks in 3D bioprinting [9]. Source: Shao et al. [9]. Reproduced with permission of Springer Nature.

activity, and promoting molecular or mechanical signaling systems, which are critical for the biological functioning of the graft. The biocompatibility in bioprinting has gradually changed from bioink not causing any adverse local or systemic reactions in the host to passively allowing or actively producing the desired effect in the host. Therefore, the study of the biocompatibility of bioink is essential to determine whether it can be used in tissue engineering. The biocompatibility test of bioink mainly includes live/dead staining to observe cell activity, CCK-8 reagent test of cell proliferation, and other characterization of cellular functionalization.

3.4 Strategies for Enabling the Printability

3.4.1 Optimization of Cross-linking Sequence

To enhance the printability of bioink, optimization of cross-linking sequence can be done. At present, the main strategies used can be summarized as two-step cross-linking strategy [10], pre-cross-linking strategy [11, 12] and in situ cross-linking strategy [12]. For example, two-step cross-linking strategy was used to print alginate/gelatin bioink. In this process, the bioink was temporarily shaped by the temperature cross-linking of gelatin, and then permanently shaped by ion cross-linking of alginate. Pre-cross-linking strategy was used to print GelMA-based bioink, in which the viscosity of GelMA was improved by temperature cross-linking of GelMA, so that it can reach the printable state. In situ cross-linking strategy was achieved by using a transparent glass tube connected at the end of the nozzle. The light-cured bioink with low viscosity and good fluidity was extruded through the glass tube and microfiber was formed.

3.4.2 Support Material-Assisted Bioprinting

For low-viscosity bioink such as collagen and fibrin, support material-assisted bioprinting (also called suspension printing) [13–18] can be applied by using a suspension medium to support the printed structure. The realization of suspension printing has a great relationship with the characteristics of suspension media, which shows solid characteristics when no external force is applied or the external force is very small, to realize the self-support of the printing structure. The yield stress caused by the movement of the printing nozzle causes the flow of the suspended medium, and the suspended medium shows the characteristics of the liquid. After the movement of printing nozzle, due to the self-healing of the suspension medium, the microstructure of the printing nozzle can be recovered spontaneously to ensure the realization of printing. Support material-assisted bioprinting makes up for the shortcomings of traditional extrusion printing in the construction of complex and high-resolution structures.

3.4.3 Microgel-Based Bioink

Another strategy to improve the printability can be realized by designing a microgel-based bioink [9]. Microgel is a kind of hydrogel in the form of micron particles (Figure 3.1), which is valuable as a rheological improver and can improve the printability of bioink. Some hydrogel can be processed into microgels, including alginate, hyaluronic acid, PEG diacrylate, and agarose gel. At present, most studies choose gelatin hydrogel as microgel material, because filled gelatin microgel not only has the same cellular responsiveness as gelatin itself, but also can improve the viscosity and yield stress behavior of bioink.

3.5 Strategies for Bioink Reinforcement

3.5.1 Composite Bioink Design

In general, bioink is composed of low-viscosity natural or synthetic biocompatible hydrogels for biocompatibility. However, pure hydrogels usually have poor mechanical and structural stability, and they often degrade too quickly, which limits their use for long-term culture or transfer to the body [19]. Therefore, composite bioink [20] is designed by adding different nanoparticles or anisotropic fillers to improve its mechanical properties, such as graphene, graphene oxide, carbon nanotubes (CNTs), hydroxyapatite (HAP) and other calcium phosphates, bioactive glass, silica nanoparticles, and clay.

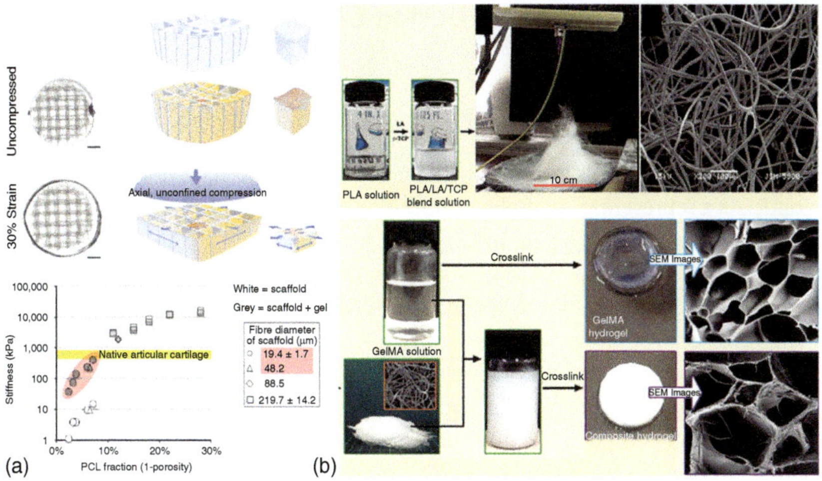

Figure 3.2 (a) The mechanism of hydrogel reinforcement with organized high-porosity scaffolds [21]. Source: Gilbert et al. [21]. Reproduced with permission of Elsevier.
(b) Schematic illustration showing the fabrication of pure gelatin methacrylate hydrogel scaffold and nanofiber-assembled gelatin methacrylate composite hydrogel scaffolds [22]. Source: Joshi et al. [22]. Reproduced with permission of Elsevier.

3.5.2 Microfiber-Assisted Reinforcement

In addition to mixing some reinforced materials, the mechanical properties of the bioink can be improved by adding some reinforced structures inside the hydrogel. In general, microfibers were chosen as the reinforced structures, which can not only improve their mechanical properties but also guide cell migration and affect cell differentiation (Figure 3.2). Because of its versatility and flexibility, electrospinning technology is widely used to prepare polymer microfibers with different diameters. At present, there are two strategies to enhance mechanical properties by using microfiber-assisted reinforcement, one is to coat hydrogel on the electrospun film to form a multilayer structure. The other is to cut the ultrafine fibers into micron sizes and then mix them into hydrogel scaffolds [21, 22].

3.6 Strategies for Improving the Biocompatibility

In the process of bioprinting, it is necessary to ensure the biocompatibility of bioink while ensuring the mechanical properties. dECM is a natural component in human body, it can affect cell differentiation, proliferation, adhesion, and morphological changes. Therefore, the biocompatibility of bioink can be improved by mixing dECM with hydrogel. Over the last few years, some researches about dECM have been reported. The related applications includes skin, vascular tissue, heart valves, and bladder.

3.7 Representative Bioink Design Case: GelMA-Based Bioinks

3.7.1 Property Characterization of the GelMA Bioink

Recently, natural hydrogels modified with methacryloyl (MA) have been widely used as bioink. As a representative of photo-cross-linking modified hydrogel, GelMA is a double bond modified from gelatin (Figure 3.3a), which has the characteristics of both natural and synthetic biomaterials. It can be seen from the nuclear magnetic spectrum (Figure 3.3b) that there are two more absorption peaks between the chemical shift 5.23–5.65 ppm, which proved that MA was successfully grafted onto the gelatin molecule, and the signal intensity of the peak of the double bond increased with the increase of the degree of substitution. To develop an appropriate printing strategy for GelMA-based bioink, it is necessary to understand the relevant characteristics of GelMA. In these figures, we summarize several important properties of GelMA, including viscosity, sol–gel transition temperature, sol–gel transition time, mechanical strength, and biocompatibility.

Figure 3.4a–d shows the effect of substitution rates and concentrations on the viscosity of GelMA. Although the value of viscosity increases with increasing the substitution rate and concentration of GelMA, it is still at a low level.

Figure 3.5 shows the effect of substitution rates and concentrations on the sol–gel transition temperature of GelMA. It can be seen that the gel point temperature of GelMA decreases with increasing the substitution rate, whereas increases with increasing the concentration.

Figure 3.6 shows the effect of substitution rates and concentrations on the photo-cross-linking process of GelMA. It can be seen that storage modulus (G′) increases with increasing the substitution rate and concentration, and sol–gel point is within five to ten seconds.

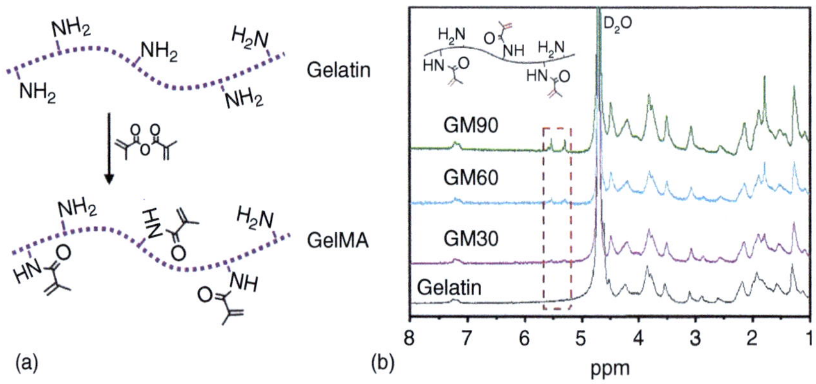

Figure 3.3 (a) The synthesis principle of GelMA. (b) The nuclear magnetic spectrum of GelMA at different substitution rates (30%, 60%, and 90%).

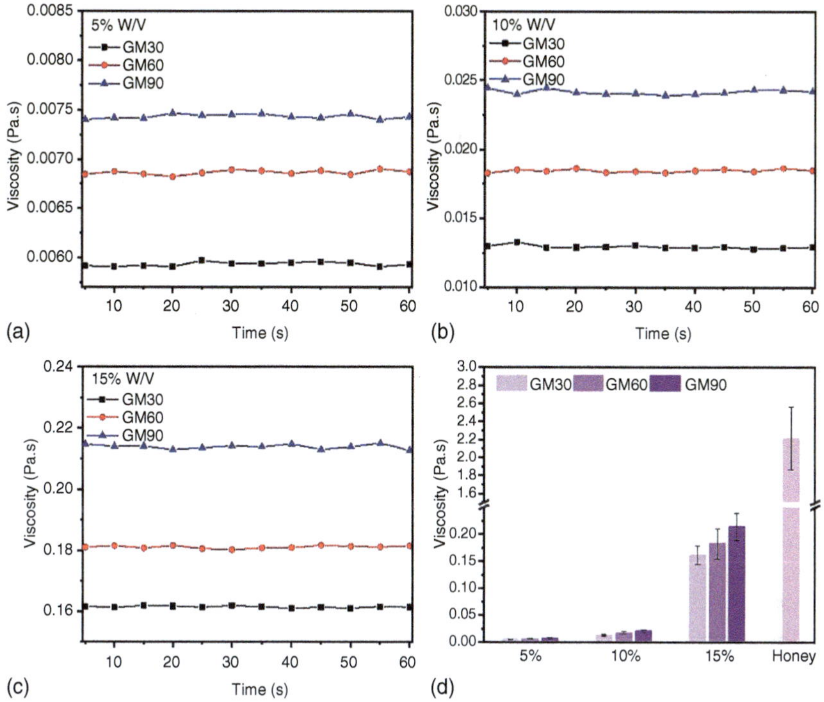

Figure 3.4 The effect of substitution rate and concentration on the viscosity of GelMA. (a) The viscosity of 5% W/V GelMA increases with increasing substitution rate. (b) The viscosity of 10% W/V GelMA increases with increasing substitution rate. (c) The viscosity of 15% W/V GelMA increases with increasing substitution rate. (d) The viscosity of GelMA increases with increasing concentration.

Figure 3.7 shows the effect of substitution rates and concentrations on the compression modulus of GelMA. It can be seen that the compression modulus increases with increasing the substitution rate and concentration.

Because GelMA has a similar bioaffinity to gelatin, it can be used for the culture of most cells, including stem cells, cancer cells, endothelial cells, and nerve cells, as shown in Table 3.1. Figure 3.8 shows that BMSCs maintained a good growth state when seeded on the surface of GelMA and encapsulated within GelMA.

3.7.2 3D Bioprinting of GelMA Bioinks with Dual Cross-linking Strategy

GelMA has both temperature behavior and photosensitive behavior. In the process of bioprinting, GelMA is a soft hydrogel and is not conducive to forming. To solve this problem, the double cross-linking strategy was presented (Figure 3.9). Specifically, after completing the 3D bioprinting of the GelMA hydrogel, the printed hydrogel structure can be temporarily shaped by the low-temperature environment, but once the low-temperature environment stops working, the printed hydrogel

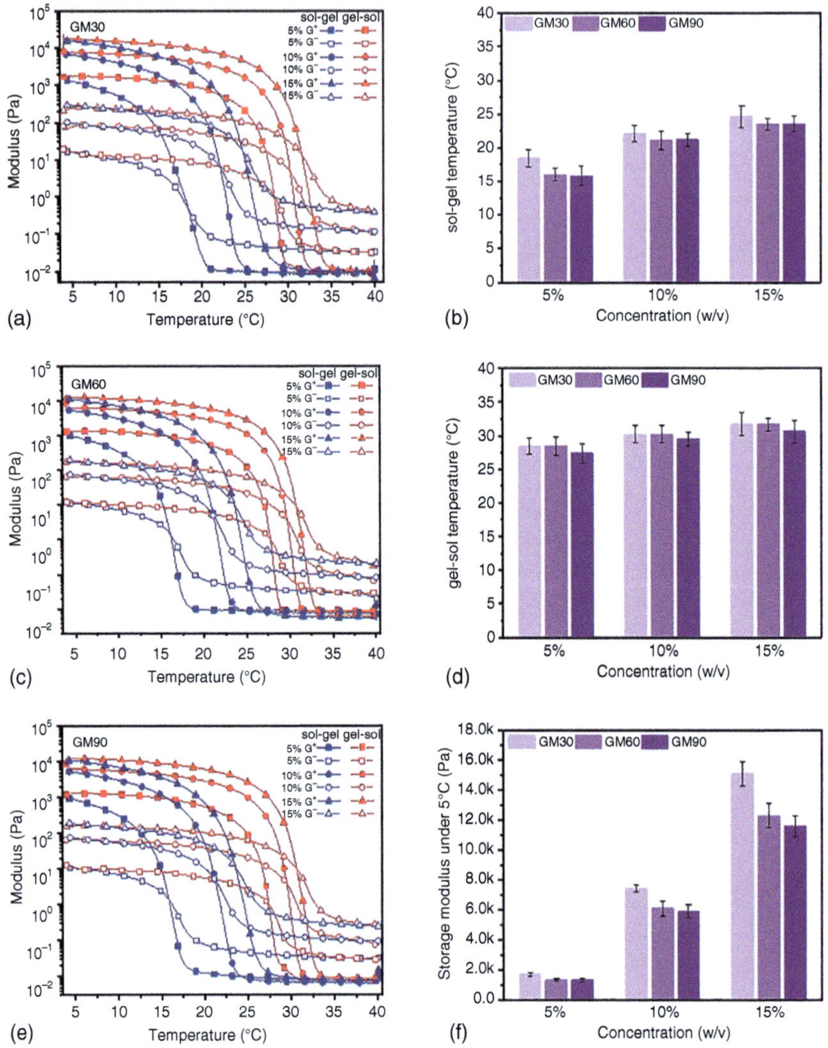

Figure 3.5 The effect of substitution rates and concentrations on the sol–gel transition temperature of GelMA. (a) The sol–gel transition temperature of GM30 increases with increasing concentration. (b) The sol–gel transition temperature of GM30/GM60/GM90 increases with increasing concentration. (c) The sol–gel transition temperature of GM60 increases with increasing concentration. (d) The gel-sol transition temperature of GM30/GM60/GM90 increases with increasing concentration. (e) The sol–gel transition temperature of GM90 increases with increasing concentration. (f) The storage modulus under 5 °C of GM30/GM60/GM90 increases with increasing concentration.

structure will gradually be destroyed because of its own temperature-sensitive cross-linking reversibility. Therefore, the printed structure needs to be cross-linked in the second step to achieve the permanent setting of the hydrogel printing structure. The structure of the hydrogel not only contains the triple helix formed by low-temperature cross-linking, the chemical cross-linking bond formed by the

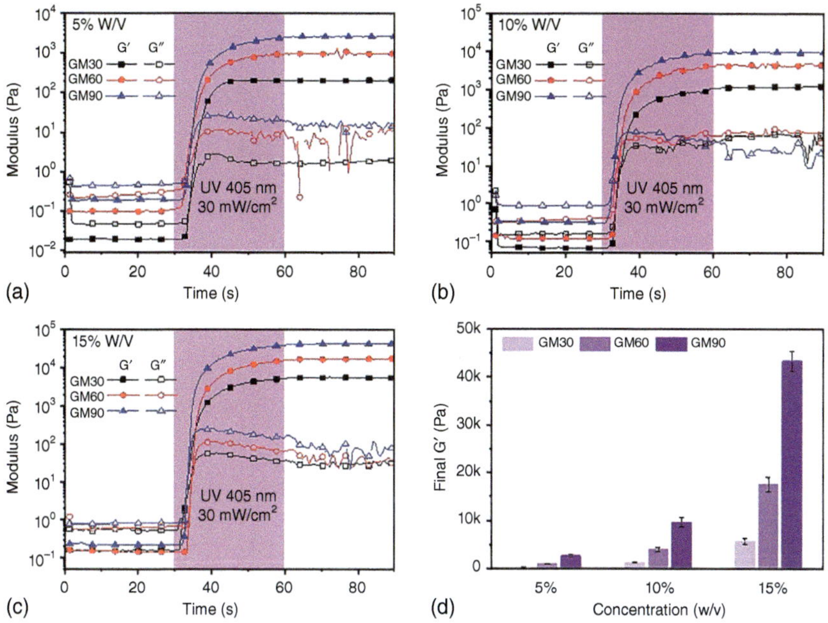

Figure 3.6 The effect of substitution rates and concentrations on the photo-cross-linking process of GelMA. (a) The modulus of 5% W/V GelMA increases with increasing substitution rate. (b) The modulus of 10% W/V GelMA increases with increasing substitution rate. (c) The modulus of 15% W/V GelMA increases with increasing substitution rate. (d) The final storage modulus (G′) of GelMA increases with increasing concentration.

second step of light cross-linking and the triple helix are combined to further enhance the strength of the printing structure based on the original. Using this strategy, the strength of the GelMA structure is greatly improved, which maintained a relatively stable energy storage modulus and mechanical strength in the culture process.

3.7.3 3D Bioprinting of GelMA Bioinks with Nanoclay as Support

Because of the limitations of low viscosity and long cross-linking time, it is still a challenge to use extrusion-based 3D printing to print GelMA directly. To balance printability and biocompatibility, this can be achieved by using bioink composed of GelMA and nanoclay [23]. The composite scaffold with bionic structures can be easily printed by using the thixotropy of nanoclay. Bioinks made of GelMA and nanoclay are specially designed to balance the printability and biocompatibility because nanoclay is a phase change material related to shear stress and has no additional sol–gel transition conditions (Figure 3.10), making it ideal for extrusion-based bioprinting.

In these studies, GelMA/nanoclay were characterized and their printing properties were studied. By testing the printing window, porosity, mechanical strength, and biocompatibility of GelMA/nanoclay bioink, it can be found that the addition of nanoclay improves the porosity of the material, increases the mechanical strength

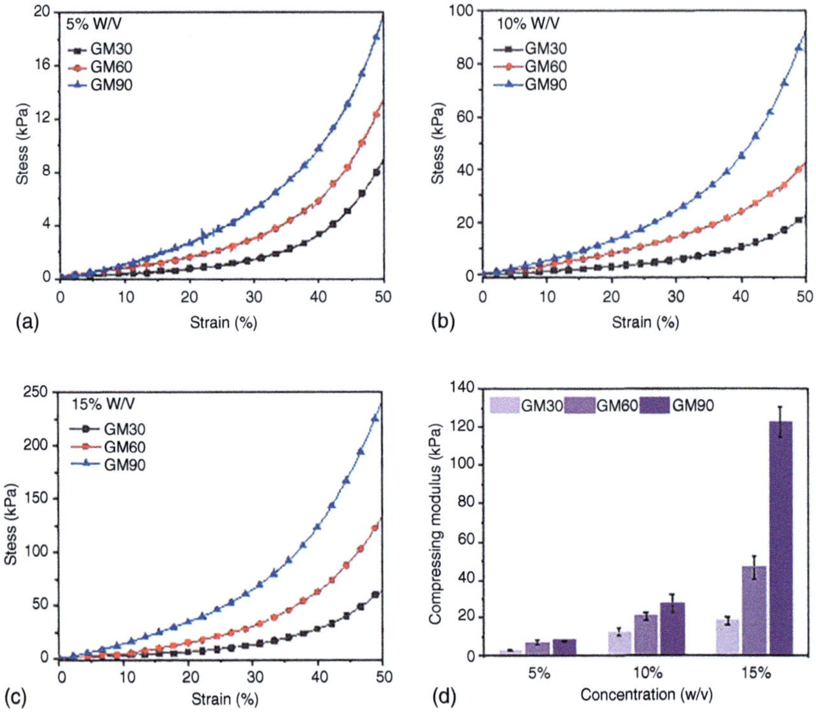

Figure 3.7 The effect of substitution rates and concentrations on compression modulus of GelMA. (a) The stress of 5% W/V GelMA increases with increasing substitution rate. (b) The stress of 10% W/V GelMA increases with increasing substitution rate. (c) The stress of 15% W/V GelMA increases with increasing substitution rate. (d) The compression modulus of GM30/GM60/GM90 increases with increasing concentration.

Table 3.1 Representative cell type that can be cultured with GelMA.

Cell category	Name
Stem cells	Bone marrow mesenchymal stem cells (BSMC)
	Adipose mesenchymal stem cells (ADSC)
	Dental pulp stem cells (DPSC)
	Neural crest stem cells (NCSC)
Primary normal cell	Cartilage progenitor cells
	Human umbilical vein endothelial cells (HUVEC)
	Cardiac myocyte
Normal cell line	Rat cerebral vascular endothelial cells (RBE4)
	Porcine hip artery endothelial cells (PIEC)
	Mouse embryonic osteoblast progenitor cells (MC3T3-E1)
	Rat hepatocytes (BRL.3A)
	Mouse chondroblasts (ADTC5)
Cancer cell line	Breast cancer cells (MDA-MB-231)
	Rat adrenal chromaffin cells (PC-12)
	Human lung cancer cells (A549)

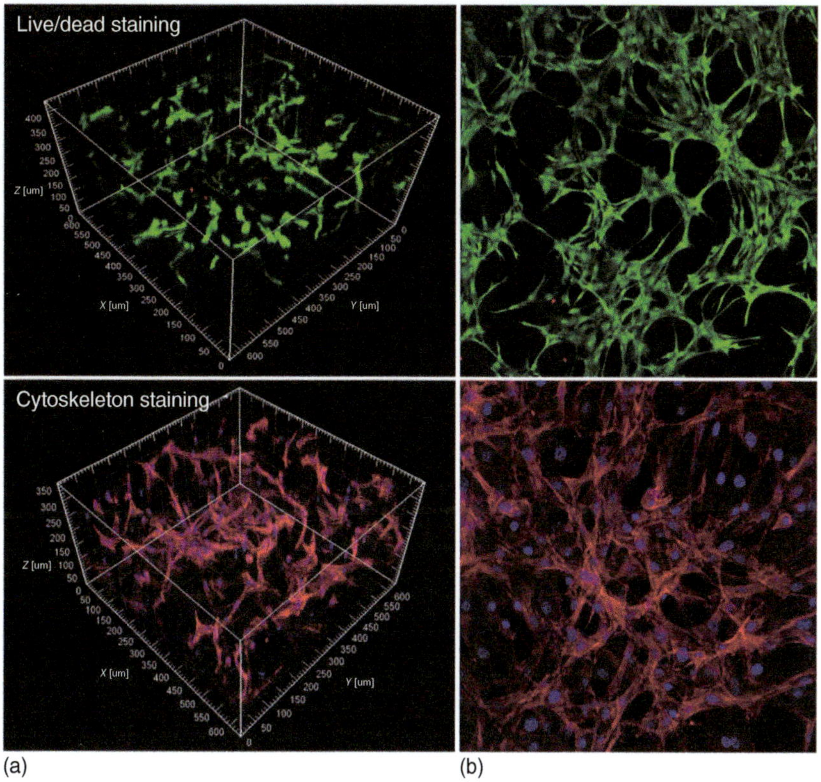

Figure 3.8 (a) The growth state of BMSCs encapsulated within GelMA. (b) The growth state of BMSCs seeded on the surface of GelMA.

of the material, reduces the degradation rate, and maintains good biocompatibility. Therefore, GelMA/nanoclay bioink provides a simple method for the preparation of composite scaffolds with realistic shape and good biological properties and opens up a new application prospect for individual treatment of tissue defects. These studies made the nanoclay-based printing strategy more practical to fabricate 3D GelMA-based scaffolds with good shape fidelity, which will have potential applications for the customized therapy of tissue defects.

3.8 Commercial Bioink

At present, the bioink used in scientific research is usually synthesized in small batches, its performance between batches is unstable due to which the output is small, and the repeatability of the experiment used for 3D bioprinting is not high, which brings a lot of inconvenience to researchers. With the growth of demand for bioinks, commercial bioink appears to grow. As photo-cross-linking bioink has the advantages of convenient operation, adjustable physical and chemical properties, and good biocompatibility, it is widely used in cell 3D culture, 3D bioprinting, tissue

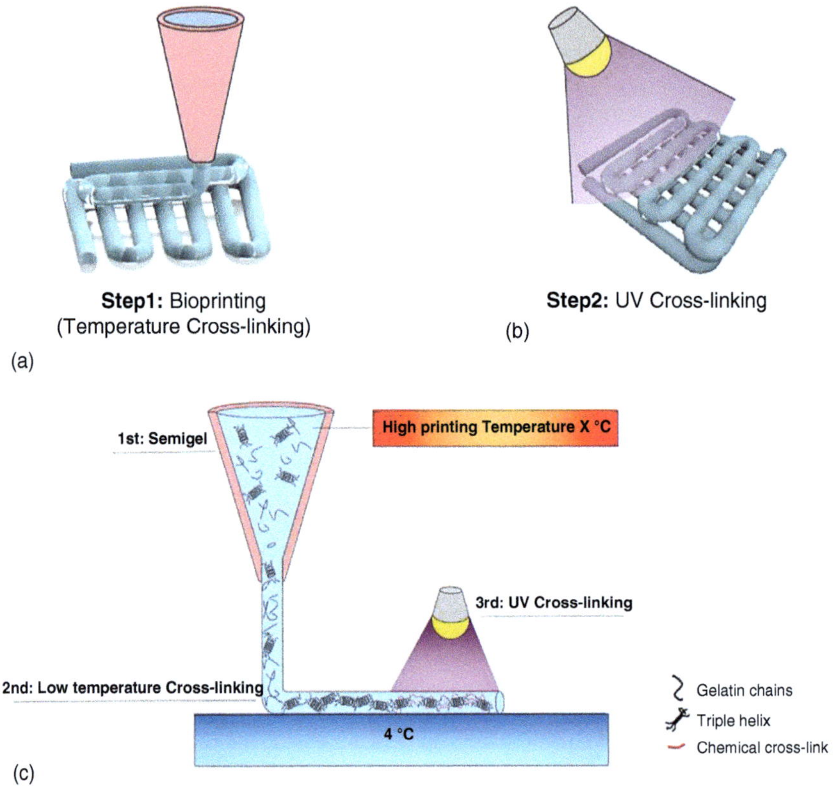

Figure 3.9 (a) Cross-linking steps in the process of bioprinting GelMA bioinks. (b) The double cross-linking strategy of bioink.

repair, and other fields. At present, there are many commercial photo-cross-linking bioink in the market, among them, photo-cross-linking bioink of EFL brand is widely used, as shown in Table 3.2. The following sections introduce the product information and applications of several typical photo-cross-linking bioink.

3.8.1 GelMA (EFL-GM Series)

GelMA of EFL can be photo-cross-linked into a gel in 10 seconds under visible light irradiation of 405 nm, and they can be applied to various bioprinting technology. Additionally, tissue structures with different strengths can be printed by choosing different product types (Figure 3.11).

3.8.2 Fluorescent GelMA (EFL-GM-F Series)

Fluorescent GelMA is achieved by grafting fluorescent molecules on GelMA and has a specific fluorescent color due to the different fluorescent molecules. Fluorescent GelMA not only has the excellent properties of GelMA, but also the characteristic of autofluorescence. Therefore, it has a broad application prospect

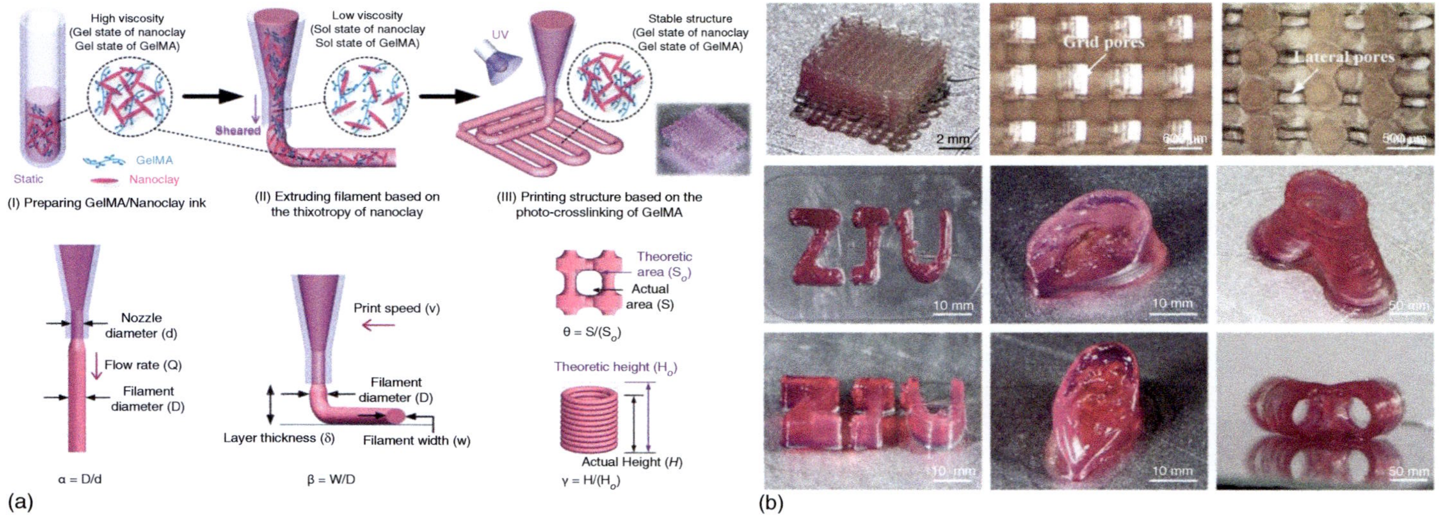

Figure 3.10 (a) 3D printing strategy of complex scaffolds with GelMA/nanoclay ink. (b) Photograph of the printed scaffolds with various shapes [23]. Source: Gao et al. [23]. Reproduced with permission of IOP Publishing.

Table 3.2 Photo-cross-linking bioink of EFL.

Name	Product type
Methacryloyl gelatin (GelMA)	EFL-GM-30
	EFL-GM-60
	EFL-GM-90
Red fluorescent methacryloyl gelatin (GelMA)	EFL-GM-RF-30
	EFL-GM-RF-60
	EFL-GM-RF-90
Green fluorescent methacryloyl gelatin (GelMA)	EFL-GM-GF-30
	EFL-GM-GF-60
	EFL-GM-GF-90
Blue fluorescent methacryloyl gelatin (GelMA)	EFL-GM-BF-30
	EFL-GM-BF-60
	EFL-GM-BF-90
Porous methacryloyl gelatin (GelMA)	EFL-GM-PR-001
	EFL-GM-PR-002
Methacryloyl silk fibroin (SilMA)	EFL-SilMA-001
Methacryloyl hyaluronic acid (HAMA)	EFL-HAMA-150K
	EFL-HAMA-400K
Methacryloyl chondroitin sulfate (ChSMA)	EFL-ChSMA-001
Methacryloyl sodium alginate (AlgMA)	EFL-AlgMA-50K
	EFL-AlgMA-300K
Methacryloyl chitosan (CSMA)	EFL-CSMA-100K
	EFL-CSMA-300K
Methacryloyl carboxymethyl chitosan (CMCSMA)	EFL-CMCSMA-200K
Methacryloyl dextran (DexMA)	EFL-DexMA-70K
	EFL-DexMA-200K
Methacryloyl polycaprolactone (PCLMA)	EFL-PCLMA-3200

in the biosensor, disease diagnosis, fluorescence tracer, bionic drive, and other research fields (Figure 3.12).

3.8.3 Porous GelMA (EFL-GM-PR Series)

Except for GelMA (EFL-GM series) and fluorescent GelMA (EFL-GM-F series), the EFL team also provided porous GelMA (EFL-GM-PR series). Porous GelMA can improve the nutrient exchange efficiency and provide more space for cell growth and proliferation (Figure 3.13).

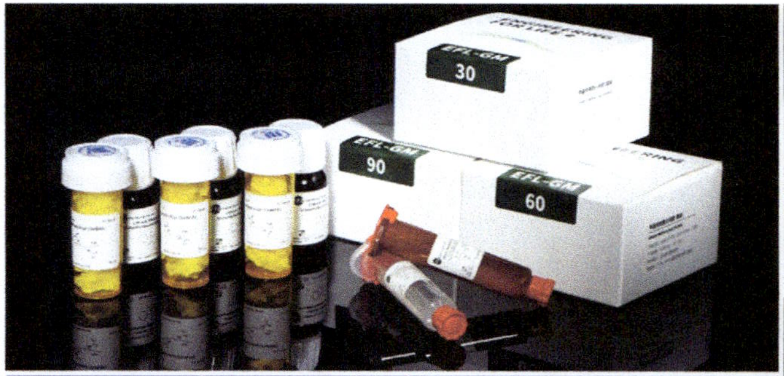

Type	Degree of substitution	Molecular weight	Turbidity
EFL-GM-30	30±5%	100 ~ 200 kDa	≤ 20 NTU
EFL-GM-60	60±5%		
EFL-GM-90	90±5%		

(a)

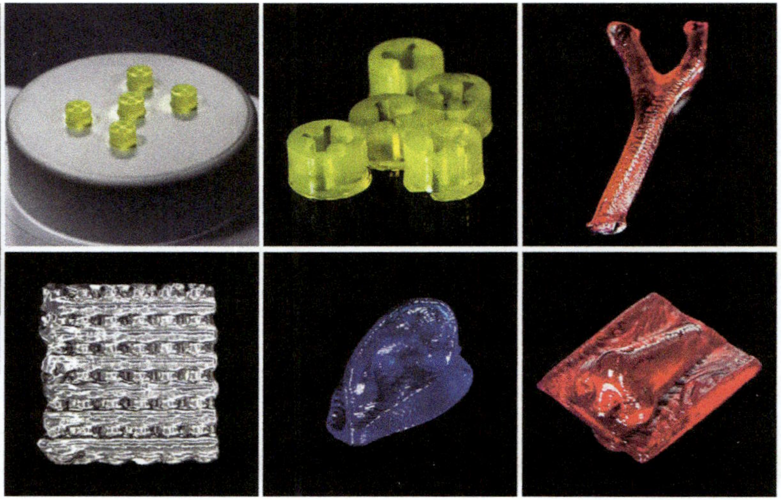

(b)

Figure 3.11 Information of GelMA (EFL-GM series). (a) Product type. (b) Printed structures.

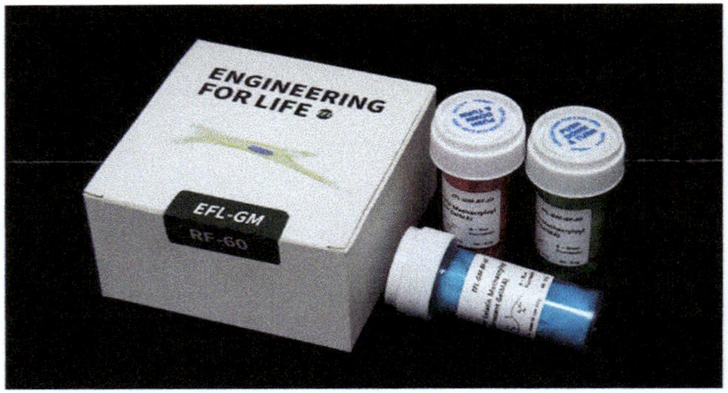

Type	Maximum excitation wavelength	Maximum emission wavelength	Color
EFL-GM-BF	429 nm	495 nm	Blue
EFL-GM-GF	492 nm	568 nm	Green
EFL-GM-RF	552 nm	618 nm	Red

(a)

(b)

Figure 3.12 Information of fluorescent GelMA (EFL-GM-F series). (a) Product type. (b) Printed structures.

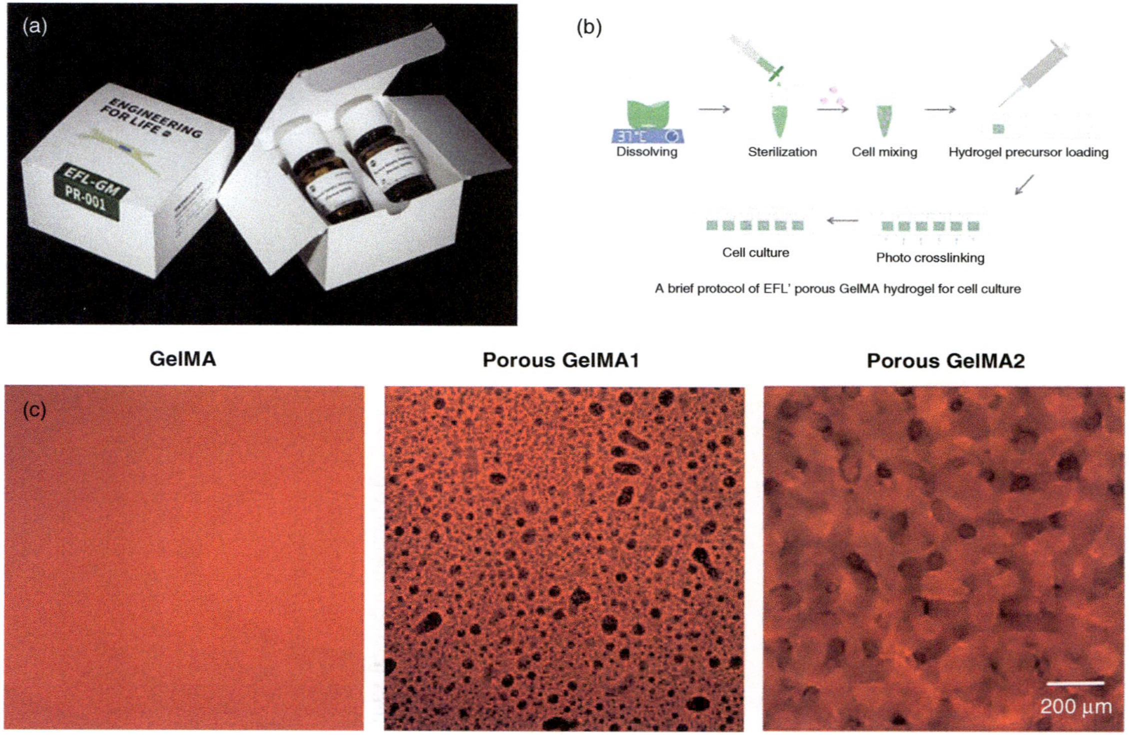

Figure 3.13 Information of porous GelMA (EFL-GM-PR series). (a) Product type. (b) Use steps. (c) GelMA with different pores.

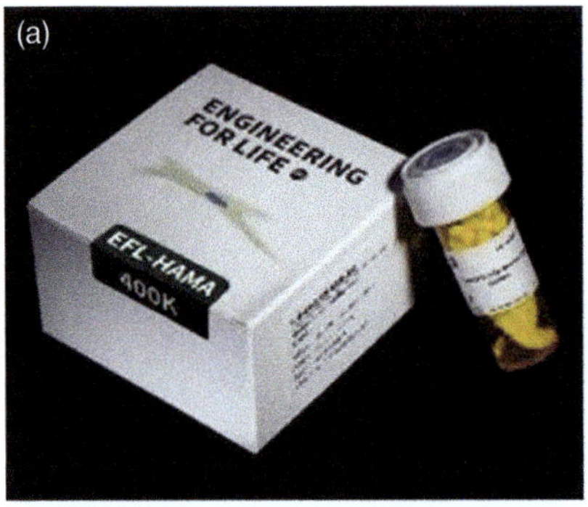

Type	Molecular weight	Viscosity
EFL-HAMA-150K	100-150 kDa	Medium
EFL-HAMA-400K	300-400 kDa	High

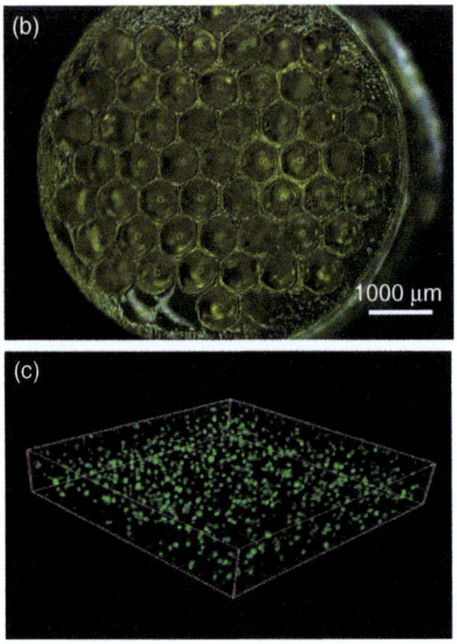

Figure 3.14 Information of porous HAMA (EFL-HAMA series). (a) Product type. (b) Printed structure. (c) The growth state of cells encapsulated within HAMA.

3.8.4 HAMA (EFL-HAMA Series)

HAMA is obtained by introducing the methacrylic group into the molecular chain of hyaluronic acid. Because of its good water-locking and moisturizing properties, it is widely used in many areas, such as cartilage regeneration, tumor model construction, microneedle preparation, wound dressings, and so on (Figure 3.14).

3.8.5 SilMA (EFL-SilMA Series)

Methacryloyl silk fibroin (SilMA) is obtained by modifying the double bond into the molecular chain of SF. It can be photo-cross-linked in 10 seconds under visible light irradiation and has good biocompatibility and strong expandability. Additionally, it can also provide a variety of viscoelastic properties for different applications (Figure 3.15).

3.8.6 PCLMA (EFL-PCLMA Series)

Methacryloyl polycaprolactone (PCLMA) is obtained by modifying double bond into the molecular chain of PCL. PCLMA has good fluidity at 40~50 °C, and has rapid photo-cross-linking properties and good biocompatibility. It can be used to print complex structures by digital light processing-based 3D bioprinting (Figure 3.16).

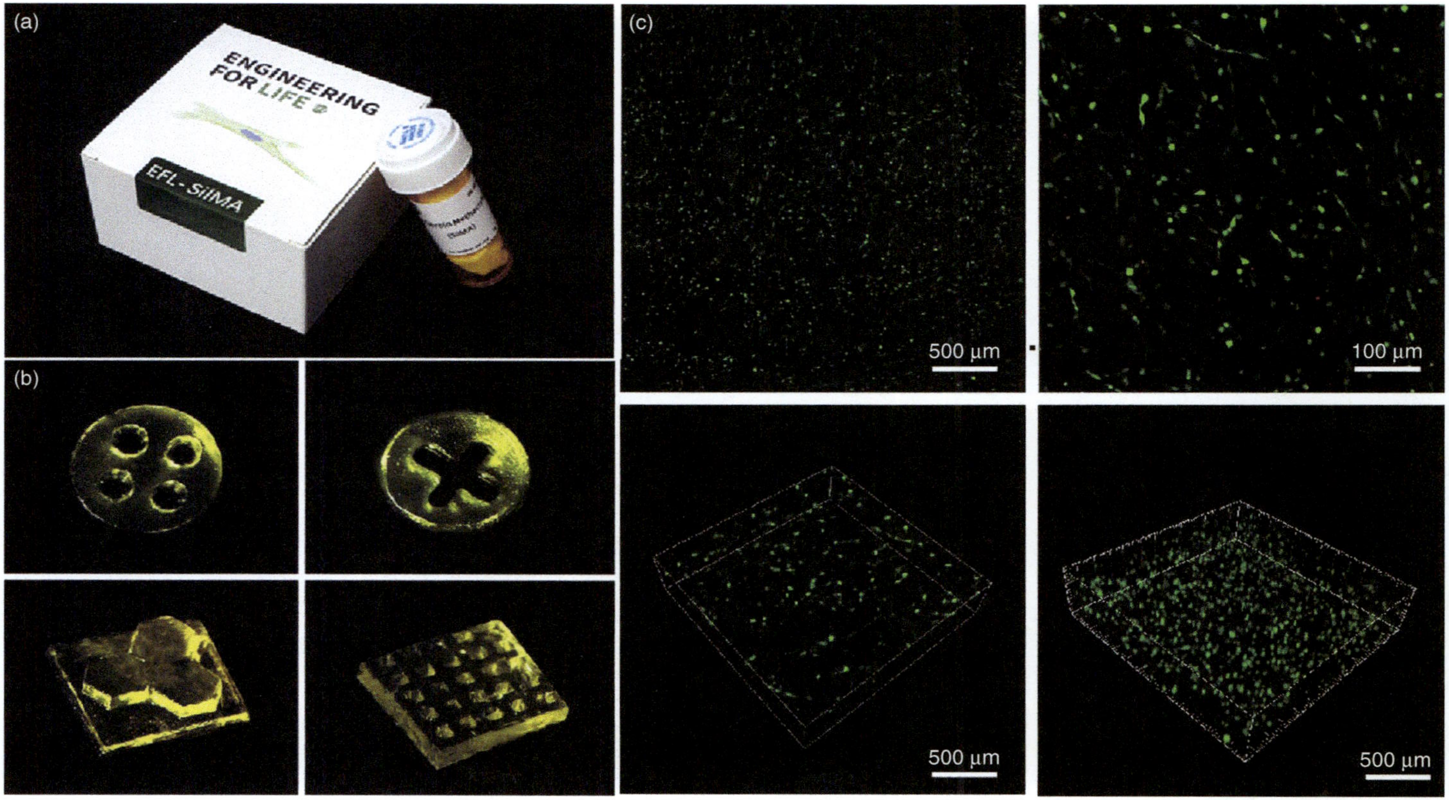

Figure 3.15 Information of SilMA (EFL-SilMA series). (a) Product type. (b) Printed structure. (c) The growth state of cells encapsulated within SilMA.

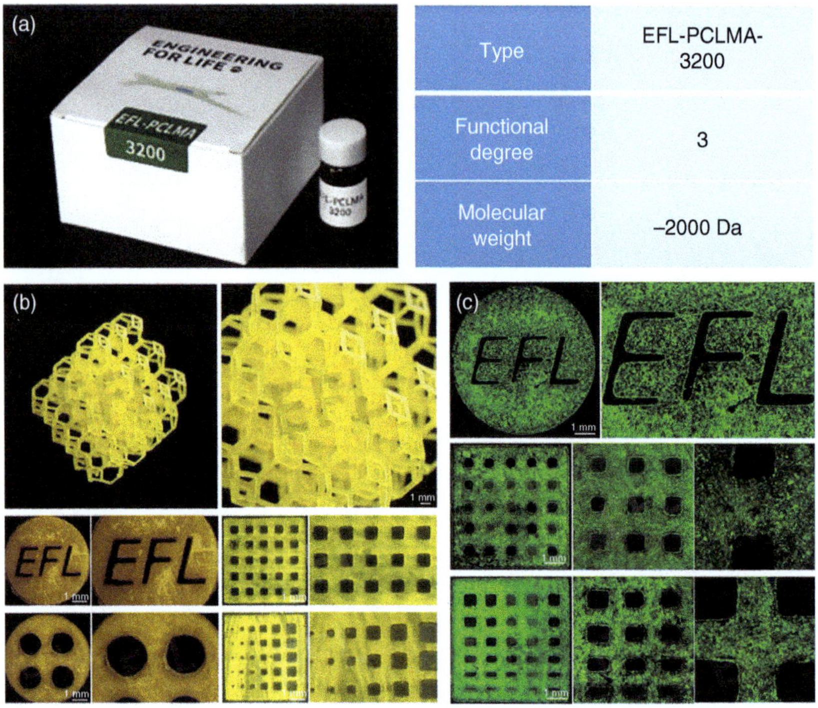

Figure 3.16 Information of PCLMA (EFL-PCLMA series). (a) Product type. (b) Printed structure. (c) The growth state of cells seeded on the surface of PCLMA scaffolds.

References

1. Hölzl, K., Lin, S., Tytgat, L. et al. (2016). Bioink properties before, during and after 3D bioprinting. *Biofabrication* 3: 32002.
2. He, Y., Yang, F.-F., Zhao, H.-M. et al. (2016). Research on the printability of hydrogels in 3D bioprinting. *Scientific Reports* 29977, 6.
3. Hospodiuk, M., Dey, M., Sosnoski, D., and Ozbolat, I.-T. (2017). The bioink: a comprehensive review on bioprintable materials. *Biotechnology Advances* 2: 217–239.
4. Chung, J.-H.-Y., Naficy, S., Yue, Z. et al. (2013). Bio-ink properties and printability for extrusion printing living cells. *Biomaterial Science* 7.
5. Ferris, C.-J., Gilmore, K.-J., Beirne, S. et al. (2013). Bio-ink for on-demand printing of living cells. *Biomaterial Science* 1.
6. Chimene, D., Kaunas, R., and Gaharwar, A.-K. (2020). Hydrogel bioink reinforcement for additive manufacturing: a focused review of emerging strategies. *Advanced Materials* 1.
7. Gungor-Ozkerim, P.-S., Inci, I., Zhang, Y.-S. et al. (2018). Bioinks for 3D bioprinting: an overview. *Biomaterials Science* 6.
8. Groll, J., Burdick, J.-A., Cho, D.-W. et al. (2019). A definition of bioinks and their distinction from biomaterial inks. *Biofabrication* 1: 13001.

9 Shao, L., Gao, Q., Xie, C. et al. (2020). Sacrificial microgel-laden bioink-enabled 3D bioprinting of mesoscale pore networks. *Bio-Design and Manufacturing.*
10 Colosi, C., Costantini, M., Barbetta, A., and Dentini, M. (2017). Microfluidic bioprinting of heterogeneous 3D tissue constructs. *Methods in Molecular Biology* 1612.
11 Liu, W., Heinrich, M.-A., Zhou, Y. et al. (2017). Bioprinting: extrusion bioprinting of shear-thinning gelatin methacryloyl bioinks. *Advanced Healthcare Materials*: 12.
12 Liliang, O., Highley Christopher, B., Wei, S., and Burdick Jason, A. (2017). A generalizable strategy for the 3D bioprinting of hydrogels from nonviscous photo-crosslinkable inks. *Advanced Materials* 29.
13 Bhattacharjee, T., Zehnder, S.-M., Rowe, K.-G. et al. (2015). Writing in the granular gel medium. *Science Advances* 1.
14 Moxon, S.-R., Cooke, M.-E., Cox, S.-C. et al. (2017). Suspended manufacture of biological structures. *Advanced Materials* 13.
15 Skylar-Scott, M.-A., Uzel, S.-G.-M., Nam, L.-L. et al. (2019). Biomanufacturing of organ-specific tissues with high cellular density and embedded vascular channels. *Science Advances* 9: w2459.
16 Lee, A., Hudson, A.-R., Shiwarski, D.-J. et al. (2019). *3D bioprinting of collagen to rebuild components of the human heart*, 482–487. New York: Science.
17 Noor, N., Shapira, A., Edri, R. et al. (2019). Tissue engineering: 3D printing of personalized thick and perfusable cardiac patches and hearts. *Advanced Science*: 11.
18 Ahlfeld, T., Cidonio, G., Kilian, D. et al. (2017). Development of a clay based bioink for 3D cell printing for skeletal application. *Biofabrication* 3, 9.
19 Isarawut, P., Sorada, K., Tanom, B. et al. (2008). Development of acellular dermis from porcine skin using periodic pressurized technique. *Journal of Biomedical Materials Research Part B Applied Biomaterials* 85.
20 Reing, J.-E., Brown, B.-N., Daly, K.-A. et al. (2010). The effects of processing methods upon mechanical and biologic properties of porcine dermal extracellular matrix scaffolds. *Biomaterials* 33: 8626–8633.
21 Gilbert, T.-W., Sellaro, T.-L., and Badylak, S.-F. (2006). Decellularization of tissues and organs. *Biomaterials* 19: 3675–3683.
22 Joshi, M.-K., Lee, S., Tiwari, A.-P. et al. (2020). Integrated design and fabrication strategies for biomechanically and biologically functional PLA/β-TCP nanofiber reinforced GelMA scaffold for tissue engineering applications. *International Journal of Biological Macromolecules* 164.
23 Gao, Q., Niu, X., Shao, L. et al. (2019). 3D printing of complex GelMA-based scaffolds with nanoclay. *Biofabrication* 11.

4

Coaxial 3D Bioprinting

4.1 Introduction

3D bioprinting has emerged as an exciting new technique in tissue engineering. This method has been applied for the fabrication of 3D tissues and organs including vessel-like structures, where layer-by-layer freeform fabrication is used to deposit a mixture of cells and biologically compatible hydrogel. However, vascularized structure has always been a difficulty in traditional 3D bioprinting. Then coaxial 3D bioprinting emerged to solve this problem.

In this section, the significance and two categories of coaxial bioprinting are demonstrated.

4.1.1 Significance

Bioprinting is an exciting new technique that has been applied to the field of tissue engineering. In 3D bioprinting, a layer-by-layer additive fabrication technology is used to directly deposit cells mixed with biologically compatible hydrogel for the fabrication of 3D tissues and organs. With its excellent scalability, repeatability, and accurate deposition positioning, 3D bioprinting reflects its significant advantages in manufacturing tissue engineering structure, which cannot be achieved by traditional manufacturing technologies [1–3]. Various printing technologies, such as inkjet bioprinting, laser-assisted bioprinting, and fiber-based extrusion bioprinting have been used to create 3D living structures.

However, each approach has its own advantages and limitations. Regardless of which bioprinting technique is used, there are always two main goals of 3D bioprinting: the easy formability for bioinks and the high viability of the cells mixed in the hydrogel for the printed structures. However, it seems that you cannot have your cake and eat it. At present, bioinks with fast cross-linking and excellent printability are often deficient in cell compatibility. More importantly, vascularization plays a significant role as the key limitation in the fabrication of large complex tissue for long-term culture (Figure 4.1). The traditional manufacturing methods are difficult to fabricate the internal microchannel network in one step. Porous structures fabricated by typical bioprinting usually result in local blockage caused by the weakness of the bioink fibrous unit (Figure 4.1a). Sacrificial bioprinting could

Cell Assembly with 3D Bioprinting, First Edition. Yong He, Qing Gao, and Yifei Jin.

fabricate multilayered vascular structures but requires a multistep printing process (Figure 4.1b). Almost all the printed large-size structures cannot solve the problem of internal nutrient delivery. And the cell survival rate in the large tissue is also reduced due to the lack of internal nutrient channels. To solve these two problems, a coaxial printing method was proposed among traditional extrusion-based and sacrificial 3D bioprinting. This approach has a simple setup of complex vascular networks and provides a feasible strategy for channel endothelialization (Figure 4.1c).

The concept of coaxial printing comes from the electrospinning technique, in which two immiscible liquids are electrospun through a coaxial, two-capillary spinneret to fabricate several kinds of hollow nanofilaments [5]. Later, researchers applied the concept to 3D bioprinting. The main feature of coaxial bioprinting is that different materials (hydrogels) are extruded separately through nozzles that are nested in another with the same axial. After extrusion, different materials contact each other, but due to the laminar flow characteristics in hydromechanics, they remain stable at the contact interface and form a multi-material core–shell structure. At the same time, bioinks with different cross-linking properties can be cross-linked at different stages, thus ensuring the rapid prototyping of printing fibers and the slow integration of the overall structures.

In the field of bioprinting, different coaxial nozzles were designed to produce multilayered hydrogel fibers. Ozbolat et al. used dispensing machine nozzles to make internal and external nested coaxial nozzles [6]. This method is simple and easy to operate, but its integrated design is easy to cause nozzle blockage with difficulty to clean. In addition, the coaxiality of this nozzle is hard to be ensured. Khademhosseini et al. used a coaxial nozzle based on the technology of a microfluidic chip to manufacture hollow gel fibers [7]. In this kind of nozzle, the coaxial flow is based on fluid laminar effect during the printing process. This method can realize micron scale tube manufacturing with high accuracy, but the process of making the microfluidic chip is complex with high cost.

With the characteristics of mixing multi-material and fabricating tube structures in one step, coaxial printing can play a great role in the field of bioprinting in the aspects of multi-material deposition and forming a special structure. To solve the problems in the process of bioprinting and the nutrition transport of the printed tissues, two branches are derived from coaxial 3D bioprinting, namely solid fiber-based coaxial bioprinting and hollow fiber-based coaxial bioprinting. By assembling bioinks with different properties to a core–shell structure, the solid fiber-based coaxial bioprinting improves the forming ability of bioprinting, promoting the feasibility of 3D bioprinting at the technological level without reducing the biocompatibility. By using sacrificing materials, the hollow tube is formed in one step in the hollow fiber-based coaxial bioprinting, which realizes the nutrition supply of the printed entity structure, improving the practicability of 3D bioprinting in the application level. Representative coaxial nozzle design and printing strategies are illustrated in Figure 4.2.

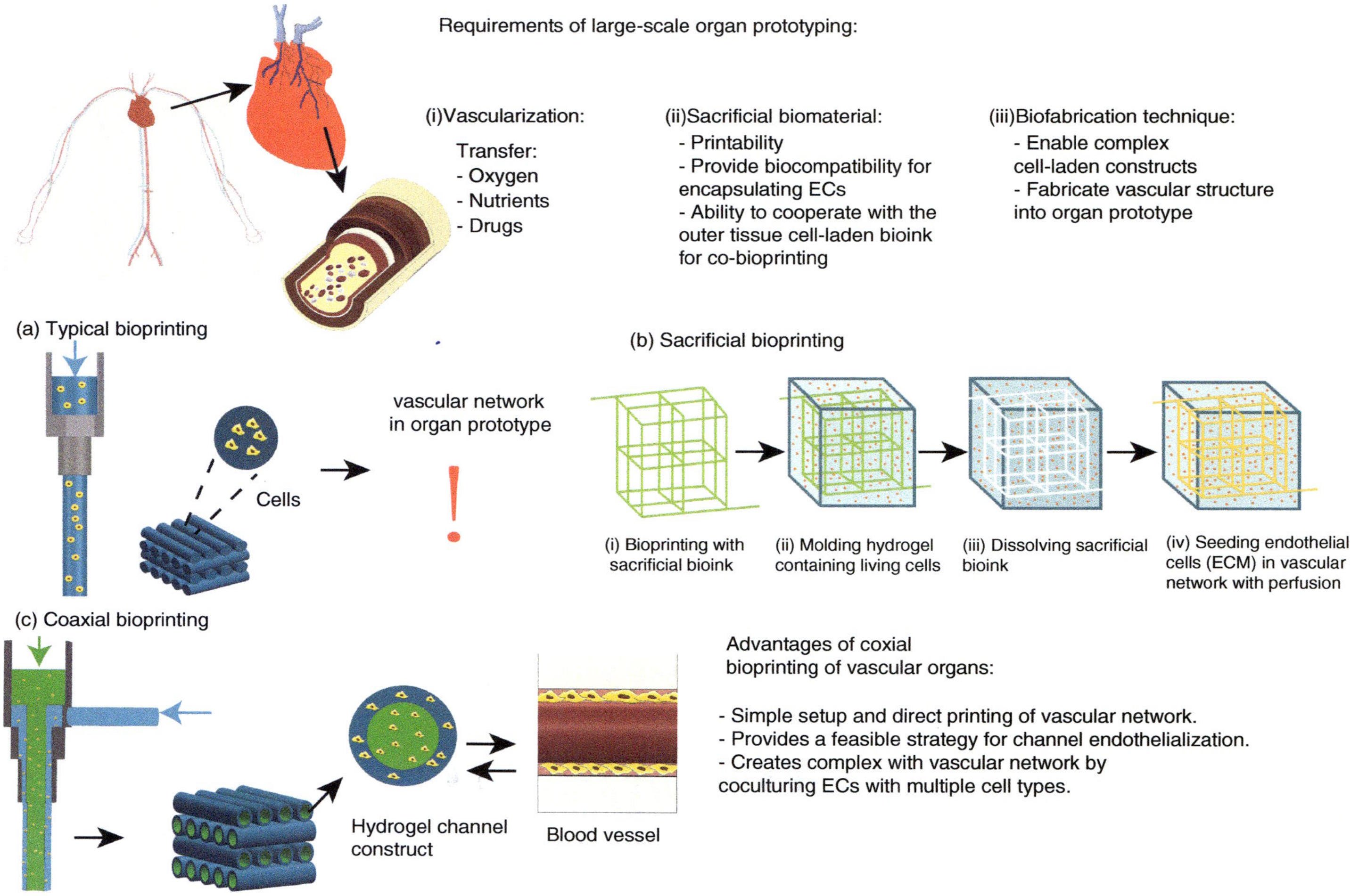

Figure 4.1 Requirements to fabricate vascular organ prototypes by coaxial bioprinting [4]. (a) Typical bioprinting. (b) Sacrificial bioprinting. (c) Coaxial bioprinting. Source: Ramezani et al. [4]. Reproduced with permission of Springer Nature.

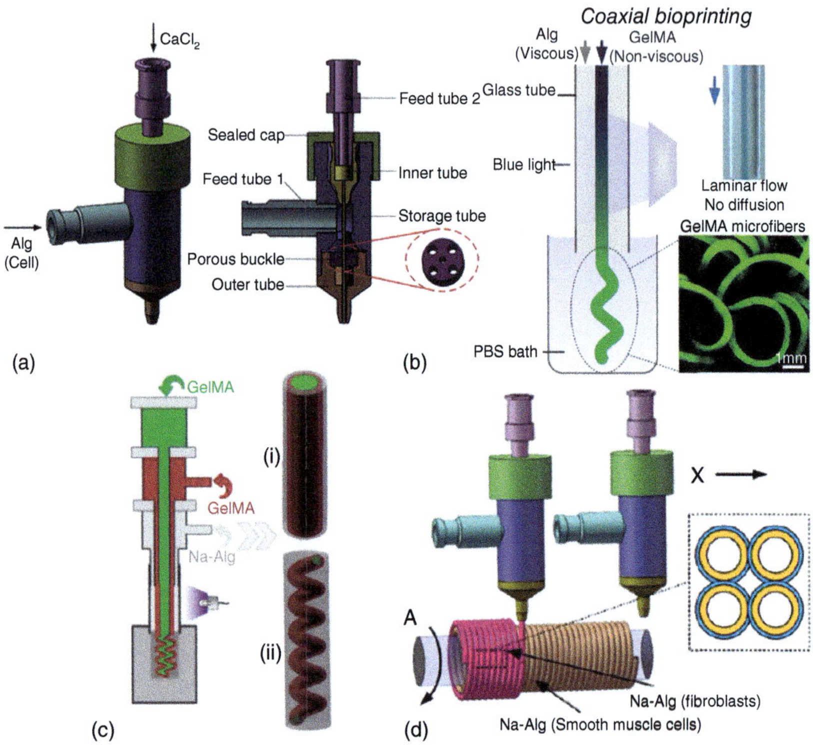

Figure 4.2 Schematic illustration of fabrication method of coaxial bioprinting. (a) Design of a kind of coaxial nozzle [8]. Source: Gao et al. [8]. Reproduced with permission of Elsevier. (b) New strategy of coaxial bioprinting used for continuously generating microfibers [9]. Source: Shao et al. [9]. Reproduced with permission of Wiley. (c) Multiple coaxial nozzles to fabricate multilayered fibers [10]. Source: Shao et al. [10]. Reproduced with permission of Wiley. (d) Fabrication process of vessel like channel by multiple scale coaxial nozzle bioprinting over a rod [11]. Source: Gao et al. [11]. Reproduced with permission of American Chemical Society.

4.1.2 Two Categories

4.1.2.1 Solid Fiber-Based Coaxial Bioprinting

A fibrous or reticular structure consists of muscle fibers, nerve fibers, blood vessels, etc. In tissue engineering, because of its biomimetic geometry and easy nutrient diffusion, the cell-loaded microfibers can be used to construct the fiber-shaped microstructures in vitro. To create fiber-based micro-tissue that approximates the real tissue, cell-filled micro-fibers must mimic the biological, morphological, and mechanical properties of natural tissue. The physical properties (rigidity, strength, viscosity, adhesion), chemical properties (degradation rate, toxicity of degradation products), and immunogenicity of bioink materials have significant influences on the biological functions of cell activity, migration, differentiation, and even gene expression. Some effects are beneficial to cell adhesion, cell intrinsic function, and its application, and some are not.

However, it is very difficult for a kind of biomaterial to have excellent performance in all these aspects. Although some biomaterials have good biocompatibility, they are poor in printability. Improper rheological properties, slow cross-linking speed, and other defects make it impossible to be used in 3D bioprinting based on fiber units. The commonly used strategy to overcome this dilemma is to add additives to these gel materials to improve printability. But these additives often have a negative impact on the biocompatibility of the original materials and affect cell growth. Simply mixing printing materials cannot solve this problem fundamentally.

In this case, solid fiber-based coaxial bioprinting was proposed. Bioinks with different properties are extruded through the coaxial nozzle at different layers. Generally, the bioink in the inner nozzle has better biocompatibility, while the bioink in the outer nozzle has better printability. In this core–shell structure fiber, the outer shell provides a good forming capability and sufficient mechanical strength to support the entire 3D-printed structure. The inner nucleus, on the other hand, is usually mixed with living cells and provides a good in-vivo like environment. With the development of technology, the structure of coaxial printing becomes more and more complex, and the multi-layer coaxial printing structure appears gradually. In a one-step extrusion process, different layers are alternately mixed with the hydrogel used to form and the hydrogel encapsulated with cells. A complex multi-cell co-culture model was obtained to better mimic the natural in-vivo environment.

4.1.2.2 Hollow Fiber-Based Coaxial Bioprinting

Under the condition of in vitro culture, the supply of nutrients should be ensured. Due to the limited distance between nutrients and oxygen diffusion, the cells cannot survive in the area beyond 200 μm away from blood vessels. In natural tissues, there is an intricate network of blood vessels to ensure the transport of nutrients. Therefore, to construct complex and functional thick tissue in vitro, the construction of a tissue engineering scaffold simulating a natural vascular network is an important problem that needs to be solved urgently.

One of the most intuitive methods for creating a network of gel-based channels is to directly print a tubular structure similar to a blood vessel. At present, there are many different bioprinting methods to realize the manufacturing of single or branch gel tubes, mainly including ink-jet printing and laser direct writing printing based on gel sphere fusion, extrusion printing based on gel fiber fusion, and lamination assembly and cell tissue microsphere fusion printing based on gel module assembly. Ink-jet printing has high accuracy; laser direct writing can print materials with high viscosity. However, both methods are based on gel particles by drop fusion forming, in which the printing efficiency is low. The method based on gel module assembly requires traditional micro-manufacturing technology to fabricate the forming module, in which the cost is high. The method based on cell tissue microsphere fusion has low strength and requires auxiliary support to realize layer-by-layer assembly. Thus, because of its simple principle and device, extrusion forming is often used in bioprinting. There are two kinds of methods to print complex structures, namely by direct print tubular structure and using the method of support. However, compared to the print method based on particle fusion, extrusion molding efficiency of printing had a very big enhancement; tubular structure is still based on a layer-by-layer manner, with complex process steps.

By coaxial printing, hollow fibers can be fabricated conveniently and quickly with appropriate technology and biomaterials. Its basic principle is that hydrogels, such as sodium alginate, which form a shell, are extruded from the outer nozzle and sacrificial materials, such as $CaCl_2$, which would be removed and form a hollow structure, are extruded from the inner nozzle. Generally, the sacrificial material also has the function of starting hydrogel for cross-linking. The printed hollow fibers can continue to stack with each other and form a 3D structure with internal channels. Nutrients can penetrate between the walls of hollow fibers, which are often a few 100 μm thick. Thus, the microchannel system consisting of coaxial printed hollow fibers could deliver nutrients for cell growth.

4.2 Printable Ink Materials

4.2.1 Forming Mechanism

The forming ability of hydrogels is very important in bioprinting. Here, printability is used as a concept to describe and analyze the forming ability. During the bioprinting process; hydrogels can be processed into different basic units through a certain device and then accumulated into a designed 3D biological structure [12]. Throughout the printing process, hydrogel would go through three different stages: (i) Hydrogel is processed into basic units, including microparticles, microfibers, and micro planes. In this process, hydrogel should be in a liquid or sol state, so its rheology will strongly affect the stability of printing. (ii) Phase transition stage. The hydrogel transforms from liquid to solid (sol to gel). In this stage, the hydrogel is triggered to begin phase transition due to specific external conditions. For ion-sensitive hydrogels, the trigger condition is to contact the corresponding specific ions. For temperature-sensitive hydrogels, the trigger condition is the temperature change. For PH-sensitive hydrogels, the trigger condition is the change of PH value. For light-sensitive hydrogels, the trigger condition is the change of light wavelength or intensity. (iii) The basic units of the hydrogel are fused with each other during the phase transition process. And the three-dimensional structure is formed by the layer-by-layer assembly. In this process, it is necessary to control the velocity of hydrogel phase transition to ensure the fusion of basic units in the process of phase transition. It can be seen from these three stages that the printability of hydrogels requires that hydrogels have controllable rheological properties and gel properties.

Coaxial printing is based on traditional extrusion printing. Thus, the forming mechanism is broadly similar but still different. Besides the extrusion process and cross-linking mechanism, the main principle of coaxial printing is to form a core–shell flow. In this approach, one or multiple materials are based on the laminar flow that can be used in parallel stream form. In the printing process, when two kinds of materials are loaded and distributed from inner and outer capillaries through a coaxial nozzle, the core-shell filaments are obtained by coaxial distribution of the different materials. In coaxial printing, the cross-linking of the material begins to react from the contact of the polyphase material after it has been extruded from the nozzle. Therefore, it reduces the requirement of rheology and fast curing property for inner material. Some bioinks with low printability can be formed passively by encapsulating in the shell. The outer material should be capable of immediate cross-linking to ensure the formation of shape fidelity. The fast cross-link ability allows the outer material to help maintain structural integrity after filament deposition. The different distribution of inner and outer material properties can also achieve multiple cross-linking order from inside to outside or from outside to inside.

4.2.2 Categories of Printable Bioinks

Bioinks for coaxial bioprinting should possess not only good printability, i.e. rapid prototyping ability, but also be able to provide high biocompatibility during the printing process. Specifically, an ideal bioink would need to meet the following requirements: (i) Bioinks could provide a similar extracellular matrix (ECM) environment; (ii) Bioinks are nontoxic with controllable degradation kinetics for cells and tissue in short and long term function; (iii) Bioinks have good printability performance, which contains a proper shear-thinning behavior, rapid crosslinking time, and mechanical strength. A proper shear-thinning behavior with desire viscosity can maintain the shape fidelity after printing and protect printed cells. As well as shear-thinning reduces the shear force required of flow during extrusion material from nozzle. The extruded hollow or solid coaxial fibers not only need to maintain coaxial shape, but it also need to be able to support the printed entire structure. It also prepares a suitable environment for the encapsulated cells.

Hydrogels are often considered as suitable biomaterials for coaxial bioprinting, which has structures similar to ECM, and with required nutrient and oxygen diffusion capabilities. Hydrogels are commonly used in coaxial bioprinting, made from natural or synthetic polymers such as alginate, gelatin, and gelatin of methacrylate anhydride (GelMA) etc. In Table 4.1, examples of coaxial bioprinting using various bioinks and crosslink mechanisms are listed.

4.2.2.1 Alginate

Sodium alginate is a sodium salt of polyanion polysaccharide (alginate) extracted from natural brown algae, which is connected by β-D-mannituronic acid (M unit) and α-L-guluronic acid (G unit) [21]. The two basic units of sodium alginate constitute homopolymer of GG and MM and alternate polymer of GM, respectively.

Sodium alginate will form a viscous sol solution after dissolving in water. Many factors will affect the liquidity of sodium alginate solution, including relative molecular weight, concentration, temperature, shear rate, PH value, solvent, and extra added substances [22]. The relative molecular weight has an important influence on the viscosity of polymer solution. Usually, a higher relative molecular weight leads to higher viscosity. The viscosity of sodium alginate solution is exponential with the solution concentration. And with the increase of temperature, the viscosity

Table 4.1 Different bioinks in coaxial bioprinting.

Bioink composition	Cell type	Cross-link mechanism	References
GelMA–alginate	HUVECs, MDAMB-231,MCF7 breast cancer cells, NIH/3T3 mouse fibroblast	$CaCl_2$ and UV crosslink	[13]
GelMA, alginate, 4-arm poly(ethyleneglycol)tetra-acrylate (PEGTA)	Human umbilical vein endothelial cells (HUVECs) and human mesenchymal stem cells (MSCs	Dual-stage cross-linking ionically cross-linked by calcium ions (Ca^{2+} ion) covalent photo cross-linking of GelMA and PEGTA	[14]
Alginate	Bovine cartilage progenitor cells (CPCs) were	2–5% $CaCl_2$ solution	[15, 16]
Alginate/ PEG-fibrinogen/	HUVECs/iPSC-CMs	Dual stage cross-link by fast $CaCl_2$ and for PF UV cross-link	[17]
Alginate/GelMA/ polyethylene glycol tetra-acrylate	HUVECs/hMSCs	Calcium ions	[18]
Alginate/GelMA/PEG	Human urothelial cells/human bladder smooth muscle cells/human umbilicalvein endothelial cells/human smooth muscle cells	Situ cross-link: $CaCl_2$ Post cross-link: UV exposure	[19]
Gelatin methacryloyl (GelMA) and gelatin	Osteoblast, human umbilical vein endothelial cells	Photo-cross-linking mechanism	[20]

of sodium alginate solution decreases. In addition, sodium alginate solutions can be miscible with a variety of substances, including thickeners, surfactants, or other hydrogels. Sodium alginate/gelatin and sodium alginate/GelMA mixtures are often used in bioprinting.

Sodium alginate is a typical ionic-triggered cross-linkable hydrogel [23]. Sodium alginate solution can react with divalent cations to form hydrogels, such as Ca^{2+}, Cu^{2+}, and Fe^{2+}. Taking Ca^{2+} as an example, Ca^{2+} cross-linking occurs on G unit, and the cross-linking process is mainly obtained by Na^{+} and Ca^{2+} exchange on G unit. Ion replacement of Ca^{2+} occurs at carboxyl sites, and the alginate on the other side of the chain can also be connected with Ca^{2+}. The carboxyl groups of two G units are combined with Ca^{2+} to form a coordination structure. Ca^{2+} replaces Na^{+}, aggregates alginate ions, and binds sodium alginate more tightly. At the same time, the polymerized molecules also constrain Ca^{2+}, and this synergistic effect ultimately leads to hydrogels with a three-dimensional network structure of high strength and viscoelasticity.

In this 3D network structure, alginate ions are like a box, and Ca^{2+} is enclosed like an egg, which is often referred to as the "egg-box structure." The strength of the cross-linked hydrogel depends on the efficiency of synergy, i.e. the concentration of Ca^{2+} and alginate ions. The higher the ion concentration, the stronger is the synergy process. And there are more "eggs" wrapped in a "box;" at the same time, there are more layers of "boxes" outside "eggs." As a result, the cross-linked calcium alginate hydrogel is dense with high strength.

Here are some commonly used methods to prepare sodium alginate hydrogel with Ca^{2+} cross-linking [24]. (i) Direct droplet addition: add the aqueous solution of sodium alginate drops to the aqueous solution containing Ca^{2+}, and Ca^{2+} starts an outside-in diffusion. In this method, the outer cross-linking is sufficient, resulting in a higher crosslink density. (ii) Reverse droplet addition: add an aqueous solution containing Ca^{2+} to an aqueous solution containing sodium alginate. Ca^{2+} diffuses from the inside to the outside, with sufficient inner cross-linking and higher cross-linking density. (iii) In situ release method: Calcium salts with low solubility or Ca^{2+} coordinated with other materials are used. After fully mixed with sodium alginate solution, a weak acid with a slow-release effect is added. Then Ca^{2+} is released and combines with sodium alginate to form colloids under the action of acid.

The process of sodium alginate and Ca^{2+} cross-linking is the same among the above three methods of cross-linking. Bolan used the surface gelling theory to elaborate on the timing process of its reactions. When sodium alginate meets Ca^{2+}, the edge of sodium alginate in contact with Ca^{2+} first conducts ion exchange, and Na^{+} is replaced by Ca^{2+}, forming a dense hydrogel shell. The pores of this hydrogel shell only allow the passage of Ca^{2+} and Na^{+} with small molecular weight, but not alginate ions with large molecular weight. Therefore, Ca^{2+} passes through this hydrogel shell and continues to react with the internal alginate ions, while the replaced Na^{+} dissociates to the outside of the gel shell. Finally, calcium alginate hydrogel is formed. The gelation rate of sodium alginate is very important in bioprinting, which affects the fusion process of the molding unit and the stability of the printed 3D structure.

However, alginate is gradually declining in bioprinting applications because it could not provide a good microenvironment for cells, nor can it simulate the internal morphology and function required for fibrous tissue in vivo. However, due to its excellent forming ability and ionic cross-linking characteristics, it is a good material for the "shell" structure in coaxial printing [25, 26]. In the core/shell method, which is mostly applied in coaxial bioprinting, the bioink-based alginate and crosslink solutions simultaneously extrude through the nozzle; the gelation mechanism is at the end of the dispensing head. The rapidly-formed shell can provide good mechanical support, and the cross-linking ions diffusing from the inside to the outside also enable the outer wall of the shell to have sufficient fusion time before curing. If the core structure is mainly composed of biological ink encapsulating cells, the tissue entity will be formed; if the core structure is mainly composed of sacrificed materials, the complex internal microchannels will be established to transport nutrients.

4.2.2.2 Gelatin

Gelatin is a hydrolyzed product of collagen, and the gelatin used in tissue engineering or in drug-carrying systems is often derived from collagen of porcine, fish, or bovine tissues. Gelatin has high biocompatibility and biodegradability in a physiological environment. Although it is of animal origin, the digestive process gives it a very low antigenicity, while the biodegradation process in vivo does not produce biotoxic products. The amino acid sequence ARG-gly-ASP (RGD) inside the molecule can improve the biological performance of gelatin and has the function of cell recognition.

In the process of collagen treatment, different types of gelatins can be obtained due to the differences in PH, temperature, and treatment time. Type A gelatins can be obtained by collagen treatment under the alkaline condition of PH 8–9, while type B gelatins can be obtained under the acidic condition of PH 4–5.

As a yellowish or translucent grain, gelatin is slightly soluble in cold water and soluble in hot water. Gelatin is a temperature-sensitive hydrogel with a thermally reversible property. Gelatin molecules contain polar groups such as amino and carboxyl groups. In its aqueous solution, the affinity for water is greater than its own affinity. In the gelatin solution, the main body of linear polymers is attached to each other, and the thermal motion of molecules decreases with the decrease in temperature. Van der Waals forces (intermolecular forces) between molecular chains make the molecular chains cross-linked to form the same structure so that the molecules cannot move with each other, and the molecules lose their fluidity. And the polar groups on the molecular chain firmly adsorb a large number of water molecules filled between the network structure, become tough, elastic gelatin.

When the hot gelatin solution is cooled, its viscosity gradually increases. If the concentration is high enough and the temperature is low enough, the gelatin solution will be converted into gelatin. Gelatin gels resemble solid substances that retain their shape and are elasticity [27]. Gelatin gels can reversibly revert to a solution state when heated. Gelatin solution gradually freezes under 40 °C. When the temperature is above 40 °C, it can be dissolved in water to form a low-viscosity solution, and its temperature-sensitive property can be used as condensate for printing scaffold to help rapid shaping [28–30].

Owing to its extremely low mechanical strength and similar cross-linking temperature with human body temperature, gelatin is not suitable for bioprinting on its own [31, 32]. As for the sacrificial bioink, due to its reversible temperature-sensitive curing property and superior biocompatibility, gelatin is the optimal candidate. Seung-Schik Yoo et al. reported a method to print hydrogel scaffolds containing fluidic channels using gelatin printed between collagen layers and then liquefied and drained at 37 °C to form a hollow channel within the collagen scaffold. Dai et al. tested the printing of a cell–gelatin mixture onto collagen layers and then liquefied the gelatin to form a dynamically perfused vasculature inside a thick hydrogel scaffold.

In addition, gelatin is usually chemical modified to obtain other special properties and to be used in corresponding scenarios.

4.2.2.3 GelMA

As mentioned above, though gelatin has great biocompatibility, as the temperature of the human body is near 37 °C, gelatin alone is not suitable for bioprinting under this condition. The gel state is easy to turn back to liquid in an in vivo like environment. Another method of cross-linking natural gelatin is chemical cross-linking. The commonly used reagents include formaldehyde, glutaraldehyde, etc. However, their application is often limited due to the strong toxicity of aldehydes. Therefore, it is of great significance to improve the processing property and property stability by modifying gelatin.

There are abundant active groups such as hydroxyl group, carboxyl group, and amino group on the molecular chain of the limb, which can be used as effective sites for grafting modification [33, 34]. GelMA is the product of introducing methacrylate groups on the basis of gelatin under the action of methacrylate anhydride. The preparation process of GelMA was first proposed by Van Den Bulcke in 2000. There are specific reaction matrix metalloproteinases peptide sequences such as aspartic acid, arginine, glycine (RGD) sequence, etc., on natural gelatin to promote cell adhesion. And because in the process of synthesis of the above sequence will not react with methyl acrylic acid anhydride, GelMA is of similar biocompatibility with gelatin; cells could attach to the various structures of GelMA with high efficiency.

The introduction of methacrylate groups makes GelMA photosensitive, which can cross-link rapidly under the action of a certain wavelength of light and initiator. In the process of reaction, the carbon–carbon double bond of GelMA is broken under the action of free radical produced by the initiator. In the process of rebonding, if the two broken double bonds are close to each other, a new covalent single bond is formed. Multiple reactions above form a network cross-linking structure.

The selection of cross-linking reagent has a great influence on the subsequent process. The selection of exposure time, intensity, and other parameters and the effect of cell culture are closely related to the selection of the cross-linking reagent. At present, there are two mainstream cross-linking methods. One is the ultraviolet cross-linking system, which requires the addition of photoinitiator 2-hydroxy-4′-(2-hydroxyethoxy)-2-methylpropiophenone (I2959). The disadvantage is that the ultraviolet system does great damage to cells, and I2959 has high cytotoxicity. Another way of cross-linking is a relatively safe blue-ray cross-linking in the visible, the initiator of lithium phenyl-2, 4,6-trimethylbenzoylphosphinate (LAP). According to the research of Benjamin et al., the newborn fibroblasts in LAP water gel cross-linking system survival rate is more than 95%, so the current trend is using LAP to cross-link GelMA.

Light cross-linking properties improve GelMA processing properties substantially. By micro manufacturing means such as lithography, it is feasible to fabricate highly complex structures under mild conditions. And its physical and chemical properties can be accessed by reliable controlling of methyl acrylic acid groups introduced quantity, concentration, light intensity, and illumination time. Thus, GelMA has a wide range of applications in drug-controlled release, tissue regeneration, tumor model building, and other fields.

However, in extrusion-based 3D bioprinting , limited by its generally low initial viscosity and relatively slow gelation speed, it is a significant challenge to produce GelMA microfibers efficiently and stably. In addition, due to the relatively poor mechanical property of GelMA, the GelMA microfibers are difficult to meet the need of long-term culture in vitro. By coaxial bioprinting, GelMA can be cross-linked during or after the printing process with the assistance of other materials. GelMA has both the characteristics of irreversible cross-linking and good biocompatibility, which means that it can be used as shell material to form tube walls, and also as core material to act as infill [9, 10]. In our early works, both methods were reported. Various shapes of GelMA/alginate microfibers could be obtained by adjusting the flow rates based on the coflow rope-coil effect. GelMA encapsulated cells to realize the function in the inner structure. Moreover, GelMA/gelatin was coaxial printed to generate a solid structure. In the subsequent culture process, the core fibers (gelatin) are dissolved to generate effective hollow channel networks, which facilitate nutrient/oxygen transport and improve the bioactivity of the entire structure.

4.3 Representative Biomedical Applications

The construction of organoids has been a challenge in tissue engineering for a long time. On the one hand, few existing bioinks could have both good printability and excellent biocompatibility simultaneously. On the other hand, large volume printed in vitro tissue needs a complex internal nutrient transport network to ensure the growth of cells. Coaxial 3D bioprinting, as a new technology, can exactly solve these two challenges to some extent. Here, we illustrate some representative biomedical applications for coaxial 3D bioprinting in morphology control and complex nutrient transport network construction, respectively.

4.3.1 Morphology-Controllable Microfiber-Based Organoids

In 3D bioprinting, cell responsive bioink with low viscosity and rapid gelation is a key requirement. Low viscosity bioink allows thinner nozzles and increased distribution speed, resulting in higher print resolution and shorter printing time. It can also reduce the possible shear stress in the process of cell deposition, improving cell viability. A fast gelation process could generate shape fixation, improving the printing morphology control ability. By extruding bioinks of different properties in a coaxial nozzle in different layers, different parts of the core–shell structures are responsible for shape forming and encapsulating cells. Assisting the moving printing platform, morphologically controllable organoids could be fabricated based on microfibers.

In our early works, Gao et al. presented a new bioprinting method based on hollow alginate filaments using a coaxial nozzle [8]. The sodium alginate solution was dispensed through the outer tube of the coaxial nozzle, while the calcium chloride solution was dispensed through the inner of the coaxial nozzle. By controlling the concentration and flow rate of the two solutions, partial cross-linking was produced.

Two sections of the alginate filament were obtained, which were fast gelled internally and temporarily ungelled externally. This allows the hollow fiber to combine with the adjacent new fiber. With appropriate time, the external bond can also be completely gelled. By moving the printing stage, the adjacent hollow fibers were deposited at precise locations and formed a 3D structure. In this study, high-strength cell-laden hydrogel 3D structures with built-in microchannels were fabricated with a fusion of adjacent hollow filaments by controlling the cross-linking time sequence. The hollow alginate filaments shown in this work serve as a scaffold to support the mechanical integrity of the cellular environment in 3D tissue constructs and can also act as built-in microchannels to deliver nutrients for cell growth. By a similar principle, Jia.et al. reported a blended ink that consisted of GelMA, alginate, and PEGTA [14]. And with a coaxial bioprinting system, complex 3D vasculature was fabricated to support the proliferation and early maturation of vascular cells. Colosi et al. printed 3D tissue constructs by loading mixed GelMA and alginate in core nozzle and calcium chloride in shell nozzle [35]. The alginate-based bioink promoted cell migration and alignment within each fiber, leading to the organization of the encapsulated cells.

In a more microscopic perspective, mixed fiber itself by coaxial extrusion can also form microtissues. Microscale fibers could facilitate nutrient diffusion in fiber-based tissue engineering and improve cell survival. Shao et al. used GelMA as the core and sodium alginate solution as the shell, fabricated morphology-controllable fiber-based mini tissue [10]. In a transparent coaxial nozzle, GelMA was first starting cross-linking under UV light. When the shell alginate was extruded and contacted with calcium chloride solution, it quickly reacts with the gel to form the shell layer. The function of alginate is to quickly shape and fix the GelMA that has not been completely cured. After GelMA is fully cured, the alginate can be digested and removed. By precisely controlling the printing parameters, the size and shape of GelMA microfiber can be adjusted, such as the Janus structure, the multi-layer GelMA structure, and the double-parallel and double-helix GelMA structure. In addition to the flexible processability of this method, GelMA microfibers also promoted encapsulated cells to realize functions. The human umbilical vein endothelial cells (HUVECs) were uniformly encapsulated in GelMA and gradually migrated toward the peripheries of the GelMA microfibers to form a lumen of confluent endothelium embedded in the Ca–Alg shell. Vascular microstructures were constructed, and these structures were cultured in vitro for a long time, illustrating the potential value of these microfibers for the construction of artificial blood vessels. Figure 4.3 illustrates the current progress of morphology-controllable microfiber-based organoids.

4.3.2 Vessel-on-a-Chip

Cardiovascular disease is a serious disease that troubles the whole world. As a typical chronic disease, its mechanism of action is complex, which is mostly related to vascular performance degradation induced by long-term abnormal biochemical and mechanical effects of hyperlipidemia, hypertension, hyperglycemia, and blood flow

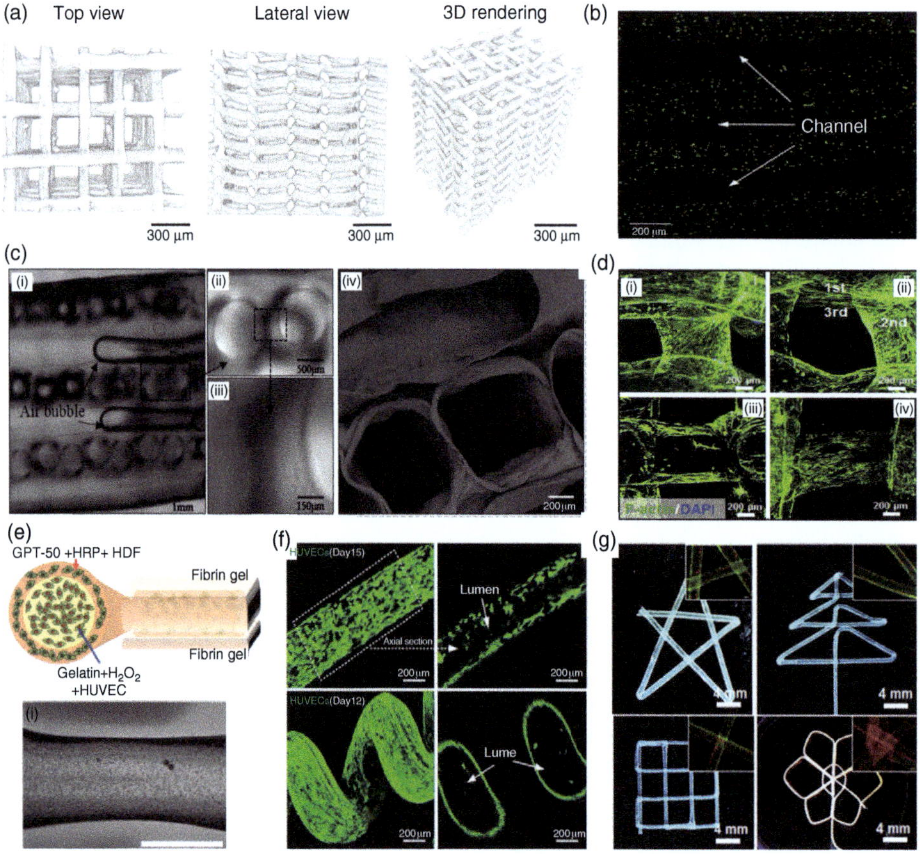

Figure 4.3 Schematic illustration of current progress of morphology-controllable microfiber-based organoids. (a) Printed structure of GelMA and alginate at low viscosity [35]. Source: Colosi et al. [35]. Reproduced with permission of Wiley. (b) Printed alginate hollow filaments [8]. Source: Gao et al. [8]. Reproduced with permission of Elsevier. (c) The fusion phenomenon between adjacent alginate hollow filaments [8]. Source: Gao et al. [8]. Reproduced with permission of Elsevier. (d) The HUVEC cell culture of perfusable hollow fibers with different layers [14]. Source: Jia et al. [14]. Reproduced with permission of Elsevier. (e) Schematic of GPT-50 bioprinting with HUVEC-core and HDF-sheath configuration [36]. Source: Hong et al. [36]. Reproduced with permission of Royal Society of Chemistry. (f) Blood vessel structures of GelMA microfibers [10]. Source: Shao et al. [10]. Reproduced with permission of Wiley. (g) Fluorescence photographs showing coaxial bioprinted perfusable tubes with various shapes [14]. Source: Jia et al. [14]. Reproduced with permission of Elsevier.

disorder. Therefore, an in-depth study of the mechanism of vascular degradation is required to provide direct guidance for the development of targeted drugs for cardiovascular diseases. At present, the research on the pathogenesis of cardiovascular diseases mainly focuses on animal model experiments and two-dimensional cell culture experiments. However, both of these two methods have certain limitations. The low success rate of animal models leads to a great demand for animals, and the ethical and moral problems of animal models impose more and more constraints on animal experiments. At the cell culture level, the current research on the mechanism of mechanical force is still dominated by two-dimensional cultured cells. The main research method is to coat the cells or adhere them to the elastic membrane and apply dilation tension through the stretching device to simulate the stretching effect of blood vessels. Since Donald reported his study of the lung micro-fluidic chip on Science in 2010, organ-chip technology has attracted more and more attention as a new analytical tool. Subsequently, some studies on vascular microarrays have been reported, but cells are still grown in the 2D environment on conventional chips.

By coaxial printing, it is easy to fabricate 3D vascular network structures that can be loaded with both mechanical and chemical stimuli. In our early work, based on the concept of organ weaving, the coaxial extrusion technique was applied to print hollow gel fibers as weaving units to the surface of a cylinder template [11]. In this research, calcium chloride solution in the inner needle played the role of a sacrificial layer to form a hollow microchannel. Sodium alginate in the outer needle wrapped around the cylinder and formed the blood vessel chip. By controlling the cross-linking velocity of sodium alginate and calcium chloride, some of the cross-linked hollow gel fibers were coaxially extruded to the cylinder. Due to the fusion of adjacent fibers, two kinds of flow channels were formed: the intermediate macroscopic flow channel formed after the removal of the cylinder template and the microscopic flow channel surrounding the hollow gel fibers. Due to progressive cross-linking, this vascular structure exhibits strong mechanical strength and can meet mechanical force loading. L929 mouse fibroblasts were more than 90% active for one week after being cultured in the printed vascular structure, demonstrating the biocompatibility of this manufacturing method.

Due to the gradual cross-linking characteristics of sodium alginate and calcium chloride, hollow gel fiber will merge into a multi-scale flow structure. The coaxial printed multilevel vascular structure has a macro–micro channel, which can load both mechanical and chemical stimuli. When the printed vascular structure is placed in a bioreactor, the microscopic channels can be filled with cell growth factors, proteins, and drugs, and the macroscopic channel can be used for the study of cell mechanics such as stretching, bending, and shearing. In a further study, separate coaxial print nozzles loaded with different cells were used to print multicellular vessel structures. And in the cell co-culture experiments, L929, MOVAS, and HUVEC were simultaneously printed in the vascular structure. After three days of culture, the endothelial cells extended and formed cellular connections.

Other research groups have also used coaxial printing for the construction of in-vitro chips. Zhang et al. constructed vascularized heart chips with coaxial printing [37]. They injected GelMA and alginate solution into the inner shaft nozzle and added calcium chloride solution to the outer shaft nozzle. At the outlet of the coaxial nozzle, the inner and outer axial solutions contact, resulting in rapid cross-linking of the inner axial solutions. During the cross-linking process, the mixed solution of the inner and outer axis is printed into a microfiber scaffold, which is irradiated by ultraviolet light to achieve complete cross-linking and the final shape of the structure. Finally, cardiomyocytes were inoculated on the scaffolds, and the cultured myocardium tissue was cultured in a two-layer PMMA chip to prepare a mature vascularized heart chip. Gao et al. engineered an EPC/atorvastatin-loaded poly(lactic-*co*-glycolic) acid (PLGA) microspheres (APMS)-laden bio-blood-vessel (BBV) by combining 3D coaxial cell printing [38]. In a nude mouse ischemic model, the EPCs/atorvastatin-laden BBVs promoted the survival and differentiation of EPCs, the neovascularization rate, and salvage of ischemic limbs. Zhang et al. investigated the manufacturability of printable microfluidic channels, where microfluidic channels support mechanical integrity as well as enable fluid transport in 3D [15]. Cells maintained good viability after bioprinting and during prolonged in vitro culture. Zhanga et al. encapsulated HUVSMCs in the sodium alginate and printed in the form of vasculature conduits using a coaxial deposition system [39]. Deposition of smooth muscle matrix and collagen was observed around the peripheral and luminal surface in long-term cultured cellular vascular conduit through histology studies. Representative applications of vessel-on-a-chip by coaxial bioprinting are shown in Figure 4.4.

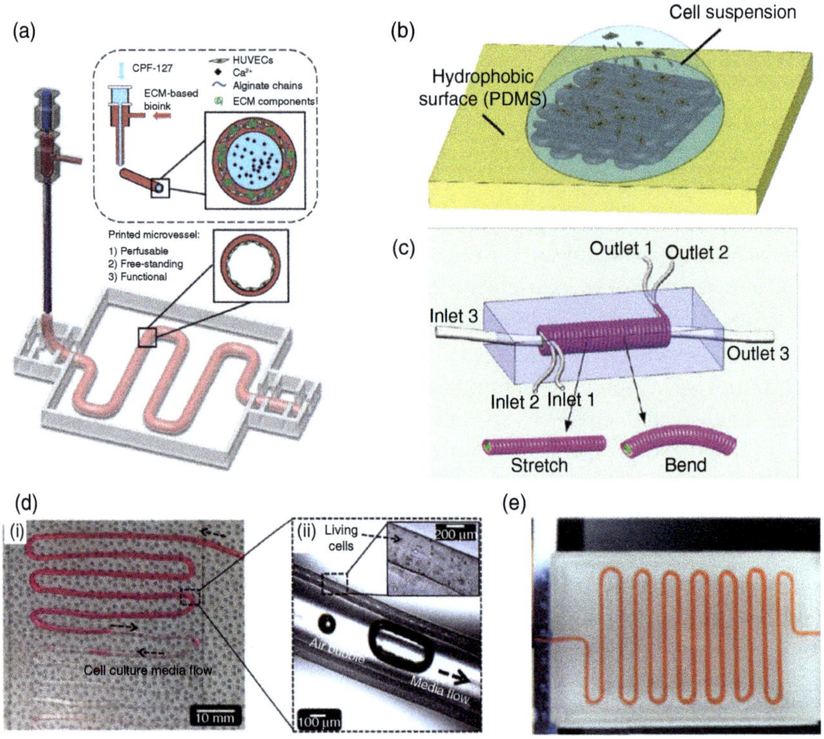

Figure 4.4 Schematic illustration of current progress of vessel-on-a-chip by coaxial bioprinting. (a) Construction of a freestanding in vitro vascular model [38]. Source: Gao et al. [38]. Reproduced with permission of Wiley. (b) Seeding procedure of cardiomyocytes onto the bioprinted scaffolds [37]. Source: Zhang et al. [37]. Reproduced with permission of Elsevier. (c) In vitro load system to simulate the microenvironment of blood vessel in vivo [11]. Source: Gao et al. [11]. Reproduced with permission of American Chemical Society. (d) Tubular channels with perfused cell type [15]. Source: Zhang et al. [15]. Reproduced with permission of IOP Publishing. (e) A long vasculature conduit printed in a zigzag pattern [39]. Source: Zhanga et al. [39]. Reproduced with permission of Royal Society of Chemistry.

4.4 Future Perspective

Vascularization is a critical factor for the biofabrication of volumetric tissue. A vascular network must be prefabricated to prevent cell death, and this is a major impediment, not only in tissue engineering but also in bioprinting. A novel technique in bioprinting, coaxial bioprinting, enabling directly depositing biomimetic vascular structure, will likely solve this problem to fabricate complex vascularized tissue constructs. Coaxial bioprinted cellular fibers provide the appropriate support and microenvironment for cell survival and intercellular interactions. The classic coaxial bioprinting is using core/shell nozzle (two-layer coaxial nozzle), where crosslinker solution and cell-laden hydrogel are pumped from inner/outer needle. Certainly, to obtain a more biomimetic multi-layered vascular structure, a multi-layer coaxial nozzle can be used. And due to the fast ionic cross-link mechanism, the alginate-based hydrogels are the most commonly used hydrogels for encapsulating cells; however, the presence of alginate inevitably affects the behavior of the encapsulated cells. Thus, novel bioink combinations need to be developed to promote cell functionalization. Currently, due to their superior biological performance and photo-/thermo-cross-link mechanisms, Gelatin/GelMA is the optimal core/shell hydrogel candidate for engineering vascularized tissue constructs.

Prospectively, to engineer vascularized tissue constructs, there are two crucial factors, (i) feasible strategy for channel endothelialization; (ii) perfusable channels and perfusion long-term culture for angiogenesis. The new strategy of coaxial bioprinting-enabled self-seeding endothelial cells is a promising approach, yet perfusion culture needs to be further performed to promote angiogenesis. Later, bioprinted tissues can slowly vascularize via angiogenesis for biomedical applications in vitro. Additionally, nerve cells can be easily introduced into coaxial bioprinting to fabricate vascularized and neurotized tissue constructs.

Coaxial bioprinting has great potential in the manufacture of vascular tissue structures. However, current coaxial methods of bioprinting have some limitations, such as (i) restricting the creation of branching vascular structures in different ranges. Since a vessel is a structure containing several layers of proteins and cells and is complex with a branching system throughout the body, the physiological function of the vessel shows the number of layers and thickness of the vessel. Therefore, coaxial bioprinting requires the creation of a branching network of blood vessels capable of neovascularization. (ii) Another limitation of coaxial bioprinting is the inability to print long vascular networks. It is important that coaxial bioprinting methods have the ability to create large diameter structures with shape fidelity and without shrinkage during printing. Therefore, we expect that in the near future, coaxial biopsies could focus on producing rapidly vascularized tissue that is clinically volumetric and used according to patient needs. In addition, it is difficult to bioprint submicron capillaries with coaxial bioprinting technology. Therefore, being able to print microvascular networks simultaneously with other large tissues may be the goal of future efforts.

References

1 Lee, J.M. and Yeong, W.Y. (2016). Design and printing strategies in 3D bioprinting of cell-hydrogels: a review. *Advanced Healthcare Materials* 5: 2856–2865. https://doi.org/10.1002/adhm.201600435.

2 McBeth, C., Lauer, J., Ottersbach, M. et al. (2017). 3D bioprinting of GelMA scaffolds triggers mineral deposition by primary human osteoblasts. *Biofabrication* 9 https://doi.org/10.1088/1758-5090/aa53bd.

3 Mandrycky, C., Wang, Z., Kim, K., and Kim, D.H. (2016). 3D bioprinting for engineering complex tissues. *Biotechnology Advances* 34: 422–434. https://doi.org/10.1016/j.biotechadv.2015.12.011.

4 Ramezani, H., Zhou, L.y., Shao, L., and He, Y. (2020). Coaxial 3D bioprinting of organ prototypes from nutrients delivery to vascularization. *Journal of Zhejiang University. Science A* 21: 859–875. https://doi.org/10.1631/jzus.A2000261.

5 Ji, X., Wang, P., Su, Z. et al. (2014). Enabling multi-enzyme biocatalysis using coaxial-electrospun hollow nanofibers: redesign of artificial cells. *Journal of Materials Chemistry B* 2: 181–190. https://doi.org/10.1039/c3tb21232g.

6 Zhang, Y., Yu, Y., and Ozbolat, I.T. (2013). Direct bioprinting of vessel-like tubular microfluidic channels. *Journal of Nanotechnology in Engineering and Medicine* 4: 1–7. https://doi.org/10.1115/1.4024398.

7 Oh, J., Kim, K., Won, S.W. et al. (2013). Microfluidic fabrication of cell adhesive chitosan microtubes. *Biomedical Microdevices* 15: 465–472. https://doi.org/10.1007/s10544-013-9746-z.

8 Gao, Q., He, Y., Fu, J.Z. et al. (2015). Coaxial nozzle-assisted 3D bioprinting with built-in microchannels for nutrients delivery. *Biomaterials* 61: 203–215. https://doi.org/10.1016/j.biomaterials.2015.05.031.

9 Shao, L., Gao, Q., Xie, C. et al. (2019). Bioprinting of cell-laden microfiber: can it become a standard product? *Advanced Healthcare Materials* 8: 1–8. https://doi.org/10.1002/adhm.201900014.

10 Shao, L., Gao, Q., Zhao, H. et al. (2018). Fiber-based mini tissue with morphology-controllable GelMA microfibers. *Small* 14: 1–8. https://doi.org/10.1002/smll.201802187.

11 Gao, Q., Liu, Z., Lin, Z. et al. (2017). 3D bioprinting of vessel-like structures with multilevel fluidic channels. *ACS Biomaterials Science and Engineering* 3: 399–408. https://doi.org/10.1021/acsbiomaterials.6b00643.

12 He, Y., Yang, F., Zhao, H. et al. (2016). Research on the printability of hydrogels in 3D bioprinting. *Scientific Reports* 6: 1–13. https://doi.org/10.1038/srep29977.

13 Liu, W., Zhong, Z., Hu, N. et al. (2018). Coaxial extrusion bioprinting of 3D microfibrous constructs with cell-favorable gelatin methacryloyl microenvironments. *Biofabrication* 10 https://doi.org/10.1088/1758-5090/aa9d44.

14 Jia, W., Gungor-Ozkerim, P.S., Zhang, Y.S. et al. (2016). Direct 3D bioprinting of perfusable vascular constructs using a blend bioink. *Biomaterials* 106: 58–68. https://doi.org/10.1016/j.biomaterials.2016.07.038.

15 Zhang, Y., Yu, Y., Chen, H., and Ozbolat, I.T. (2013). Characterization of printable cellular micro-fluidic channels for tissue engineering. *Biofabrication* 5 https://doi.org/10.1088/1758-5082/5/2/025004.

16 Yu, Y., Zhang, Y., Martin, J.A., and Ozbolat, I.T. (2013). Evaluation of cell viability and functionality in vessel-like bioprintable cell-laden tubular channels. *Journal of Biomechanical Engineering* 135: 1–9. https://doi.org/10.1115/1.4024575.

17 Maiullari, F., Costantini, M., Milan, M. et al. (2018). A multi-cellular 3D bioprinting approach for vascularized heart tissue engineering based on HUVECs and iPSC-derived cardiomyocytes. *Scientific Reports* 8: 1–15. https://doi.org/10.1038/s41598-018-31848-x.

18 Wu, Z., Su, X., Xu, Y. et al. (2016). Bioprinting three-dimensional cell-laden tissue constructs with controllable degradation. *Scientific Reports* 6: 24474. https://doi.org/10.1038/srep24474.

19 Pi, Q., Maharjan, S., Yan, X. et al. (2018). Digitally tunable microfluidic bioprinting of multilayered cannular tissues. *Advanced Materials* 30: 1–10. https://doi.org/10.1002/adma.201706913.

20 Shao, L., Gao, Q., Xie, C. et al. (2019). Synchronous 3D bioprinting of large-scale cell-laden constructs with nutrient networks. *Advanced Healthcare Materials* 1901142: 1–9. https://doi.org/10.1002/adhm.201901142.

21 Pawar, S.N. and Edgar, K.J. (2012). Alginate derivatization: a review of chemistry, properties and applications. *Biomaterials* 33: 3279–3305. https://doi.org/10.1016/j.biomaterials.2012.01.007.

22 Axpe, E. and Oyen, M.L. (2016). Applications of alginate-based bioinks in 3D bioprinting. *International Journal of Molecular Sciences* 17 https://doi.org/10.3390/ijms17121976.

23 George, M. and Abraham, T.E. (2006). Polyionic hydrocolloids for the intestinal delivery of protein drugs: alginate and chitosan - a review. *Journal of Controlled Release* 114: 1–14. https://doi.org/10.1016/j.jconrel.2006.04.017.

24 Boland, T., Xu, T., Damon, B., and Cui, X. (2006). Application of inkjet printing to tissue engineering. *Biotechnology Journal* 1: 910–917. https://doi.org/10.1002/biot.200600081.

25 Onoe, H., Okitsu, T., Itou, A. et al. (2013). Metre-long cell-laden microfibres exhibit tissue morphologies and functions. *Nature Materials* 12: 584–590. https://doi.org/10.1038/nmat3606.

26 Costantini, M., Colosi, C., Święszkowski, W., and Barbetta, A. (2019). Co-axial wet-spinning in 3D bioprinting: state of the art and future perspective of microfluidic integration. *Biofabrication* 11 https://doi.org/10.1088/1758-5090/aae605.

27 Liu, F., Chen, Q., Liu, C. et al. (2018). Natural polymers for organ 3D bioprinting. *Polymers (Basel)* 10: 1–26. https://doi.org/10.3390/polym10111278.

28 Zhang, Y., Zhou, D., Chen, J. et al. (2019). Biomaterials based on marine resources for 3D bioprinting applications. *Marine Drugs* 17 https://doi.org/10.3390/md17100555.

29 Zhang, L., Li, G., Shi, M. et al. (2017). Establishment and characterization of an acute model of ocular hypertension by laser-induced occlusion of episcleral veins. *Investigative Ophthalmology and Visual Science* 58: 3879–3886. https://doi.org/10.1167/iovs.16-20807.

30 Suntornnond, R., An, J., and Chua, C.K. (2017). Bioprinting of thermoresponsive hydrogels for next generation tissue engineering: a review. *Macromolecular Materials and Engineering* 302: 1–15. https://doi.org/10.1002/mame.201600266.

31 Sekine, H., Shimizu, T., Sakaguchi, K. et al. (2013). In vitro fabrication of functional three-dimensional tissues with perfusable blood vessels. *Nature Communications* 4: 1–10. https://doi.org/10.1038/ncomms2406.

32 Suntornnond, R., An, J., Yeong, W.Y., and Chua, C.K. (2015). Biodegradable polymeric films and membranes processing and forming for tissue engineering. *Macromolecular Materials and Engineering* 300: 858–877. https://doi.org/10.1002/mame.201500028.

33 Xing, Q., Yates, K., Vogt, C. et al. (2014). Increasing mechanical strength of gelatin hydrogels by divalent metal ion removal. *Scientific Reports* 4: 1–10. https://doi.org/10.1038/srep04706.

34 Van Den Bulcke, A.I., Bogdanov, B., De Rooze, N. et al. (2000). Structural and rheological properties of methacrylamide modified gelatin hydrogels. *Biomacromolecules* 1: 31–38. https://doi.org/10.1021/bm990017d.

35 Colosi, C., Shin, S.R., Manoharan, V. et al. (2016). Microfluidic bioprinting of heterogeneous 3D tissue constructs using low-viscosity bioink. *Advanced Materials* 28: 677–684a. https://doi.org/10.1002/adma.201503310.

36 Hong, S., Kim, J.S., Jung, B. et al. (2019). Coaxial bioprinting of cell-laden vascular constructs using a gelatin-tyramine bioink. *Biomaterials Science* 7: 4578–4587. https://doi.org/10.1039/c8bm00618k.

37 Zhang, Y.S., Arneri, A., Bersini, S. et al. (2016). Bioprinting 3D microfibrous scaffolds for engineering endothelialized myocardium and heart-on-a-chip. *Biomaterials* 110: 45–59. https://doi.org/10.1016/j.biomaterials.2016.09.003.

38 Gao, G., Park, J.Y., Kim, B.S. et al. (2018). Coaxial cell printing of freestanding, perfusable, and functional in vitro vascular models for recapitulation of native vascular endothelium pathophysiology. *Advanced Healthcare Materials* 7: 1–12. https://doi.org/10.1002/adhm.201801102.

39 Zhanga, Y., Yub, Y., Akkouchc, A. et al. (2015). In vitro study of directly bioprinted perfusable vasculature conduits. *Journal of Investigative Dermatology* 135: 612–615. https://doi.org/10.1038/jid.2014.371.

5

Digital Light Projection-Based 3D Bioprinting

5.1 Introduction

5.1.1 Printing Process

The prototyping principle of DLP can be summarized as a layer-by-layer forming method: a model is divided into several slices by computer and the forming of each layer is based on a photo-cross-linking reaction (Figure 5.1a) [1]. The simplest DLP printing system consists of three parts: a liftable printing platform, an ink tank with a transparent bottom, and a monochromatic light projector. As the printing process begins, the platform first goes down, submerged under the liquid surface, leaving a thin gap with the tank bottom. This gap, filled with materials, has the same thickness with each digital slice. Then, the projector sends the image of the first slice through the transparent bottom, illuminating the materials within the image boundary. In a short period, the photocurable liquid materials cross-link, and becomes solid. After that, the platform rises to leave another gap between the first print and the tank bottom. Then the image of the next slice is projected to become solid. The special design of the hydrophobic bottom guarantees that the new print will stick to the platform or the former slice. Finally, as it cycles, the whole model will be printed layer by layer.

5.1.2 Significance

The feasibility of 3D printing applied in tissue engineering has been proved by many extrusion printing cases. As an engineering issue, the further medical application requires large numbers of samples with good consistency to prove its effectiveness and security. However, conventional printing strategies can hardly provide adequate samples and finish the printing stage before the cells looses their viability. Meanwhile, as a scientific issue, advanced research requires a higher printing accuracy to mimic the micro–nano tissue features. However, because of the existence of gravity and interfacial tension, it is difficult to achieve an extrusion fiber with less than 100 μm diameter, especially using the biocompatible materials with living cells [2–4].

DLP-based bioprinting can solve the above problems to some extent, for it has high printing accuracy, mass production with consistency, and the ability to construct complex features (Figure 5.1b,c) [5, 6].

Cell Assembly with 3D Bioprinting, First Edition. Yong He, Qing Gao, and Yifei Jin.

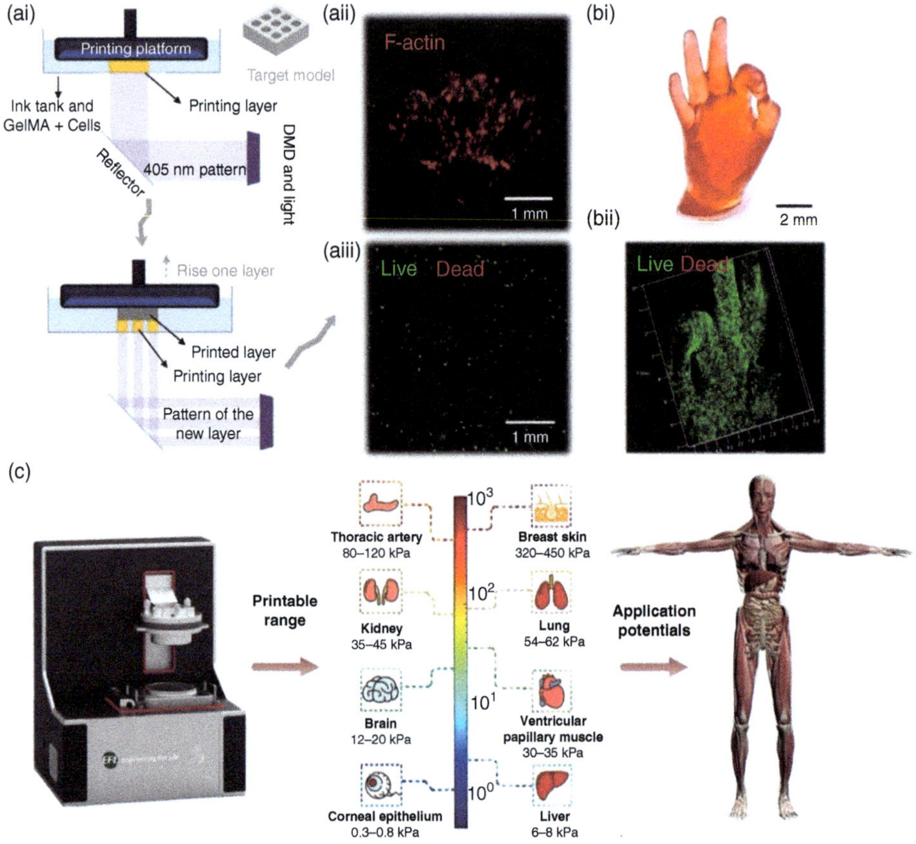

Figure 5.1 Complex DLP printed models with high precision, superior biocompatibility, and well-controlled physical properties. (ai) Printing principle: slicing model into thin layers and projecting their patterns layer by layer. (aii) Confocal images of F-actin and (aiii) LIVE/DEAD staining of the BMSCs grid. (bi) A human hand-like GelMA scaffold printed with optimized parameters. (bii) Confocal image of LIVE/DEAD staining of epithelial cells seeded on the hand scaffold. (c) DLPBP provides an efficient method to produce mimic tissues and organs with high precision, covered tensile modulus ranging from 1 kPa to 1 mPa corresponded to some typical organs and tissues like skin, artery, lung, kidney, muscle, brain, and liver. Source: Reproduced with permission of IOP Publishing.

High printing accuracy: theoretically, the accuracy of DLP printing can reach the same level as the resolution of the projector. By using a digital micromirror device (DMD) and precise objective lens, the commercialized projector has reached a resolution of less than 2 μm.

Mass production with consistency: the unique printing strategy of DLP gives it a significant advantage, which is the printing duration is only related to the height of the model. This allows us to print duplicates with the same ingredients in one process.

Constructing complex features: submerged forming process provides buoyancy force and breaks the surface tension. Solidification only causes a very slight

Table 5.1 How to change these parameters to improve prints.

When parameter increases	Printability of material	Printing precision	Mechanical properties	Biocompatibility	Recommended value
Concentration of GelMA	++	Rise first then fall	++	3D: −− 2D: Ns	3D Cell: 7%–10% Scaffold: 10%–20%
Substitution degree of GelMA	++	Rise first then fall	Harder but less flexible	3D: −− 2D: Ns	3D Cell: 30%–90% $-NH_2$ 0% -OH
Exposure time	Irrelevant	−	+	3D: −−− 2D: Ns	10–40 s per layer
Light intensity of the light source	Irrelevant	++	+	Ns	More than 1 mW/cm^2
Concentration of photoabsorber	−−	++	−	Ns	0.1–0.5% w/v

Notes: "+" refers to increasing, "−" refers to decreasing, while "Ns" refers to "no significance." And the number of "+" or "−" indicates its degree.

Table 5.2 The selections of parameters in each printing strategy.

Structure type	Exposure time	Light intensity	Layer thickness	Concentration of GelMA	Concentration of light absorbent
Solid structure	+++	+	++	+	++
Channel structure	+	+++	+	++	+++
Conduit structure	+	+++	+++	++	+
Thin-walled wall	++	++	++	++++	++
Microcolumn structure	+++	+++	++	++++	+

Notes: The number of "+" means the degree of each item, i.e. the printing strategy of solid structure is choosing low concentration of GelMA and middle of light absorber with long exposure time, small light intensity, and middle layer thickness.

change in its density, enabling buoyant to almost completely offset gravity. Meanwhile, solidified material has the same chemical units as the liquid, which leads to little surface tension.

5.2 Photocurable Biomaterials

Materials used for DLP bioprinting must have three basic capabilities: biocompatibility, formability, and mechanical property. Biocompatibility is the fundamental requirement in bioprinting. The material must be nontoxic and provide a suitable cell living environment. Formability, in DLP bioprinting, specifically refers to the ability to transform into a solid phase under exposure conditions [7]. The faster the phase transition is, the better the print will be achieved. Fine mechanical property prevents the print from rupturing. Cells can sense the stiffness of the surface, so it also regulates cell growth. To sum up, the requirements for DLP printing material are stricter than those for extrusion, and few natural materials can meet them. Recently, the synthetic method provides some new strategies to break through the limitation of photocurable biomaterials.

As mentioned above, photocurable biomaterials should be photocurable and biocompatible at the same time. A common synthetic strategy is that first biomaterials are modified to be cross-linkable and then a photoinitiator is added to trigger the reaction while exposing.

The nature of the solidification phenomenon is polymerization. Various types of biomaterials can be modified to gain this ability: peptides such as gelatin and silk fibroin, polysaccharides such as alginic acid and chitosan, and even artificial biocompatible materials such as polycaprolactone and polylactic acid. Each type of these molecules has one type or more active functional groups, for example, amino, carboxyl, and hydroxyl, which are easy to react with other molecules under controlled conditions. So, it is possible to add other functional groups like double bond and sulfhydryl onto the biomolecules. Double bond groups, for example, can link with each other continuously to polymerize while there are anions, cations, or free radicals. So far, if there is any substance that can use light energy to generate ions or free radicals, then the photo-cross-linking strategy is feasible, and that substance is a photoinitiator.

5.2.1 Photo-Cross-Linking Mechanism

Photo-cross-linking reaction can be divided into two steps: the cracking of the initiator to form free radicals and the polymerization of biomolecules. The first step is a photochemical reaction in which photoinitiator absorbs light energy and converts it into chemical energy. The second step is a simple chemical reaction in which biomolecules use the chemical energy to bond to each other and form a three-dimensional network.

5.2.1.1 Conversion of Light Energy to Chemical Energy: Photoinitiator

Photoinitiator a class of small molecule compounds that can generate high-energy active species, like ions or free radicals, when cured with a specific wavelength of light. Every substance, every molecule that constitutes a substance, and every chemical bond between molecules has its own natural frequency of vibration. When a vibration that is the same as or close to its natural frequency is input on chemical bonds, it will resonate and absorb the energy. If the vibration energy is large enough, the chemical bond will be broken, and the original molecule will break into two parts. Different states of the broken parts determine the types of initiators: those who gain an unbonded electrons are free radical initiators, while those who gain a charged group are ion initiators. There are some chemicals for which the natural frequency of some weak bonds is within the ultraviolet and visible spectrum, so the bond breaking can be caused by light. And those are so-called photoinitiators.

A variety of photoinitiators have been discovered or invented, but very few of them are suitable for biological applications. The first problem is the wavelength of initiating light. Most of the initiators can only be excited by ultraviolet light, especially those whose wavelength is shorter than 320 nm. It harms the cells. A sufficient dosage of exposure to excite the initiator can cause serious damage to the cells' DNA. The second problem is cytotoxic products. During the bioprinting, the cells have to stay in the bioink along with the initiator, and the dilution of toxic chemicals takes time. So, the photochemical reaction products along with the initiator itself should be nontoxic, or at least, the dosage of toxic chemicals must be safe.

Here, lithium acylphosphinate (LAP) and 2-hydroxy-4′-(2-hydroxyethoxy)-2-methyl-propiophe (I2959) are two safe photoinitiators for bioprinting. Both of them can provide free radicals when exposed to 365 nm light and result in cross-linking within one minute (Figure 5.2). Comparing these two initiators, LAP has a faster reaction rate to cross-link within seconds. Moreover, LAP has a wider absorption spectrum in the visible light band. It can be excited by 405 nm visible light, which would hardly cause any damage to the cells. Therefore, LAP has become the most widely used photoinitiator in bioprinting due to its rapidity and safety.

5.2.1.2 Formation of Molecular Network: Monomer Polymerization

Cross-linking is a reaction that separates molecules bond to each other and forms a whole molecular network (Figure 5.3). It is what we observed, solidification of biomaterials at the macro level. The molecular site that can form new bonding among molecules is called a monomer. During this process, the average molecular mass gains rapidly, and its solubility gradually decreases. When the molecular weight aggregates to a certain extent, it will no longer be soluble and precipitates. The changes in rheology can quantitatively describe what happens in this progress.

Storage modulus and loss modulus are two important rheological parameters in cross-linking study. Storage modulus describes the ability of materials to store elastic deformation energy whereas loss modulus is another index that describes the rate of energy is converted into heat when the material deforms. Cross-linking process can be divided into three stages, and each of them can be reflected in the rheological diagram. The first stage is side reaction, where impurities in materials like oxygen react first, and two moduli remain unchanged. The second stage is pre-polymerization, where separated molecules bond with each other to form a network, and both storage and loss modulus increase. But the storage modulus increases much faster than

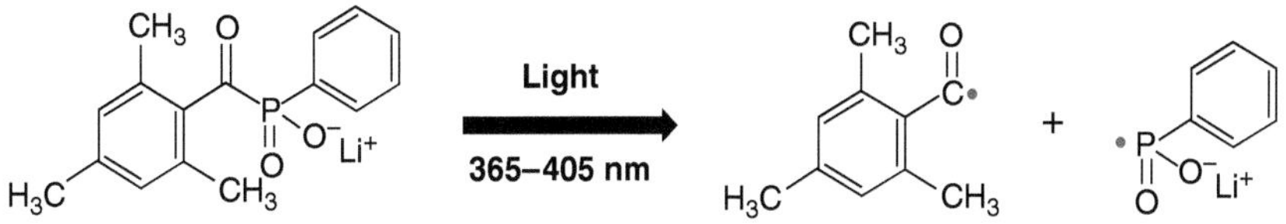

Figure 5.2 The process that lithium acylphosphinate (LAP) initiated by 405 nm light to form free radicals.

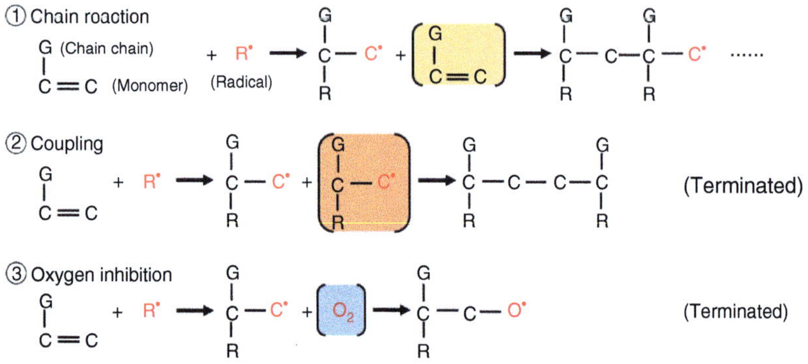

Figure 5.3 In the process of forming a cross-linked molecular network, there are three main reactions involving double bonds and free radicals.

the loss modulus does, and at one point, the value of the former exceeds the latter. That is the gelation point, where the material just becomes solid. As the reaction goes on, the third stage post-polymerization comes. When the reaction reaches the third stage, most of the molecules have already bonded together, but there are still many unbonded sites that can further cross-link to strengthen the molecular networks. So, the storage modulus keeps growing, until few sites remain in the reaction system.

A problem is that, what kind of monomer is suitable for cross-linking reaction. It should be inert enough when not excited but cross-link rapidly when excited. And there must be a sufficient amount to form a stable network. Carbon–carbon double bond is such an ideal functional group for cross-linking. It is inert at room temperature, and can form carbon–carbon single bonds for connecting different molecules rapidly when meeting with free radicals. Unluckily, few natural biomaterials come with enough double bonds. Such materials must be artificially synthesized.

5.2.2 Typical Materials: Gelatin Methacryloyl (GelMA)

Gelatin methacryloyl (GelMA) is a typical type of photocurable biomaterial successfully applied in DLP bioprinting. GelMA is modified gelatin by adding a double bond on the peptide chain to give it cross-linking abilities. Gelatin is the product of collagen hydrolysis and has the same chemical composition as the extracellular matrix, promising its good biocompatibility.

5.2.2.1 Composition and Synthesis

After hydrolysis, gelatin remains the same primary structure as collagen, composed of 20 kinds of essential amino acids. Among these, three kinds of basic amino acids have amino groups on their R-base (lysine, arginine, and histidine), and another three kinds have hydroxyl (serine, threonine, and tyrosine). Both these two groups can react with carboxyl under milder conditions. So, methacrylate, a dibasic organic acid with olefin groups, can add a double bond to the gelatin molecules by substitution reaction (Figure 5.4).

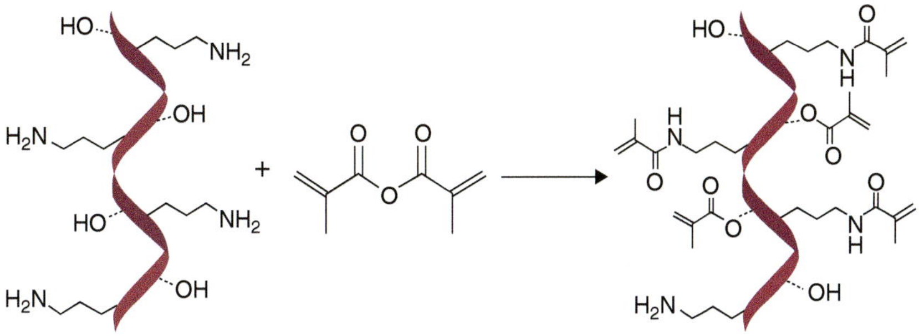

Figure 5.4 The reaction principle of using gelatin and MA to synthesize GelMA.

A brief synthesis protocol is given in this section. GelMA is synthesized by the reaction of gelatin and methacrylate in phosphate buffer (pH 7.4). Methacrylate should be added into the solution very slowly to prevent solidification. Temperature is kept around 50 °C, to avoid decomposition of reactants while maintaining a high reaction rate. For the same purpose, pH is controlled between pH 8 and 9 by adding sodium hydroxide solution. The reaction is completed when the pH no longer changes. Then the mixture is diluted by pure water (5×) and transferred to the dialysis unit. During dialysis, unreacted methacrylate (MA) and side reaction products are fully removed and the process is stopped automatically until an ion sensor indicates that the products are qualified. Finally, the products are freeze-dried to remove water and stored at −20 °C.

5.2.2.2 Substitution Degree

The substitution degree of the double bond is an important parameter for GelMA. It has a significant influence on its biocompatibility and formability. A higher substitution degree means more R-base of amino acid groups are replaced, making the gelatin lose more of its original biocompatibility. But it also means more cross-linkable double bonds are added, leading to faster and easier cross-linking. Under the same reaction conditions, different substitution degrees can be achieved by controlling the ratio of gelatin to MA.

Amino acid with the amino group has a higher pK_a (9–10) than that with a hydroxyl group (5–6). During a GelMA substitution reaction, MA tends to react with amino groups first. Approximately, MA would react with hydroxyl groups only when amino groups have been totally substituted. So, GelMA substitution degrees can be further divided into amino and hydroxyl substitution.

Several methods can be used to test substitution degrees, for example, the trinitrobenzene sulfonic acid (TNBS) method and the ^{1}H nuclear magnetic resonance (^{1}H NMR) method. TNBS is a quantitative method based on optical absorbance. TNBS acid can react with amino groups to produce products with specific light absorption at 430 nm. This can be used to quantify the number of unsubstituted amino groups and further calculate the substitution degrees. However, this method cannot be used to test hydroxyl substitution. So it is only applicable to GelMA with a low substitution degree. ^{1}H NMR is another common method to quantitatively

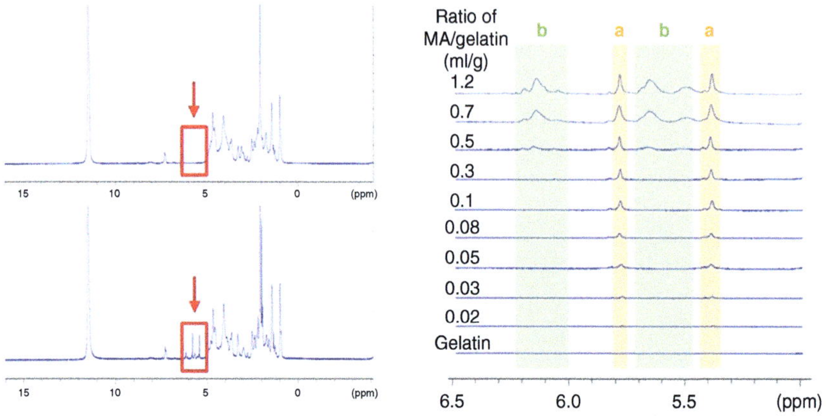

Figure 5.5 ^{1}H NMR spectroscopy of gelatin and GelMA, and characteristic peaks of different substitution degrees. Source: Reproduced with permission of IOP Publishing.

analyze the number of functional groups. GelMA, compared with gelatin, has a characteristic peak at 2.2–3.9 and 5.6–6.5 ppm on a ^{1}H NMR spectroscopy. And the area of characteristic peak is related to substitution degrees. But GelMA is a rather complex mixture composed of 20 amino acids, which can hardly get clear spectroscopy for accurate peak area analysis. So it is a qualitative method for both amino and hydroxyl substitution degrees (Figure 5.5).

5.3 Printing Equipment

A DLP bioprinting equipment has the whole basic units of a traditional resin DLP. But many necessary improvements are needed to adapt bioprinting. Overall, a DLP bioprinting system can be divided into three parts: optical units, material storage units, and mechanical movement units.

5.3.1 Optical Units

The images, used to expose the material to cross-link, are provided by a monochromatic light projector, the core of the entire optical units. It is similar to a common office projector, but has a much smaller projection format. A comprehensive analysis of the projector structure helps to better understand the principle of high-precision printing. The simplest projector consists of a monochromatic light source, dodging and collimation units, a DMD, and the objective lens system. Among them, light source and dodging unit provide the initial uniform monochromatic light. DMD processes the light into a target-shaped figure. While the objective lens system transforms the figure to a suitable size. Printing precision depends on both DMD and objective lens.

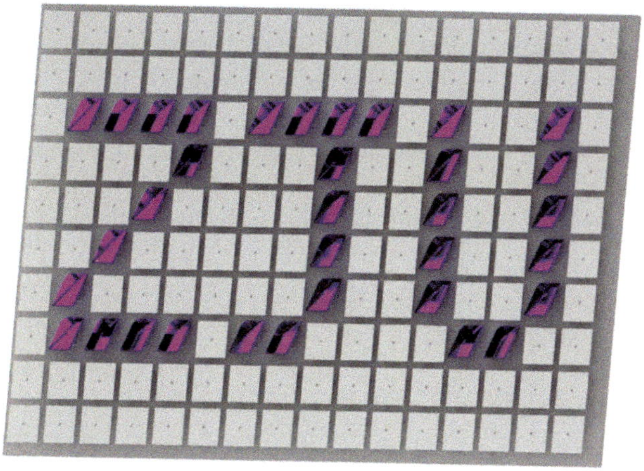

Figure 5.6 The structure of digital micromirror device: how to form a digital resolution.

5.3.1.1 Image Forming: Digital Micromirror Devices

The resolution of a DMD determines how detailed the image could be divided.

The DMD is a microoptoelectromechanical system with several hundred thousand microscopic mirrors arranged in an array on its surface corresponding to the pixels in the image to be projected. Each microscopic mirror can rotate within a small angle (around −12° to +12°) individually, corresponding to an on or off state. In the on state, light is reflected into the objective lens to make the pixel appear bright. The shape of the monochrome image consists of these on-state pixels (Figure 5.6). Some advanced printing applications require greyscales to form gradient cross-linking. This is realized by controlling the duty cycle of each micromirror between on and off state. It is worth noting that the whole DMD chip is smaller than 1 cm^2 and can provide 1280 × 800 resolutions.

5.3.1.2 Objective Lens: Focusing System

The performance of a focusing system determines how large a pixel is on the screen.

The focusing system consists of a set of spherical mirrors to zoom in or out the image reflected from the DMD chip. Scale down the image can result in a higher printing precision but a smaller format and longer printing time. Furthermore, an infinitesimal pixel is impossible due to the existence of aberration. Considering the cost and efficiency, 50–100 μm scale of each pixel is a reasonable choice for a common DLP bioprinting equipment. It can fulfill most of the requirements of micro features in tissues while finish printing a centimeter-scale model within an hour. In addition, there is already some commercial equipment for resin printing that can reach a resolution around 1 μm. It has a very powerful but expensive focusing system that might adapt to bioprinting, only if the biomaterials could respond as good as resin.

5.3.1.3 Material Storage Units

As mentioned before, cross-linking process happens in the materials during DLP printing. The design of material storage units needs to consider many aspects to support projection-based printing.

Unlike traditional resin, the volume of biomaterials needed in printing is much smaller. Biomaterials also need a controllable environment to maintain the viability of cells. Meanwhile, an assisting system for a soft and weak biomaterial is of the same significance.

5.3.1.4 Environment Controlling Systems

Sterile design: The sterile strategy of the first-generation bioprinter is to cover the whole printing area with a sealed box. An operation window allows the researcher to put materials in and print them out. UV light and ozone generator will disinfect the area before and after printing. The disadvantages of this design are obvious, because the sealed box occupy a large space and the whole equipment could be bulky and expensive.

New designs bring us desktop-level printers. DLP printer can be made as small as a basketball for it does not have an extra system like an air pump to occupy large space. As a result, the small equipment can be directly placed in a biological safety cabinet, and then all sterile problems are solved.

Heated system: temperature is an important parameter during bioprinting. Excessive temperature is fatal to cells. However, low temperature will increase the viscosity of the material, or even condensation, which greatly affects the forming ability. Therefore, the printing temperature is generally controlled between 25 and 37 °C. In addition, for scaffold printing where there are no cells in the biomaterial, the temperature can be raised to improve fluidly and formability. It is worth reminding that all pipelines for conveying materials should be heated. Otherwise, it is easy to cause blockage due to condensation.

Humidity control: upright forming type has no humidity requirements because the whole print is submerged in liquid material during the printing process, and the release film effectively prevents the evaporation of the material.

The upside-down forming type should work in a high humidity environment. When a tall model is being printed, the previously formed part will likely be exposed to dry air and then shrink. This will cause the newly formed slice to increase in height, deviate from its original position, or even fail to adhere to the previous layer. Meanwhile, low humidity will accelerate the evaporation of the material solution, resulting in higher and higher material concentrations during the printing process. Therefore, a stable humidifier is necessary in this case.

Feeding system: some types of cells are very fragile, like neural stem cells, while some others cannot proliferate, like cardiomyocytes. For these types of cells, it is unwise nor uneconomical to mix them with ink and then fill the entire trough. The feeding system can incubate the cell-containing ink in a more suitable environment than printing. Then transfer a small amount of liquid to the printing area for every few layers. However, a feeding system may bring new problems like

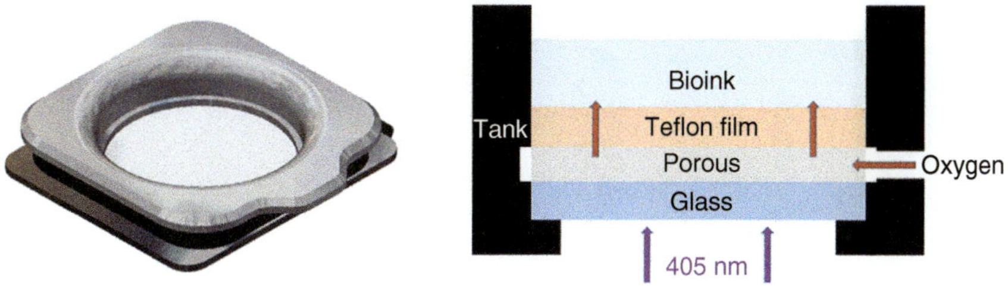

Figure 5.7 The multilayer structure of transparent ink tank bottom.

additional consumption in pipelines and cleaning matters. So whether you need this system is worth weighing.

5.3.1.5 Ink Tank: Transparent and Non-stick Bottom

For the upside-down forming type, the ink tank must be transparent for light to pass through. Hydrophobic surface guarantees that the newly formed slice will stick to the platform or previous layer instead of the tank bottom. Besides, separation of the bottom surface and printing layer cannot cause excessive deformation, which means the bottom should have enough stiffness. Therefore, a three-tier structure is employed in the tank bottom. The first layer structure is made of Teflon film, a hydrophobic transparent polymer material that can permeate oxygen. The second layer is filled with a porous material to provide passage of oxygen. The third layer is made of quartz glass to provide mechanical support. This structure allows oxygen to penetrate to the bottom of the material, forming an uncrosslinkable dead zone. Material within the dead zone, to an approximate depth of several microns, will never cross-link due to oxygen inhibition, preventing it from sticking. Meanwhile, the first two layers are tightly stretched on the glass surface to prevent them from deforming (Figure 5.7). However, long-term use will cause the oxygen permeability and transparency of the Teflon file to decrease, so it needs to be replaced frequently.

5.4 Mechanical Movement Units

Differing from extrusion printing, DLP does not require mechanical movement within the x–y plane. The simplest DLP equipment only has a degree of freedom for one-way movement, which lifts the printing platform when each layer is formed. This allows the equipment to be very compact. According to application experience, adding a separation mechanism will greatly improve the success rate of printing.

5.4.1 Lifting Mechanism: Main Movement

The lifting mechanism consists only of a lead screw and a stepper motor, realizing the lifting movement of the platform between slices (Figure 5.8). Theoretically, after one layer of printing is completed, the platform only needs to rise by one layer height. But in actual application, it will rise about a few centimeters and then drops. The

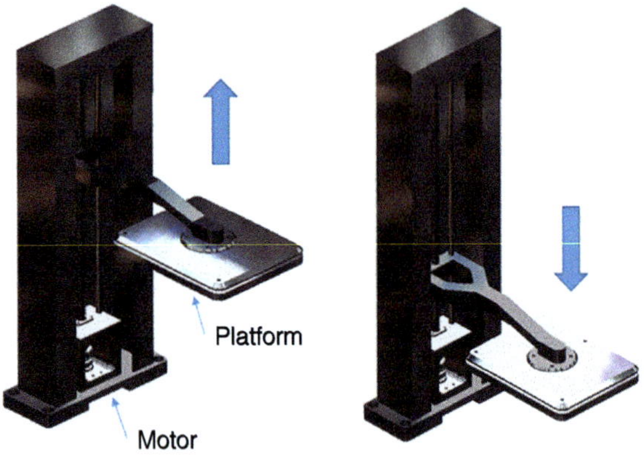

Figure 5.8 Main movement: platform rising and going down.

rising distance is one layer larger than the falling distance. During the process, the model will be more reliably separated from the bottom surface, and it will be cleaned by the liquid material.

5.4.2 Tilting Mechanism: Mixing and Separation

Vertical separation movement provided by the lifting mechanism often causes the fracture of the formed model, especially those models with a large area. Negative pressure is easily formed between the large printing surface and the tank bottom. When separating, the huge pulling force generated by atmospheric pressure will act on the printed model. To solve this problem, a tilting mechanism is designed (Figure 5.9). When the separation begins, the tank will tilt to a small angle, creating a gap between the bottom and the edge of the model by stress concentration. This gap will release the negative pressure, making the separation rather easier. However, this tilting movement must be carefully measured and calculated to coordinate with the main movement. Otherwise, it will introduce lateral shear forces which are more likely to cause damage to the model.

5.4.2.1 Printing Error Formation and Optimization Strategies

Printing precision has a larger influence on subsequent biological applications. Many factors make it impossible to print a model the same as what it was designed.

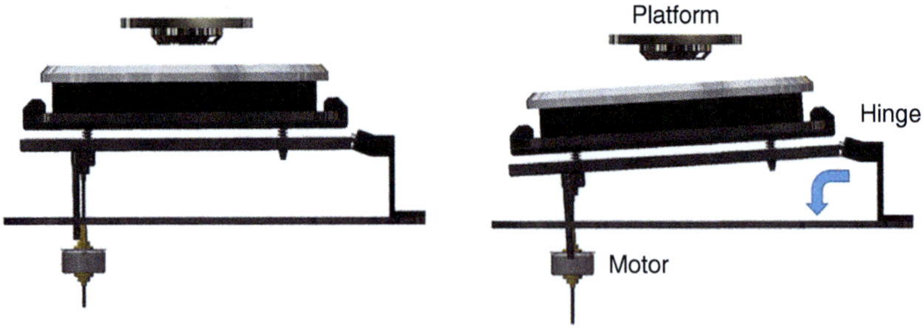

Figure 5.9 Tilting mechanism: mixing and separation.

So, it is necessary to control printing errors within a reasonable range. To realize this, we must have a deep understanding of how these errors occur, what parameters influence them and what is the law of influence.

5.4.2.1.1 Transmission and Scattering of Light

DLP printing is based on a basic concept: cross-links where the light is exposed. However, it is only in a theoretical situation that exposure occurs with a sharp and instant-fade-off boundary. Instead, the exposed area is gradual. The distribution of light field determines the print quality. Along the z axial direction, the exposed images are gradually defocused, and the GelMA has a weak absorption of the light that passes through, resulting the light to expose a deep distance (Figure 5.10). In the x–y plane, because the ink is not 100% transparent and contains countless microparticles that form a complex light reflection interface, light scattering occurs everywhere and will light up nearby materials (Figure 5.11). The illuminated materials will have different degrees of cross-linking, and those which reach the gel point will become solid. Therefore, controlling the gel point close to the model boundary is the key to decrease printing errors.

5.4.2.1.2 Analysis of the Process Parameters

Whether the material can reach gel point or not is a matter of reaction degree. Further, the reaction degree depends on the concentration of reaction substrates, reaction time, and conditions. Cross-linking reaction here in DLP is a simple

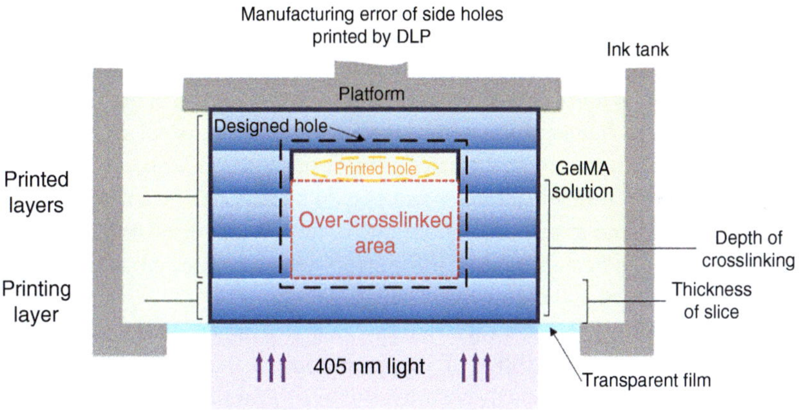

Figure 5.10 The principle of forming printing errors in the z-direction. Source: Reproduced with permission of IOP Publishing.

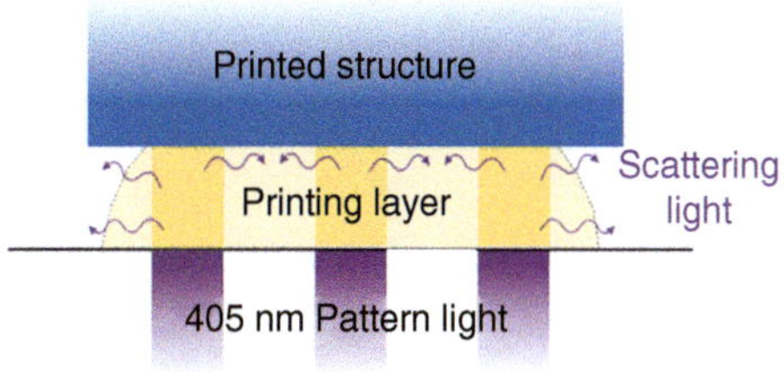

Figure 5.11 The principle of forming printing errors in the x–y plane.

chemical reaction that involves only two substrates: free radicals and double bond groups. As the half-period of free radicals is very short, the reaction time depends on how long it is exposed. And a biocompatible reaction situation is strictly fixed (37 °C, pH 7.2–7.4, salinity = 0.5%NaCl). Above, **exposure time** is a directly controlled parameter, while the concentration of substrates depends on the previous reaction.

The concentration of double bond groups is easy to control. Only **substitution degree** and **concentration of GelMA** influence it, and both of them can be directly adjusted by simple material preparation.

Free radicals are the product of the previous photochemical reaction. So, to control their concentration a further analysis of the previous reaction is needed. Same as before, the reaction situation is fixed and duration also equals to exposure time (if the amount of initiator is adequate). But the reaction rate is greatly affected by the light intensity. However, as it has been mentioned before, the distribution of light intensity is affected by both the **power of the light source** and the absorption of the solution. Absorption can be adjusted by adding a photoabsorber. Lemon yellow is a kind of food coloring that has a strong absorption of 405 nm light and is nontoxic. The **concentration of photoabsorber** can significantly control the distribution of light intensity.

5.4.2.1.3 How to Change These Parameters to Improve Prints

Concentration and substitution of GelMA can affect printability significantly. The rise of these two parameters will improve printability. However, excessive formability can lead to more printing errors. And cells can hardly survive in such a tight hydrogel network (Table 5.1).

Light intensity depends on the power of the light source. The higher the light intensity is, the easier and better print can be achieved. And it will greatly reduce printing time. But the financial cost of the large power light is expensive. The lowest light intensity that can ensure a successful printing of GelMA is around 1 mW/cm^2.

Exposure time is a flexible parameter and easy to be modified on software. Short exposure time may lead to printing failure due to insufficient cross-linking. While long exposure time will cause large printing errors, and decrease cell viability. This parameter can be tested and optimized in experiments.

The concentration of the absorber can be calculated precisely by experiment. The light absorption rate of different concentrations of the absorber to 405 nm wavelength can be measured by a spectrophotometer. From this, we can calculate the corresponding absorber concentration when reaching a certain depth and light attenuation to a certain intensity.

5.5 Optimization of Several Typical Structures

The target structures matter in controlling printing results. Even with the same material adopted, the different structure requires different selections of printing parameters. For example, the light intensity for printing a "cube" is lower than that for printing a "needle," and exposure time is longer ensuring enough cross-linking

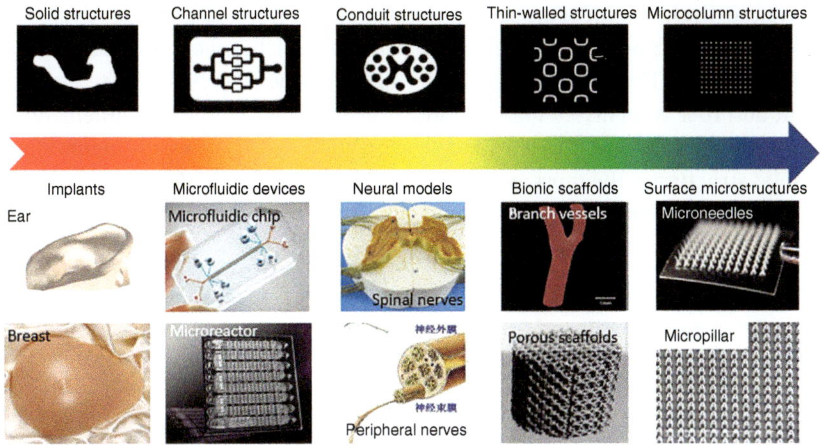

Figure 5.12 Five types of structures proposed from specific applications.

density. It is difficult to derive a unified printing strategy to select the best printing parameters and choose suited material systems when the target structures are changing. The target structures applied in biomedical fields are usually complicated with their shape which is composed of complex 3D curved surfaces, i.e. ear, heart, or liver. In other cases, the structures used in preclinic applications require self-design, like microfluidic chips, drug screening models, and tissue engineering repair. Thus, it is a big challenge to know how to design the structures with suitable size and shape corresponding to the cross-linking properties of materials, printing processing, and the biocompatibility in practical applications.

From the perspective of the digital light processing bioprinting (DLPBP) process, the printing results are closely related to the sliced patterns of the structure. Almost all of the structures applied in biomedical engineering can be divided into five basic typical structures: solid structure, channel structure, conduit structure, thin-walled structure, and microcolumn structure (Figure 5.12). Within the printing process of these five typical structures, the requirements of materials and the selections of printing parameters are different due to the different key features and the printing difficulties. Correspondingly, five printing strategies are proposed to guide the process of successful DLPBP printing (Table 5.2).

5.5.1 Printing Strategies of Solid Structures

Solid structures refer to the structures without a cavity inside, which are regarded as the easiest structures to print in projection-based printing, the shape of the slices of which are single and concentrated, as shown in Figure 5.5I. Solid structures have extensive applications in prosthesis manufacturing, implant regeneration, tissue repairing, and so on. The whole model of solid structures is usually nongeometric and looks like solid blocks. The largest printing error of solid structures is the fidelity of the dimensions caused by the internal shrinkage of GelMA hydrogel.

Therefore, the key point of the strategy is to adjust the internal shrinkage degree of controlling the photo-cross-linking density at a low level, but successful for keeping the printed shape. Thus, a low concentration of GelMA was adopted and the amount of light absorber is relatively high and low exposure dose of low light intensity or short exposing time is applied in solid structure printing. It is worth mentioning that solid structure is the most available model potential for cell-laden printing for the low concentration of hydrogel adopted due to the insurance of bio-performance of the cell.

5.5.2 Printing Strategies of Channel Structures

The channel structures are usually applanate, i.e. the scale of height is much smaller than the scale of the sliced area. The channel structures can be applied in building preclinic applications as microfluidic chips, microreactors, and a large mass of disease models with channels in direct 3D printing ways. The feature of the slice patterns of channel structures is that there are some non-exposure patterns of microfluidic or holes within large exposure area, because of which the scale of the non-exposure area is smaller than the exposure area, as shown in Figure 5.5II. The difficulty of printing channel structure lies in the need to improve the accuracy of microfluidic as much as possible under the premise of ensuring the overall forming. Due to the phenomenon of superimposed light intensity of pixel lights and the influence of the scattering situation of light, the microfluidic structure will most likely fail to print.

Therefore, the printing strategies must base on the structural features of channel structures. To ensure the forming of the microfluidic feature, a big dose of light absorber tartrazine is adopted to avoid the redundant curing of scattering light. Besides, large light intensity and short exposure time are adopted to cross-link the GelMA hydrogel, which inevitably leads to small curing depth so that small layer thickness is required in a slice set of channel structures.

5.5.3 Printing Strategies of Conduit Structures

The conduit structures are tubular with perforated holes in the structures, the height scale of which is usually bigger than the scale of the sliced area. Different from the channel structure, the channel direction of the conduit structure becomes lengthways from the crosswise of the channel structure, so the heights of conduit structures are commonly high. And the slice patterns of each layer are all same, i.e. there are no lateral features in the structures. The main difficulty in the printing of conduit structures successfully is that perforated holes get easily blocked, the reason for which is the cross-linking of GelMA hydrogel trapped in the perforated holes due to the presence of capillarity under repeated exposures during the printing process. Without the circulation of GelMA bioink, the molecular weight of GelMA hydrogel will gradually increase as the projection time goes on.

Therefore, the key to the successful printing of conduit structures is reducing the overexposing situation in the holes. The best way to improve the block situation is to decrease both the intensity of scattering light and exposure time, which means small light intensity and large layer thickness are required. To get photo-cross-linked in this situation, a high concentration of GelMA is adopted to ensure the cross-linking rate and low concentration of light absorber tartrazine is added to improve the curing depth matching the big layer thickness.

5.5.4 Printing Strategies of Thin-Walled Structures

The thin-walled structures are a common type of porous scaffolds, which means the thin and discrete exposing area in sliced pattern, the exposure areas are separated by the non-exposure areas and the scales of exposure area are much smaller than nonexposed. Analyzing the characteristics of thin-walled structures, due to the small exposure area, leads to an insufficient exposure dose, which is the main reason that causes the printing failure of this kind of structure. Thus, the strategy for overcoming this situation is keeping over-exposure of each layer to ensure the shape retention by increasing the cross-linking density. Big exposure doses are applied when printing thin-walled structures for the cross-linking rate of thin-wall feature which is more likely to cause structural damage due to underexposure.

Another issue of concern is the printing resolution of thin-walled structures, including the printing resolution of the structure and the resolution of internal pores, i.e. the minimum size of the thin-wall and internal pores. The printing resolution is closely related to the photocuring characteristics of the material. The ideal photocuring material can perfectly reproduce the size and resolution of the designed model. However, due to the low photocuring rate and low optical cross-linking density of current hydrogel materials such as GelMA, thin-walled structures must be fabricated by controlling the formation of over-curing, which inevitably sacrifices part of the printing resolution. As for the resolution of internal pores, the pores of the thin-walled structure are usually interconnected, which is conducive to the circulation of bioink and makes it less likely to cause blockage.

5.5.5 Printing Strategies of Microcolumn Structures

The microcolumn structure is composed of micropillars or microcolumns, and usually is sheet-like. This kind of structure is often adopted in manufacturing microneedles for drug release research or research on cell behavior on the microstructure surfaces. The features of microcolumn structures are similar to thin-walled structures, the exposing areas are thin and discrete, but sometimes displays as arrays and the columns are close to each other and the height of microcolumn structures are generally low.

Therefore, the key to fabricate this structure is to print out the microcolumns with the whole shape and without deformation. The printing strategies of microcolumn structures are different from the strategies of thin-walled structures because the over-exposure situation exists if large exposure doses are applied leading to the redundant curing between columns. To manufacture the microcolumn structures with good fidelity, a high concentration of GelMA is selected to improve the

Table 5.3 The summary sheet of printing strategies.

Structure type	Applications	Structure features	Technological difficulty	Printing analysis	Printing advices
Solid structure	Implants: ear, nose, beast	Solid, the exposure area is large and concentrated	Surface contour accuracy Size of solid structure	① Easy to shape and large printing window ② Easy to fall off the printing platform	① Low concentration of GelMA (5–10%) ② High concentration of photoabsorber ③ Low exposure dose
Channel structure	Microfluidic chips, microreactors	Flat structure with microchannel, unexposed area surrounded by large exposure area	Size and forming accuracy of microchannels	① Overall neat and easy to from ② Flow channel is easily blocked due to overexposure	① Appropriate concentration of GelMA (10–15%) ② High concentration of photoabsorber ③ Small layer height ④ High light intensity and short exposure time
Conduit structure	Nerve conduits	High structure with holes in the section, surrounded by large exposure area	Size and forming accuracy of holes	① Overall neat and easy to from ② Material flow is not smooth in the holes and easy to block	① Appropriate concentration of GelMA (10–15%) ② Low concentration of photoabsorber ③ Large layer height ④ Evenly distributing the holes
Thin-walled wall	Topological scaffolds, Bionic vascular	Thin structure and strips with small exposure area and uniform are distribution	Forming integrity and internal permeability of thin-walled structure	① Small exposure dose ② Hard to support its own structure ③ Increasing cross-link density to ensure the rigidity of structure	① High concentration of GelMA (12–18%) ② Appropriate concentration of photoabsorber ③ High exposure dose ④ Increasing the wall thickness appropriately

Table 5.3 (Continued)

Structure type	Applications	Structure features	Technological difficulty	Printing analysis	Printing advices
Microcolumn structure	Microneedles, microarray, micropillars	Thin structure with a base and tiny columnar structures on it.	Complete forming of the tiny exposure area	① Easy to form the base	① High concentration of GelMA (12–18%)
				② Micropillar need sufficient exposure	② Low concentration of photoabsorber
				③ Densely arranged micropillars are easy to cause printing errors	③ Large layer height
					④ High exposure dose

cross-linking density to retain the shape, correspondingly, a light absorber is added to decrease the overexposing caused by scattering light, and high light intensity is adopted to improve the cross-linking rate. Exposure time is adjustable to control the printing results.

In summary, developing a unified printing method of DLPBP with hydrogel materials is unpractical, to overcome the difficulty, five printing strategies are proposed matching five kinds of basic typical structures for practical applications. Based on the different sliced patterns of each structure, the guide of the GelMA system formulations and the selection of printing parameters are different. And the analysis of each structure is summarized in Table 5.3. We believe that the printing strategies can be extended to other hydrogel materials for further biomedical engineering applications.

5.6 Applications

5.6.1 DLPBP Structures with High Precision

DLPBP has significant advantages in forming complex structures with great precision. Here, we demonstrate some successfully printed biological models (Figure 5.13). These models are all made of GelMA and are impossible to be printed by traditional extrusion printing.

5.6.2 Customized Physical Properties Bioprinting

The mechanical performance of GelMA is related to its cross-linking degree. As analyzed above, the parameters that affect cross-linking degree are the same as what we

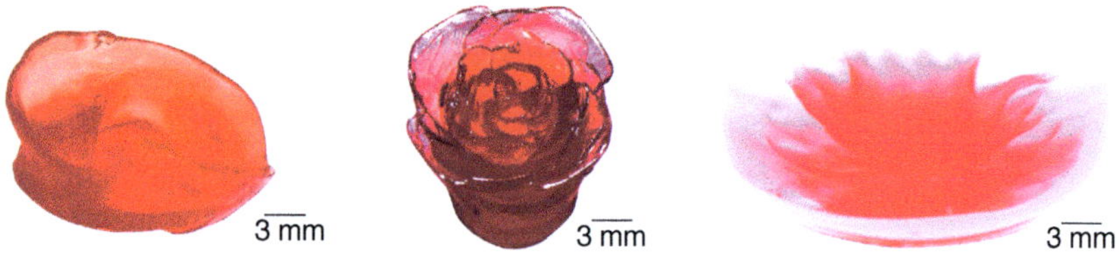

Figure 5.13 DLPBP GelMA models: ear, rose, and lotus. Source: Reproduced with permission of IOP Publishing.

used to control DLPBP printing qualities. This means we can control the mechanical properties by matching printing parameters. Here we conducted a lot of experiments and found the law that how these parameters influence mechanical properties. A parameter-mechanical properties map was fitted by mathematical methods based on experimental data. In this map, you can set your target tensile/compression modulus/strength and find corresponding parameters that will help you to achieve your goal. Furthermore, the optional range of modulus and strength is reasonably wide, from 0.1 to 400 kPa modulus and 1 kPa to 1 mPa strength (Figure 5.14). This almost covers most of the soft tissues.

5.6.3 Regenerative and Biomedical Applications

Blood vessels are the most common and basic tissues in the body. Both real tissues and artificial substitutes need vessels to bring nutrition and take metabolic waste away. Recently, cardiovascular-related diseases are increasing, because of the improvement of people's living conditions. So it is meaningful to construct artificial blood vessel substitutes.

Although the blood vessel looks simple in structure, it also needs enough strength to resist blood pressure and the entire endothelial cell layer to prevent blood clotting. Here, we choose a branching vessel model to demonstrate the application potentials of the DLPBP method. A branching vessel is a biological-functions-demanded structure with specific mechanical properties. It is possible to print an artificial branching vessel model with high precision, real mechanical properties, and good biocompatibility using the DLPBP technology.

Herein, a part of the maxillary artery was chosen to demonstrate customized organ printing. The demands for this application include three aspects. The manufacturing precision should be accurate enough to ensure structural integrity and vascular patency. The tensile modulus should be approximately 120 kPa, while tensile strength should be more than 150 kPa. Finally, HUVECs should be able to adhere to the scaffolds and proliferate in adequate numbers to cover the entire inner surface.

According to the mechanical map, we chose 100%-NH_2 30%-OH degree of substitution, while the other two corresponding parameters were a 19% GelMA concentration and 44 seconds of exposure. 0.5% photoabsorber was added to the materials to improve printing precision. As shown in Figure 5.15c, the printed branching vessel nicely presented the shape of the target model (Figure 5.15b) with unobstructed piping. The tensile test achieved a good result with acceptable error compared to the

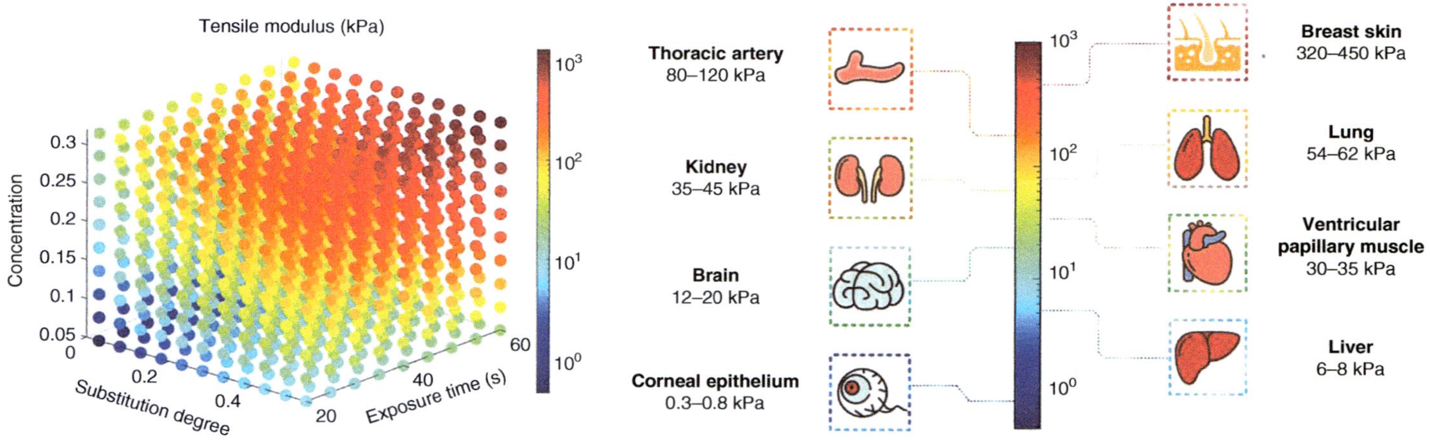

Figure 5.14 Mechanical parameter map: controllable mechanical properties. Source: Reproduced with permission of IOP Publishing.

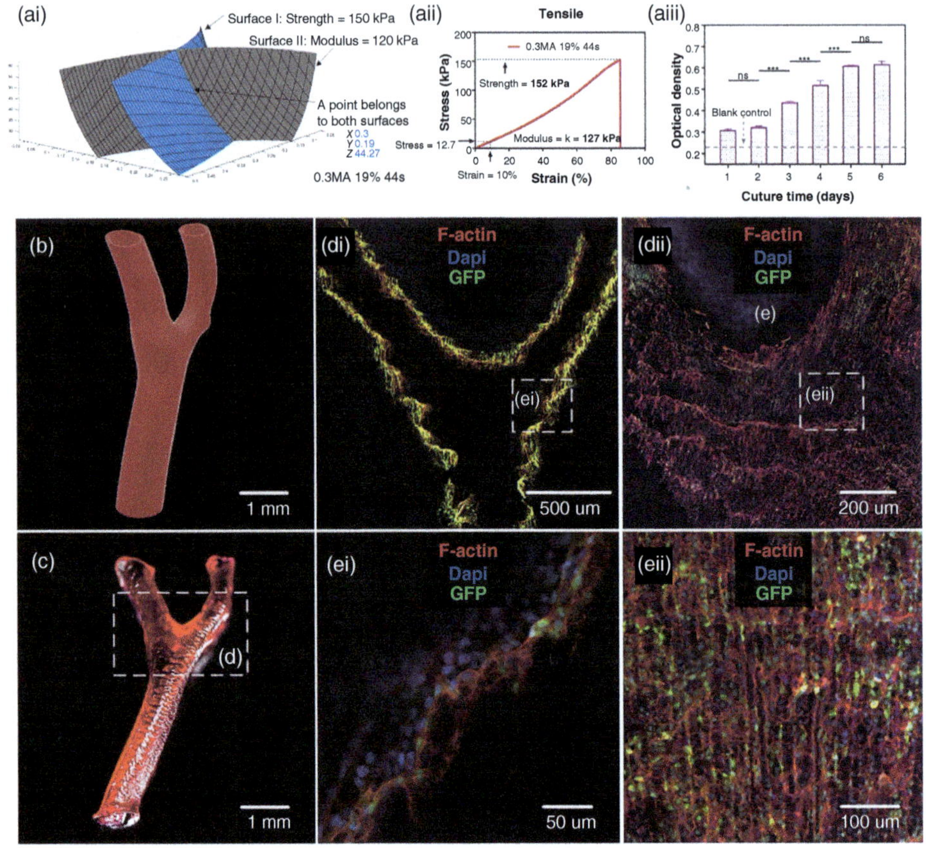

Figure 5.15 DLPBP application case: mechanical properties controllable and biocompatible branching vessels. ***$p < 0.001$, **$p < 0.01$, *$p < 0.05$, Ns $p > 0.05$. Source: Reproduced with permission of IOP Publishing.

target (Figure 5.15aii). The optical density of the samples tended to increase during the culture, indicating a normal proliferation rate of GFP-HUVECs seeded on the scaffolds (Figure 5.15aiii). GFP-HUVECs adhered to the inner surface clearly and outlined the branching vessel (Figure 5.15di). After combining the confocal images together with a half vessel, the whole lumen appeared to be fully covered by GFP-HUVECs (Figure 5.15dii). A monolayer of GFP-HUVECs was clearly identified in Figure 5.15ei, and the F-actin molecules of different cells were linked.

References

1 Sun, Y., Yu, K., Nie, J. et al. (2020). Modeling the printability of photocuring and strength adjustable hydrogel bioink during projection based 3D bioprinting. *Biofabrication* 13 (3): 035032.

2 Kolesky, D.B., Truby, R.L., Gladman, A.S. et al. (2014). Bioprinting: 3D bioprinting of vascularized, heterogeneous cell-laden tissue constructs. *Advanced Materials.*

3 Mandrycky, C., Wang, Z., Kim, K. et al. (2016). 3D bioprinting for engineering complex tissues. *Biotechnology Advances*: 422–434.

4 Zhao, H., Chen, Y., Lei, S. et al. (2018). Airflow-assisted 3D bioprinting of human heterogeneous microspheroidal organoids with microfluidic nozzle. *Small* 14: 1802630.

5 Vicente, M.F., Canyada, M., and Conejero, A. (2015). Identifying limitations for design for manufacturing with desktop FFF 3D printers. *International Journal of Rapid Manufacturing* 5 (1): 116.

6 Mu, Q., Wang, L., Dunn, C.K. et al. (2017). Digital light processing 3D printing of conductive complex structures. *Additive Manufacturing* 18:S2214860417300398.

7 Chia, H.N. and Wu, B.M. (2015). Recent advances in 3D printing of biomaterials. *Journal of Biological Engineering* 9 (1): 1–14.

6

Direct Ink Writing for 3D Bioprinting Applications

6.1 Introduction

Direct ink writing (DIW), which can deposit inks using computer-controlled nozzles to create models with specific compositions, is one of the most popular biomanufacturing methods for constructing complex three-dimensional (3D) structures at the micro and nanoscale [1, 2]. DIW, the first developed extrusion-based bioprinting technology, can be divided into two categories depending on the printing mechanisms: (i) continuous filament printing and (ii) discontinuous droplet printing [3, 4]. When bioinks are used as the build materials, aiming at the formation of tissues or organs, continuous filament printing is always preferred [5]. During the printing process, bioinks are extruded out from the dispensing nozzles and deposited according to the designed CAD models to construct complex 3D tissues/organs layer by layer. Due to its versatility, flexible printing process, and wide range of printable materials, DIW has become widely used for *in vitro* 3D printing of tissues/organs [6–8].

According to the supporting mechanism of the printed architectures, DIW technology has two strategies: (i) rapid solidification-induced 3D printing and (ii) self-supporting material-assisted 3D printing . The main difference between these two strategies lies in the cross-linking process. In the first strategy, bioinks need to be cross-linked at a certain rate during printing to ensure that printed layers have sufficient mechanical stiffness for supporting subsequently deposited layers, as shown in Figure 6.1a. Since the cross-linking step is performed simultaneously along with the printing step, this strategy is also called the "cross-linking-while-printing " approach. Rapid solidification-induced 3D printing, which was developed earlier, is the mainstream DIW technique for cell printing applications. However, the rapid cross-linking mechanism required during printing greatly limits the range of printable materials. Moreover, it is difficult to precisely control the cross-linking rate, resulting in frequent clogging of nozzles. Therefore, in recent years, researchers have developed a new printing strategy: self-supporting material-assisted 3D printing, which overcomes the main technical challenges in rapid solidification-induced 3D printing.

In this strategy, self-supporting materials are used as additives to prepare hydrogel-based bioinks, which present sol state in a dispensing nozzle, but quickly

Cell Assembly with 3D Bioprinting, First Edition. Yong He, Qing Gao, and Yifei Jin.

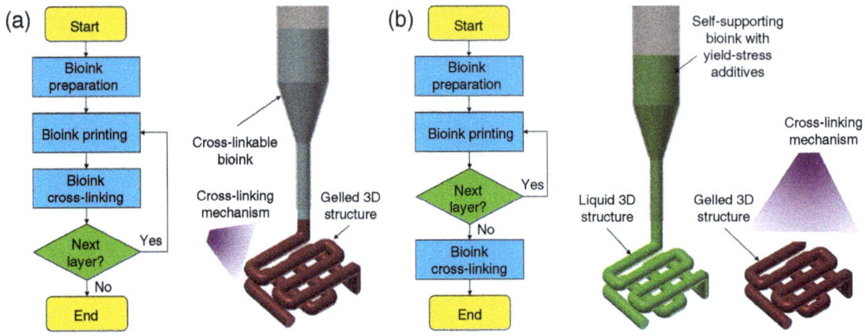

Figure 6.1 Schematic of (a) rapid solidification-induced 3D bioprinting strategy and (b) self-supporting material-assisted 3D bioprinting strategy. Source: Yifei Jin.

return to gel state after being extruded out of the nozzle. Thus, the printed filaments/layers have self-supporting property at gel state to keep the short-term structural integrity (as shown in Figure 6.1b). After the whole 3D structure is finished, the corresponding cross-linking mechanism(s) is utilized to solidify the structure. Since the cross-linking step is separated from the printing step in self-supporting material-assisted 3D printing, this new strategy is also named as the "printing-then-cross-linking " approach. During DIW, printing accuracy greatly depends on the rheological properties of bioinks and mechanical properties of the printed solid (in rapid solidification-induced 3D printing) and/or solid-like (in self-supporting material-assisted 3D printing) filaments/layers [9]. Therefore, how to achieve bioinks with suitable rheological properties as well as sufficient mechanical stiffness (after deposition) become one of the main challenges in the development of DIW technology in the future.

The purpose of this chapter is to provide an overview of the DIW technology. First, different supporting mechanisms during rapid solidification-induced 3D printing and self-supporting material-assisted 3D printing will be discussed in detail. Some representative bioinks will be reviewed. The criteria to design new bioinks for DIW will be proposed. Then, some technical issues during DIW will be discussed, including the printability of bioinks as well as various printing/cross-linking strategies to facilitate either rapid solidification or self-supporting. Finally, some biomedical applications using DIW will be summarized.

6.2 Printable Bioinks in DIW

The development of bioinks is of great significance in DIW, particularly for cell printing applications, which must possess appropriate rheological properties as well as rapid cross-linking capacity and/or self-supporting capacity. The goals of this section are to introduce different supporting mechanisms and representative bioinks as well as to further propose the design criteria of new bioink materials for DIW.

6.2.1 Supporting Mechanisms and Representative Bioinks

Rapid solidification-induced 3D printing and self-supporting material-assisted 3D printing strategies have totally different supporting mechanisms. The former relies on the rapid solidification of hydrogels to achieve sufficient mechanical stiffness during printing, while the latter mainly depends on the self-supporting property provided by yield-stress additives.

6.2.1.1 Rapid Solidification-Induced Mechanical Stiffness Improvement

Different cross-linking mechanisms are used during rapid solidification-induced 3D printing to rapidly cross-link/solidify/cure extruded filaments/layers. Based on the quantity of the cross-linking steps, it can be classified into two subcategories: single-step cross-linking and multi-step cross-linking. The cross-linking mechanisms in single-step cross-linking can further be categorized into physical cross-linking (including thermal cross-linking [10, 11] and ionic cross-linking [12]) and chemical cross-linking (including photo cross-linking [13], pH cross-linking [14], and enzyme cross-linking [15]), as shown in Figure 6.2. Bioinks in multi-step cross-linking are usually composed of multiple biomaterials. Thus, different cross-linking mechanisms are performed to cross-link the corresponding biomaterials in multiple steps [16]. In this section, various cross-linking mechanisms will be discussed and then representative bioinks, especially hydrogels, will be introduced.

6.2.1.1.1 Physical Cross-linking

Thermal Cross-linking Thermal cross-linking can realize rapid *in situ* solidification of thermal cross-linkable hydrogels by controlling different temperatures of the hydrogel bioinks during and after printing. Usually, these hydrogels have low viscosity within a certain temperature range. Once the temperature exceeds or falls below a certain temperature threshold, the hydrogels start to be cross-linked and rapidly present gel state [17], enabling the deposited hydrogel layers to have

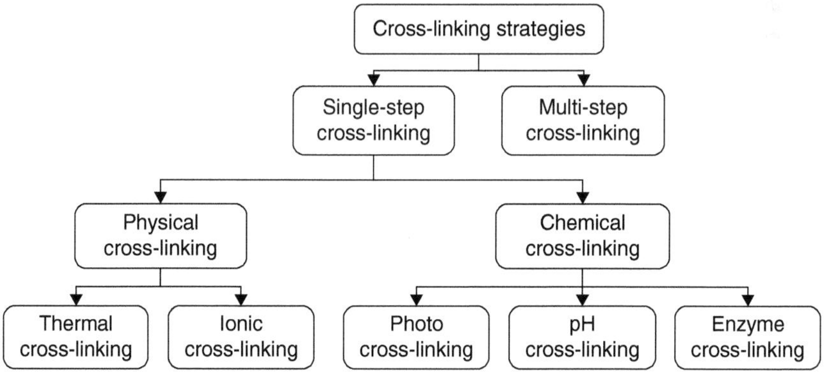

Figure 6.2 Classification of cross-linking strategies during rapid solidification-induced 3D bioprinting. Source: Yifei Jin.

mechanical stiffness for supporting the next layers. Commonly used thermal cross-linkable hydrogels include agarose [18], collagen, gelatin [19], Matrigel™, Pluronic [20], polyethylene glycol (PEG), and poly(propylene fumarate-*co*-ethylene glycol) (PPF-*co*-EG), etc. Since agarose, collagen, Matrigel, Pluronic, and PEG have been introduced in previous chapters, herein, only gelatin and PPF-*co*-EG will be discussed.

Gelatin, mainly derived from the denaturation of collagen [21], widely exists in animal bones and tendons. Gelatin undergoes a reversible thermal cross-linking at ~27 °C that can promote cell adhesion, migration, proliferation, and differentiation [22]. The main advantages of gelatin include high gelation rate, excellent biocompatibility and biodegradability, low antigenicity, and no harmful byproducts during degradation. In addition, gelatin is easy to process and inexpensive. These characteristics, especially its excellent biocompatibility, make it a universal biomaterial for tissue engineering and cell printing applications. However, due to its low transition temperature between sol and gel states, 3D structures from pure gelatin have poor structural stability at physiological temperatures. Thus, gelatin is always mixed with other hydrogels such as alginate to prepare composite bioinks, which are commonly used in 3D bioprinting with multiple cross-linking steps. During printing, the rapid gelation rate of gelatin can facilitate the fabrication of complex 3D structures, while the irreversible cross-linking of other hydrogel components can maintain the structural completeness during cell incubation at ~37 °C. For example, Duan et al. [23] used alginate-gelatin bioink for bioprinting. The extruded gelatin was cross-linked at low temperature to ensure the short-term stability of the printed structure. After printing, sodium alginate was further cross-linked by calcium chloride solution [24–26]. Another approach to print gelatin is to chemically modify gelatin into both physical and chemical cross-linkable hydrogel, such as GelMA. Shao et al. [16] printed hydrogels containing GelMA microgels onto a cold platform for an initial thermal cross-linking and the printed architecture was permanently stabilized by photo cross-linking of GelMA after printing.

PPF-*co*-EG is a block copolymer of hydrophobic polyester PPF and hydrophilic PEG [27]. When the temperature reaches 37 °C, PPF-*co*-EG gets cross-linked with the formation of copolymer chains. The increase of PEG concentration within this composite hydrogel provides better cellular responses with higher post-printing cell viability [28]. PPF-*co*-EG supports chondrocyte survival and extracellular matrix (ECM) production *in vitro* [29]. PPF-tricalcium phosphate's complex and fully cross-linked PPF-*co*-EG hydrogel has been investigated as a good substrate for the proliferation and differentiation of seeded osteoblasts and rat bone marrow osteoblasts [30, 31].

Ionic Cross-linking Ionic cross-linking is another commonly used physical cross-linking mechanism in which cross-linkers are utilized to form reversible bonds between hydrogel molecules. Sodium alginate [32–35] is a typical ionic cross-linkable hydrogel, which has mild cross-linking conditions, ease of adoption, good biocompatibility, and low cost. During printing, calcium chloride is usually used as a cross-linker. The main challenge during alginate printing is the fast cross-linking speed, which can rapidly cross-link both the deposited alginate bioinks and the bioinks in a dispensing nozzle, resulting in nozzle clogging. Thus, how to control alginate cross-linking speed becomes a critical problem.

Gellan gum, produced by bacteria, is a hydrophilic and anionic polysaccharide with a high molecular weight. Similar to alginate, gellan gum gels when mixed with monovalent or divalent cations at low temperatures [36]. In the food and pharmaceutical industries, gellan gum, which has been approved by US Food and Drug Administration (FDA) [37], is mainly used as a gelling agent and stabilizer. In the field of regenerative medicine, gellan gum is mainly used as a component of bioinks for tissue/organ printing applications. It is proved that bioinks with gellan gum have excellent rheological properties and can achieve 3D structures with improved shape fidelity [38, 39]. In addition, gellan gum is a low-cost hydrogel with tunable rheological properties, which may further broaden its biomedical applications in the future [40].

6.2.1.1.2 Chemical Cross-linking

Photo Cross-linking Photo cross-linking is one of the most commonly used cross-linking methods in rapid solidification-induced bioprinting. When a photo cross-linkable bioink is exposed to ultraviolet (UV) and/or visible light with appropriate wavelength, the bioink gets cross-linked to entrap living cells *in situ*. Due to its low energy cost, no-pollution, and easy implementation, this cross-linking method is considered to be one of the most promising methods for cell printing applications. However, photo cross-linkable hydrogels usually have low viscosities and relatively low cross-linking rates. Thus, it is difficult to rapidly cross-link deposited filaments into solid state and ensure the structural stability of 3D structures from photo cross-linkable hydrogels. The commonly used photo cross-linkable hydrogels in cell printing include *N*-isopropylacrylamide (NIPAAm), methacrylate hyaluronic acid (HAMA), GelMA, dextran derivatives, PEGDA, etc.

NIPAAm is a derivative of poly (acrylic acid), which can be cross-linked by UV radiation to form poly(*N*-isopropylacrylamide) (pNIPAAm). At a temperature higher than its low critical solution temperature (LCST) (about 32 °C), pNIPAAm behaves hydrophobically [41, 42]. The LCST of pNIPAAm can be adjusted by copolymerizing [43, 44] or adding salt [45, 46] and surfactant [47] to pNIPAAm. For example, the copolymerization of NIPAAm with hydrophilic monomers, such as acrylic acid (pNIPAAm-*co*-AAc), can adjust the AAc content to control the LCST value of pNIPAAm-*co*-AAc in a neutral pH environment at physiological temperature [48]. pNIPAAm has low toxicity, so it can be widely used in cell culture [49, 50]. Kikuchi et al. [49] used pNIPAAm as a grafted surface to culture cells and prepared cell sheets for tissue engineering applications (as shown in Figure 6.3). Kesti et al. [51] grafted pNIPAAm onto hyaluronic acid (HA) and mixed them with HAMA for cell printing applications. It is found that cells cultured on the surface of HA-pNIPAAM had a high survival rate even though a large number of encapsulated cells died. However, after HA-pNIPAAm was washed away, the cells encapsulated in the printed structure showed high cell viability after 7-day incubation. Therefore, pNIPAAm is more suitable to be used as a sacrificial and supporting material.

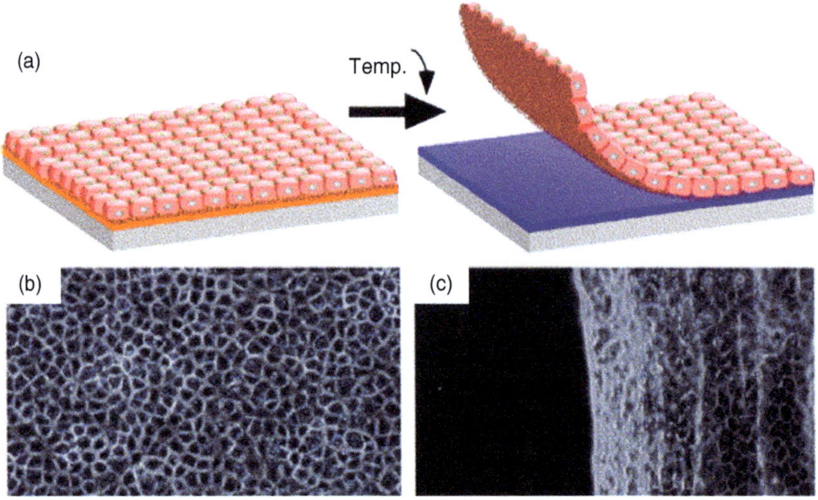

Figure 6.3 Engineered cell sheet using pNIPAAm-grafted surfaces. (a) Schematic of culturing temperature-induced recovery of intact monolayer. (b) Confluent endothelial cells on pNIPAAm-grafted dishes at 37 °C. (c) Detaching endothelial cell sheet by lowering the culturing temperature to 20 °C. Source: Kikuchi et al. [49]. Reproduced with permission of Elsevier.

In addition, the unfamiliarity of metabolic process of pNIPAAm and its activation of platelets in the blood limit the *in vivo* applications of pNIPAAm.

HAMA is a naturally derived photo cross-linkable hydrogel, which is mainly obtained by the chemical modification of hyaluronic acid with methacrylate. Hyaluronic acid is a non-sulfated linear polysaccharide with anions [52], which can be extracted from animal products or produced by bacterial fermentation (such as *Bacillus subtilis*) with controlled molecular weight. It is the main component of connective tissues, epithelial tissues, nerve tissues, joint synovial fluids, and natural ECM [53]. As a biocompatible and hydrophilic polymer, HAMA plays a key role in wound healing by promoting cell migration and proliferation. In addition, HAMA can be biodegraded by hyaluronidase and reactive oxygen species in mammals to produce low molecular weight hyaluronic acid and oligosaccharides. Thus, HAMA has been used to make heart valve-like 3D structures [54].

Dextran is a natural polysaccharide that has been widely used in tissue engineering applications due to its non-toxicity and hydrophilicity. In mammalian tissues, dextran can be degraded by glucanase. Thus, it is classified as a biodegradable material [55]. Pescosolido et al. [56] used HA and hydroxyethyl-methacrylated dextran (dexHEMA) to prepare the hydrogel with polysaccharide semi-interpenetrating polymer networks and mixed with horse chondrocytes to form the bioink. The results showed that HA-dexHEMA hydrogel had good viscoelasticity and exhibited pseudoplastic characteristics, which were very suitable for cell printing.

pH Cross-linking Chitosan is a representative pH cross-linkable biodegradable glucosamine polymer , which has been widely used for various biomedical applications. It is mainly presented in the exoskeleton of arthropods and insects and can be obtained by deacetylation of chitin [57, 58]. Acid-soluble chitosan can be cross-linked by adjusting the pH, but the gelation time is long, and the mechanical properties are poor. A high degree of deacetylation of chitin chain can increase its biocompatibility [59]. Human periodontal ligament stem cells (PDLSCs) printed with chitosan showed a good cell survival rate after printing [60]. Chitosan molecules, which have similar structures as glycosaminoglycans (GAG), can be cross-linked with GAG to promote cartilage formation [61] and the formed structures are able to be self-supported [62].

Enzyme Cross-linking Similar to ionic cross-linking, in enzyme cross-linking , enzyme and corresponding hydrogels are mixed together to cause the gelation of hydrogels. Silk is a typical enzyme cross-linkable hydrogel, which is a natural polymer that has been used for centuries as a suture material in medical applications. It has many excellent characteristics, such as high elasticity and slow degradation rate, which can provide sufficient support for cells during and after printing. In addition, its low immunogenicity and good biocompatibility also make it suitable for clinical applications. Thus, silk has been developed as a new bioink for 3D cell printing. One reported silk-based bioink was prepared by suspending mesenchymal progenitor cells derived from human subnasal turbinates in a mixture of gelatin and silk fibroin solution. This bioink can be enzymatically cross-linked with mushroom tyrosinase [15]. The study showed that the prepared bioink supported the differentiation of cartilage as well as fat and osteogenic differentiation after cross-linking.

6.2.1.2 Yield-stress Additive-Induced Self-Supporting Capacity

Different from rapid solidification-induced 3D printing, yield-stress additive-assisted self-supporting strategy is performed in the "printing-then-cross-linking" fashion. The core of this new extrusion 3D bioprinting strategy is to improve the self-supporting property of hydrogel precursors before cross-linking. Currently, two classifications of yield-stress additives have been proposed: internal scaffold additives and microgel additives.

6.2.1.2.1 Internal Scaffold Additives

Internal scaffold additives are yield-stress materials that can rapidly and frequently switch between liquid and solid-like states under different stressed conditions. The liquid state under the sheared condition makes it possible to be extruded through a dispensing nozzle, while the solid-like state under the non-sheared condition leads to the self-supporting property of deposited structures from yield-stress materials [63]. The internal scaffold additives are usually composed of nano- and/or microparticles with inherent microscale arrangements, and yield stress is the threshold force/energy to break up these microscale arrangements. When mixed with other hydrogel precursors to prepare hydrogel composite inks, internal scaffold additives can remain the inherent microscale arrangements in the composites, which makes the inks inherit this self-supporting property and enables directly printing into 3D structures at a liquid state in air. Once the printing process is completed, hydrogel precursors will be cross-linked to further stabilize the structural integrity of the printed structures.

Internal scaffold additives have been proposed for 3D bioprinting applications since 2017. Laponite® nanoclay is a representative yield-stress internal scaffold additive, which is a synthetic nanoclay widely used in industries. Laponite® nanoclay is composed of numerous nanosilicates that have a high aspect ratio with a thickness of 1 nm and diameter of 20–40 nm [64], as shown in Figure 6.4a. When Laponite® nanoclay is dispersed in an aqueous solution, sodium ions dissociate from individual nanosilicate, leaving the surface negatively charged, while the dissociation of hydroxide ions at the edge results in the edge with positive charges. This charge distribution drives Laponite® nanoclay to possess a stable "house-of-cards" arrangement (Figure 6.4a) and form a colloidal suspension with yield stress. When the applied stress on the nanoclay suspension exceeds its yield stress, the "house-of-cards" arrangement will be destroyed, and the suspension will present a liquid state. When the stress is below the yield stress, the nanoclay suspension can quickly restore equilibrium and behave like a solid, thereby providing the necessary support for the printed structure.

Laponite® nanoclay can be mixed with various hydrogel precursors such as alginate, PEGDA, gelatin, etc., to prepare hydrogel composite inks. After mixing with nanoclay, all these hydrogel precursors possess a similar yield-stress property like nanoclay suspension. Thus, during extrusion, hydrogel composite inks experience higher shear stress and behave as shear-thinning fluids, which can be easily extruded from a dispensing nozzle. After extruded out, shear stress decreases below the yield stress. The filament from hydrogel composite inks rapidly returns to a solid-like state to maintain the shape as-deposited that can support the following deposited layers without undergoing cross-linking as shown in Figure 6.4b.

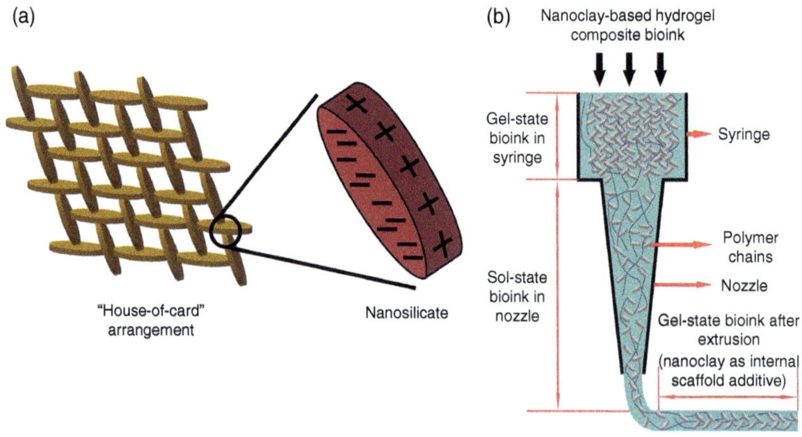

Figure 6.4 (a) Inherent "house-of-cards" arrangement of Laponite® nanoclay suspension and charge distribution on nanosilicate and (b) schematic of nanoclay-assisted 3D bioprinting in which nanoclay functions as internal scaffold additive. Source: Yifei Jin.

6.2.1.2.2 Microgel Additives

Microgels are cross-linked hydrogel microparticles, which can be prepared by various techniques, such as batch emulsions, microfluidic emulsions [65], lithography, electrohydrodynamic spraying [66], and mechanical fragmentation [67], to name a few. Microgels are made from both natural and synthetic hydrogels, and their size ranges from 1 to 1000 μm [68]. When dispersed in bioinks, these cross-linked microgels adhere densely to each other due to the mutual physical interaction and form a jammed state, which enables the bioinks to exhibit a yield-stress property and behaves like solid [69]. During extrusion, since the applied shear stress exceeds the inherent cohesive forces between cross-linked hydrogel microgels, relative motion occurs among microgels, which makes the bioinks behave like liquid [70]. When extruded out, the applied shear stress reduces below the yield stress. Thus, the relative motion disappears, and the microgel system returns to a solid-like state, enabling the deposited filaments to possess mechanical stiffness to support the subsequently printed filaments and/or layers, as shown in Figure 6.5a. After printing, the bioinks can be further cross-linked using different cross-linking mechanisms to improve the mechanical properties of the printed architectures, as shown in Figure 6.5b.

Since microgel additives can be created from hydrogels with excellent biocompatibility, the resulting bioinks usually not only have good printability and self-supporting property but also are completely cytocompatible to living cells, which significantly benefits the following cell incubation and proliferation. As a result, microgel additives are viewed as one of the most promising additives for 3D cell printing applications. Some representative microgel additives include norbornene-modified hyaluronic acid (NorHA) [69] and PEG-microgels [70].

6.2.2 Design Criteria of Bioinks for Direct Writing Applications

From the aforementioned discussions, the development of suitable bioinks has become one of the main challenges during DIW-based cell printing. The ideal

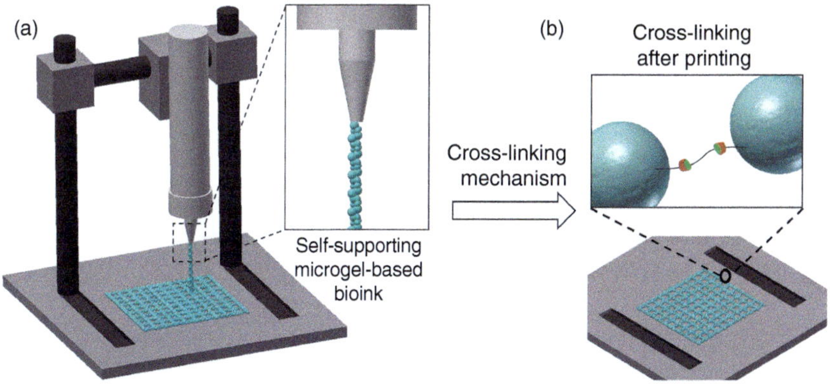

Figure 6.5 Schematic of microgel additive-assisted 3D bioprinting strategy. (a) Extruded filament from microgel-based bioink exhibits solid-like state due to the inherent cohesive forces among microgels. (b) 3D structures being cross-linked after printing. Source: Yifei Jin.

bioinks for DIW must have the following characteristics [71]: (i) suitable rheological properties; (ii) good cross-linking capacity; (iii) excellent biocompatibility and tunable biodegradation; and (iv) similar mechanical properties to natural tissues and organs after cross-linking. In this section, the design criteria of biomaterials suitable as bioinks will be discussed in detail from the aspects of rheological properties, cross-linking capacity, biocompatibility, and mechanical properties.

6.2.2.1 Rheological Properties

(1) Viscosity is one of the most important rheological properties of bioinks. The level of viscosity greatly affects the pressure required to form filaments with controllable dimensions and well-defined shapes. Usually, bioinks with higher viscosity are easier to form stable filaments after extrusion and have better self-supporting property [72]. However, highly viscous bioinks may result in nozzle clogging and high shear stress during extrusion may cause damages to the living cells in the bioinks. In addition, high viscosity has negative effects on cell growth, migration, and differentiation during the incubation process after printing. There are many factors that can affect the viscosity of a bioink, such as polymer type, concentration, molecular weight, solubility, etc. By optimizing these factors, the viscosity of the bioink can be flexibly adjusted to facilitate the printing process.
(2) Shear-thinning behavior is another very important rheological property of bioinks. During extrusion, bioinks with shear-thinning behavior can transform to a low-viscosity state so that these bioinks are able to be smoothly extruded out of a dispending nozzle and cause fewer damages to living cells in the bioinks. After extrusion, when the shear rate decreases, bioinks immediately restore to a high-viscosity state and form stable filaments.
(3) Yield stress is the most important rheological property of bioinks for self-supporting material-assisted 3D printing strategy, which is the minimum stress that makes the bioinks switch between liquid and solid-like states. Typical yield stress for printing liquid 3D structures in air is reported to be approximately 200 Pa [63].
(4) Thixotropic time needs to be considered when designing bioinks with yield-stress additives. From a microscopic aspect, this thixotropic time is the time scale bioinks take to recover the inherent microstructures after removing the applied shear stress. From the macroscopic aspect, thixotropic time is defined as the time bioinks take to finish the transition between liquid and solid-like states. Usually, short thixotropic time is required to design the bioinks for self-supporting material-assisted 3D printing. The typical thixotropic time scale falls in the range of 0.1–0.3 seconds [63, 73].

6.2.2.2 Cross-linking Capacity

Cross-linking capacity of bioinks is of great significance in rapid solidification-induced 3D printing strategy, in which bioinks must have a suitable cross-linking speed. If the cross-linking speed is too high, bioinks inside a dispensing nozzle may also be cross-linked, which may result in nozzle clogging. In contrast, if the cross-linking speed is too slow, printed filaments or layers are still in liquid state when the next filaments or layers are being deposited. Thus, the liquid filaments or layers lack sufficient mechanical stiffness to maintain the shape fidelity of printed structures. For example, during photo cross-linking, light intensity determines cross-linking speed, which further affects the stability and integrity of printed structures [74]. Lower light intensity usually leads to a low cross-linking speed. Therefore, deposited hydrogel filaments cannot be quickly and fully solidified at low light intensity. The higher light intensity can effectively improve the cross-linking speed, which makes the deposited filaments solidified before landing on the substrate. In this case, 3D structures are composed of batches of filaments without inter-filament fusion, resulting in poor mechanical properties and structural integrity. Only when the light intensity falls in an optimal range between 10 and $15\,mW/cm^2$ [74], a stable 3D structure can be successfully printed. Cross-linking method and time also need to be considered from the cellular aspect to assess the cross-linking capacity. For example, long-term UV irradiation with high intensity can significantly reduce cell bioactivity [75]. Thus, cross-linking capacity must be precisely controlled when designing a new bioink material.

In self-supporting material-assisted 3D printing strategy , cross-linking time and speed of bioinks also need to be designed. Unlike the cross-linking capacity in rapid solidification-induced 3D printing strategy, cross-linking time/speed of self-supporting bioinks are determined by the yield stress of bioink materials and the geometries of printed structure. When the printed structure has only a few layers, yield stress-induced self-supporting property can provide sufficient support. However, when multiple layers are printed, the bottom layer(s) may experience gravity-caused stress higher than the yield stress. Thus, the bottom layer(s) may switch to a liquid state, leading to the gradual collapse of the entire structure. To provide the printed structure with sufficient support at the bottom, it is necessary to carry out partial cross-linking to the bottom layer(s) during printing. As a result, the cross-linking capacity of self-supporting bioinks needs to be considered, which may affect the selection of suitable cross-linking strategies and operating conditions.

6.2.2.3 Biocompatibility and Biodegradation

Biocompatibility is an important indicator for evaluating bioinks. Biocompatible inks should also have the ability to support cell attachment, migration, proliferation, and functional expression and to remodel the ECM. In addition, 3D structures from biocompatible inks cannot cause immunogenicity after implantation [76]. Besides, bioinks should have a suitable biodegradation rate to maintain the shape of printed structure during incubation, which has a great effect on the production and reconstruction of the ECM. Ideally, the biodegradation rate needs to match the native ECM growth rate [77]. Also, the degradation products from biocompatible ink materials must be able to be absorbed or excreted by the body. Finally, the design of bioinks should meet the biological needs of the target tissue/organ. For example, in cartilage tissue engineering [78], the type of bioinks has a great influence on the phenotype of bone marrow mesenchymal stem cells. It is found that bioinks with hydrogels such as alginate and agarose can promote the formation of hyaline cartilage, while bioinks composed of GelMA and PEGMA can better support the development of fibrocartilage.

6.2.2.4 Mechanical Properties

Since different cell types require different culturing environments, printed 3D structures should have similar mechanical properties to the corresponding tissues/organs in nature and meet the functional requirements [79]. For example, bioinks with higher mechanical strength can induce stem cells to differentiate into osteoblast cell lines, while softer gelled bioinks may induce stem cells to differentiate into myoblast cell lines [80]. Thus, when designing a new bioink material, mechanical properties after cross-linking need to be considered based on its biomedical applications.

6.3 Technical Specifics in Direct Ink Writing

In this section, some technical specifics during DIW will be discussed, including printability, printing strategies in rapid solidification-induced 3D printing, and 3D structure printing using a self-supporting material-assisted 3D printing approach.

6.3.1 Investigation on Printability of Bioinks

Bioink printability is the primary requirement in DIW, which is composed of extrudability and formability. Extrudability is the capacity of extruding continuous filaments out of a dispensing nozzle. The studies on extrudability mainly focus on the droplet/filament formation behavior [81–85] (as shown in Figure 6.6a) and size [13, 85]. Formability is the capacity of forming filaments and/or filament patterns with well-defined shapes and controllable geometries after deposition, which has been investigated based on the printing resolution [13, 87–89], shape fidelity [74, 85, 90] (as shown in Figure 6.6b), and post-printing stability [10, 81, 83]. Several key factors including the bioink rheological properties [71, 87, 91, 92], bioink compositions [10, 88, 93, 94], and operating conditions [13, 90, 95] can severely affect the bioink printability.

Among these factors, the rheological properties and compositions of bioinks mainly affect the filament morphology, shape fidelity, and post-printing stability. In rapid solidification-induced 3D printing, a suitable viscosity range is required to form continuous filaments during extrusion and provide good shape fidelity after deposition [88, 95, 96].Different methods have been adopted to modify the viscosity of bioinks. For example, Chung et al. [91] mixed gelatin with sodium alginate to prepare an interpenetrating network (IPN) bioink with better printability. The extruded filaments had well-defined morphology and a controllable diameter. Thus, the printed structures also had better resolution. Mouser et al. [83] added gellan into gelatin methacrylate (GelMA) to improve filament deposition by inducing yielding behavior. In addition, cross-linking mechanisms can be used at the dispensing nozzle's exit to *in situ* adjust the viscosity of extruded bioinks. For

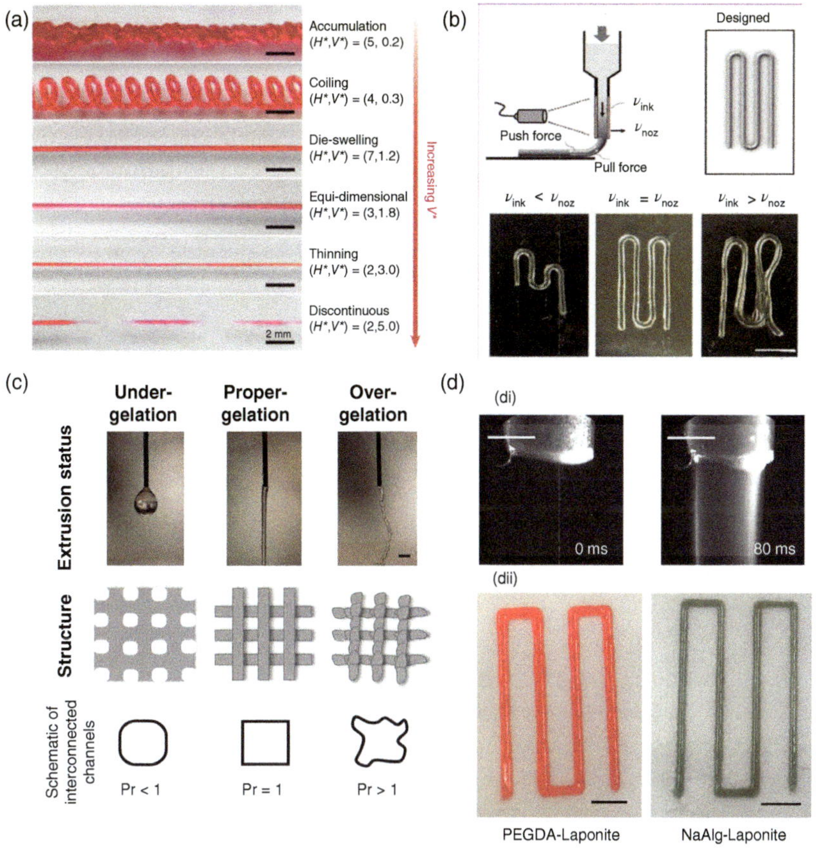

Figure 6.6 Assessment of the bioink printability. (a) Different types of filaments in DIW. Source: Yuk et al. [84]. Reproduced with permission of John Wiley and Sons. (b) Printing fidelity investigations by providing different extents of UV cross-linking at the nozzle's exit. Source: Ouyang et al. [74]. Reproduced with permission of John Wiley and Sons. (c) Printing fidelity of filaments with different gelation conditions. Source: Ouyang et al. [86]. Reproduced with permission of IOP Publishing, Ltd. (d) (i) Extrudability of yield-stress additive-based hydrogel bioinks. Source: Jin et al. [85]. Reproduced with permission of Springer Nature. (ii) Formability of yield-stress additive-based hydrogel bioinks. Source: Jin et al. [63]. Reproduced with permission of American Chemical Society.

example, Ouyang et al. [86] extruded gelatin-alginate bioink and divided the extent of gelation into three levels: under-gelation, proper-gelation, and over-gelation. When the bioink was at under-gelation state, the extruded filaments broke up into droplets due to surface tension and the printed structure had poor shape fidelity. When the bioink experienced over-gelation, it would be extruded irregularly and discontinuously. Only when the bioink was at proper-gelation state, smooth and uniform filaments can be formed, and the printed structures had excellent shape fidelity (as shown in Figure 6.6c). In self-supporting material-assisted 3D printing, the addition of yield-stress additives significantly improves the printability of bioinks. For example, Jin et al. [85] mixed nanoclay (an internal scaffold additive) with hydrogel precursors, and the resulting nanocomposite hydrogels had the yield-stress property, which presented excellent extrudability (Figure 6.6d(i)) and formability (Figure 6.6d(ii)).

Operating conditions usually determine the filament size and printing resolution. Some commonly used operating parameters in DIW include nozzle diameter, dispensing pressure, standoff distance, and path speed (nozzle speed) [82, 85, 95, 97]. Generally, smaller filament width and higher printing resolution can be achieved by decreasing the nozzle diameter and dispensing pressure while increasing the standoff distance and print speed [85, 95].

6.3.2 Different Printing Strategies in Rapid Solidification-Induced 3D Printing Approach

The core of rapid solidification-induced 3D printing approach is how to rapidly cross-link bioinks after deposition. Based on different cross-linking mechanisms, various printing strategies have been designed and implemented.

6.3.2.1 Printing of Thermal Cross-linkable Biomaterials

There are several commonly used temperature control methods for 3D bioprinting of thermal cross-linkable biomaterials. For bioinks, which present liquid state at high temperatures but solid state at low temperatures, such as gelatin and agarose, usually, heating elements need to be mounted with a bioink barrel to increase the temperature of the bioinks before extrusion, as shown in Figure 6.7a [98]. Thus, the bioinks are at sol state and can be easily extruded out of a dispensing nozzle. The temperature of a receiving substrate needs to be controlled below the glass transition temperature of the bioinks to cause the rapid solidification of the printed filaments/layers.

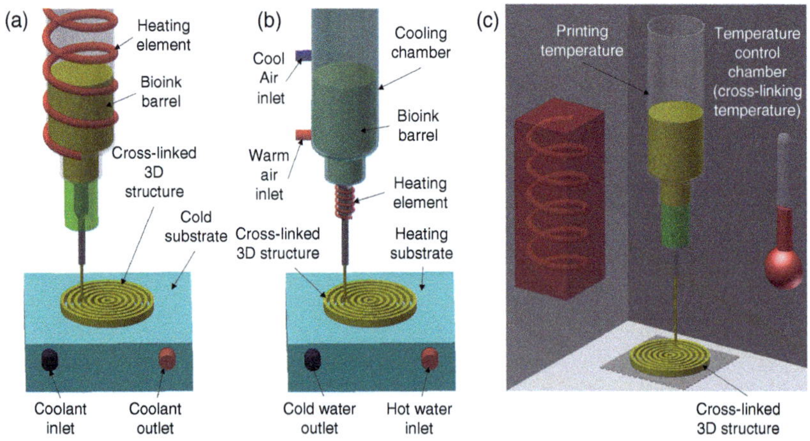

Figure 6.7 Strategies for thermal cross-linkable bioink printing. (a) Heating bioink barrel with cooling receiving substrate. (b) Cooling bioink barrel with heating receiving substrate. (c) Temperature control chamber to control the cross-linking temperature. Source: Yifei Jin.

For thermal cross-linkable bioinks with lower glass transition temperature, such as Pluronic F127 [99] and Matrigel [100], a bioink barrel should be inserted into a cooling chamber to keep the bioink at sol state during printing. In contrast, heating elements are added to both the exit of a dispensing nozzle and a receiving substrate for increasing temperature, as shown in Figure 6.7b. Thus, deposited filaments/layers can rapidly transit to solid state after deposition. However, these two temperature control methods have a main technical challenge. The temperature on the receiving substrate can be accurately controlled by either coolant or hot water, which makes it possible to rapidly cross-link the first few printed layers. With the increasing height of the printed structures, localized temperature must be different from that on the substrate. As a result, the cross-linking speed decreases, which inhibits the rapid solidification of subsequently deposited filaments/layers, severely constraining the geometries, especially the height, of achievable 3D structures.

To overcome this challenge, a temperature control chamber has been investigated. In this method, the temperature of a bioink barrel is controlled by approaches similar to the aforementioned two methods, while the environmental temperature, including the substrate temperature, is controlled by a temperature control chamber, as shown in Figure 6.7c. Using this method, scaled-up 3D structures can be printed from thermal cross-linkable biomaterials.

6.3.2.2 Printing of Ionic Cross-linkable Biomaterials

Ionic cross-linkable bioinks usually have high cross-linking speed, and nozzle clogging is one of the main challenges during printing. The commonly used strategy in 3D printing of ionic cross-linkable bioinks is separating bioinks away from cross-linker solution. Some representative methods are listed as follows:

(1) **Printing ionic cross-linkable bioinks into a cross-linker reservoir.** As shown in Figure 6.8a, a print head moves along the horizontal plane to deposit ionic cross-linkable biomaterials into 2D patterns on a receiving substrate. After printing one layer, the receiving substrate moves downward into the cross-linker reservoir. Thus, the deposited layer can rapidly solidify in the reservoir. Then, the adjacent layer of bioinks will be printed. By repeating the printing and cross-linking steps, 3D structures can be constructed layer by layer [101]. The main advantage of this method is that the *in situ* supports are not only provided by the mechanical stiffness of cross-linked layers but also from buoyancy of the surrounding cross-linker solution. As a result, this strategy is promising for scaled-up structure printing applications. However, since continuous filaments are directly deposited to the cross-linker reservoir, ions may climb up along the filaments into the dispensing nozzle and cross-link bioinks in the nozzle. Thus, nozzle clogging is the main challenge in this method.

(2) **Printing ionic cross-linkable bioinks and cross-linkers using two separated nozzles.** As shown in Figure 6.8b, two nozzles are used to deposit bioinks and cross-linkers simultaneously in this method. When these two solutions coincide with each other, cross-linking occurs on a substrate. As a result, cross-linked filaments/layers have sufficient mechanical stiffness to support the following deposited filaments/layers [102].

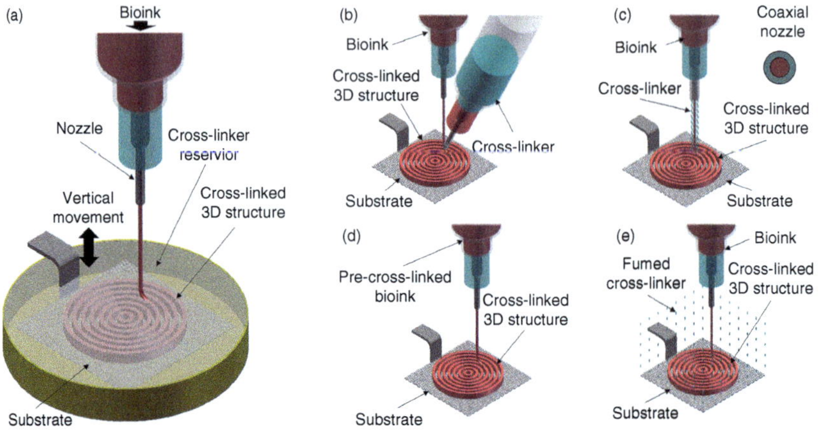

Figure 6.8 Different strategies for 3D printing of ionic cross-linkable bioinks. (a) Printing ionic cross-linkable bioinks into cross-linker reservoir. (b) Printing ionic cross-linkable bioinks and cross-linkers using two separated nozzles. (c) Printing ionic cross-linkable bioinks and cross-linkers using coaxial nozzle. (d) Pre-cross-linking ionic cross-linkable bioinks inside barrel. (e) Printing ionic cross-linkable bioinks through fumed cross-linker atmosphere. Source: Yifei Jin.

(3) **Printing ionic cross-linkable bioinks and cross-linkers using a coaxial nozzle.** Different from previous method, in this strategy, a coaxial nozzle is designed and used, which has a core channel for bioink extrusion and a sheath channel for cross-linker extrusion, as shown in Figure 6.8c. This method can not only print 3D structures but also generate hollow microtubes by injecting cross-linker into the core channel and bioink into the sheath channel [103].

(4) **Pre-cross-linking ionic cross-linkable bioinks inside barrel.** In this method, cross-linkers and bioinks are mixed together before loading into a barrel. By accurately controlling the cross-linking extent, bioinks can possess both good extrudability and self-supporting property, as shown in Figure 6.8d. However, cross-linking extent is a function of mixing time. If the bioinks are mixed with the cross-linkers for a long time, they may be completely cross-linked in the barrel and cannot be extruded anymore. Thus, this method usually has limited printing windows.

(5) **Printing ionic cross-linkable bioinks through fumed cross-linker atmosphere.** As shown in Figure 6.8e, cross-linker agent is atomized in this method. When bioink filaments are extruded out and go through a fumed cross-linker region, the atomized cross-linker can be evenly attached on the surface of the filaments to cause solidification [104]. Although nozzle clogging is a potential challenge during printing, it can be overcome by adjusting the concentration of the cross-linker. Cross-linking methods can be flexibly selected for printing different materials.

6.3.2.3 Printing of Photo Cross-linkable Biomaterials

Three strategies are commonly used for 3D printing of photo cross-linkable bioinks including (i) pre-cross-linking before extrusion, (ii) cross-linking after deposition, and (iii) *in situ* cross-linking during extrusion, as shown in Figure 6.9.

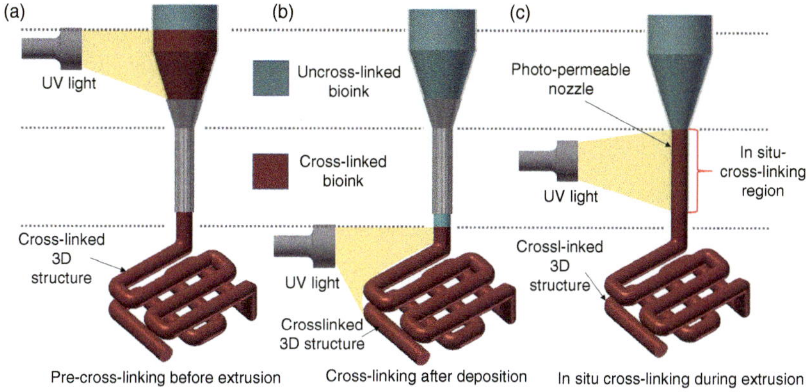

Figure 6.9 Three strategies for 3D printing of photo cross-linkable bioinks. (a) Pre-cross-linking before extrusion, (b) cross-linking after deposition, and (c) *in situ* cross-linking during extrusion. Source: Yifei Jin.

Pre-cross-linking before extrusion is suitable for photo cross-linkable bioinks with low viscosity. The cross-linking step before printing can improve the printability of bioinks to form well-defined filaments with sufficient mechanical stiffness [105]. However, the cross-linking extent in this strategy must be accurately controlled. Otherwise, over-exposure to light sources such as UV radiation may lead to the complete cross-linking of bioinks inside a barrel.

Cross-linking after deposition is the most commonly used strategy for photo cross-linkable biomaterial printing. In this strategy, sufficient light intensity should be provided to ensure the rapid solidification of deposited filaments/layers [90]. If bioinks have relatively high viscosity, the cross-linking time can be longer.

Recently, some researchers have proposed a new photo cross-linking method, called "*in situ* cross-linking." In this method, a light-permeable microtube is used to make a dispensing nozzle, replacing the traditional opaque nozzles [74]. Thus, by exposing the UV light onto the microtube nozzle, low-viscosity bioinks start to cross-link when flowing through the nozzle, which can rapidly solidify after extruded out [106, 107]. During *in situ* cross-linking, the light intensity needs to be carefully controlled to avoid bioink complete cross-linking inside the nozzle, which leads to nozzle clogging.

6.3.2.4 Printing of Enzyme Cross-linkable Biomaterials

Enzyme cross-linking , which leads to the sol–gel transition of biomaterials under mild conditions (in a neutral pH and physiological temperature environment), is an emerging cross-linking method for 3D bioprinting applications. Since it is unnecessary to use chemical agents or organic solvents to cause gelation of biomaterials such as silk fibroin [15], fibrin [108], etc., enzyme cross-linking has excellent biocompatibility. Usually, the gelation/reaction speed of enzyme cross-linking is fast. Thus, traditional cross-linking strategies do not work for enzyme cross-linkable biomaterials [71]. Therefore, a mixing-and-printing system has been investigated and proposed, as shown in Figure 6.10. In this system, multiple chambers are used to feed bioinks and enzymes into one dispensing nozzle. There is a mixing

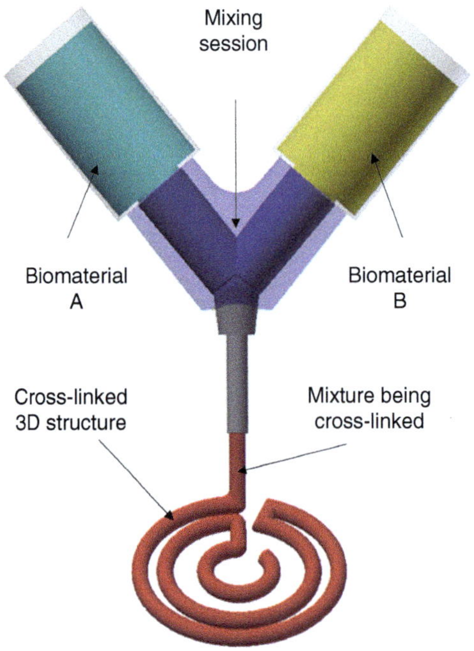

Figure 6.10 Printing-mixing system for 3D printing of enzyme cross-linkable biomaterials. Source: Yifei Jin.

session in the nozzle, in which bioinks and enzymes are mixed uniformly and they undergo enzyme cross-linking. Simultaneously, the cross-linking bioinks are extruded through the dispensing nozzle to form continuous filaments, which have sufficient mechanical stiffness to be self-supported. However, when enzymes are used as cross-linkers for *in vivo* applications, unexpected biological changes may be induced [109]. Therefore, selecting the appropriate enzyme is crucial in this new cross-linking method.

6.3.3 3D Structure Printing Using Self-Supporting Material-Assisted 3D Printing Approach

Due to the self-supporting property of bioinks, rapid solidification is not necessary in self-supporting material-assisted 3D printing approach. Thus, cross-linking step can be applied either after printing or during printing.

6.3.3.1 Internal Scaffold Additive-Assisted 3D Printing

Internal scaffold additive-assisted 3D printing was proposed in 2017. Jin et al. [63] used Laponite® RD and Laponite® XLG as internal scaffold additives to mix with different hydrogel precursors, including PEGDA, alginate, and gelatin. The resulting nanocomposite hydrogels (PEGDA-Laponite®, alginate-Laponite®, and gelatin-Laponite®) had self-supporting property and can be printed into complex 3D structures in air without cross-linking, as shown in Figure 6.11a. After printing, these 3D structures can be cross-linked using the corresponding cross-linking mechanisms to stabilize the structural integrity.

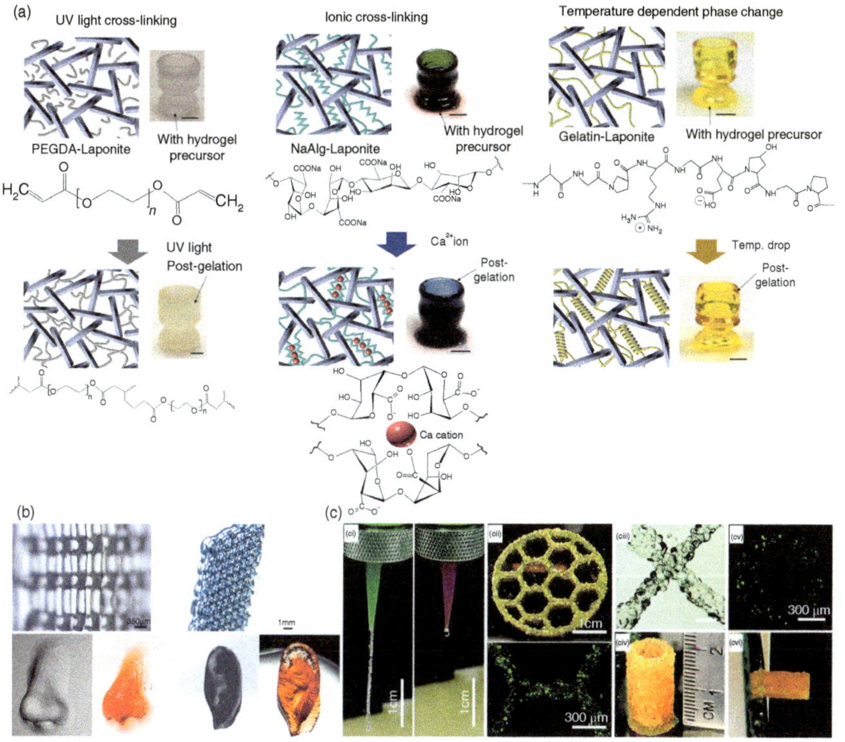

Figure 6.11 3D structures printed using self-supporting material-assisted 3D printing approach. (a) Yield-stress additive-assisted 3D printing of nanocomposite hydrogels. Source: Jin et al. [63]. Reproduced with permission of American Chemical Society. (b) 3D printing of complex structures using κCA-nanosilicate-based bioinks. Source: Wilson et al. [110]. Reproduced with permission of American Chemical Society. (c) Microgel-additive-assisted 3D printing of cellular constructs. Source: Xin et al. [70]. Reproduced with permission of Royal Society of Chemistry.

Wilson et al. [110] prepared a new self-supporting hydrogel by mixing Laponite® XLS with kappa-carrageenan (κCA), which is a linear sulfate polysaccharide from red algae that can be cross-linked both thermally and ionically. By adjusting the ratio of κCA and Laponite®, the reversible thermal gelation of the bioink can be controlled to obtain high printability and shape retention characteristics, as shown in Figure 6.11b. After printing, the cell survival rate was high, but cell diffusion and proliferation were limited.

Ahlfeld et al. [111] used Laponite® XLG, alginate, and methylcellulose to prepare a new type of bioink with self-supporting property for cell printing applications. The cells showed a high survival rate within 21 days after printing. Peak et al. [112] prepared the bioink based on the mixture of Laponite® XLG and PEG. The addition of Laponite® enhanced the viscosity, storage modulus, and network stability of PEG, and the cells encapsulated in PEG-Laponite®-based bioink showed a higher survival rate after printing. In general, nanoclay-based composite hydrogel bioinks have excellent printability, structural fidelity, mechanical properties, and good biocompatibility, which are promising for more cell printing applications in the future.

6.3.3.2 Microgel Additive-Assisted 3D Printing

Microgel additive-assisted 3D printing approach was proposed in 2019. Highley et al. [69] mixed cells with NorHA to prepare microgels and printed 3D cellular constructs. They found that the printing process did not reduce the survival rate of the encapsulated cells. In addition, Xin et al. [70] mixed human bone marrow mesenchymal stem cells (hMSCs) with PEG microgels as the cell-laden bioink and printed three-layered honeycomb structures and tubular structures (Figure 6.11c). They found that cells showed good diffusion ability and cell viability after 5-day and 10-day culturing still exceeded 90%. Although the cells were not encapsulated in the microgels in their study, the activity of the cells in the microgel-based bioink was not affected by the extrusion process.

6.4 Representative Biomedical Applications

DIW has been widely applied to fabricate 3D living tissues/organs in tissue engineering. This section aims to review several representative biomedical applications using DIW technique.

6.4.1 Aortic Valve Printing

The heart is composed of four valves: aortic, pulmonary, tricuspid, and mitral valves. The aortic valve is mostly involved in heart-related diseases such as stenosis and reflux. The aortic valve located between the ascending aorta and the left ventricular outflow tract [113] plays a major role in ensuring the proper function of the cardiovascular system [114]. Therefore, investigations on engineered heart valves mainly focus on how to create cellular aortic valves.

Hockaday et al. [90] printed the aortic valve structure using DIW. In their study, micro-CT was used to image the geometries of porcine aortic valves with sinuses and leaflets. Then, a 3D model was generated based on micro-CT data. The mixture of alginate and PEGDA was selected as the bioink, in which alginate was mainly used to improve the printability, and PEGDA served for improving the mechanical properties of the bioink. UV radiation was utilized to rapidly cross-link the printed structure and the cross-linking time ranged from 17 to 1140 seconds depending on the concentration of photoinitiator. Thus, aortic valves with different sizes (diameter ranging from 12 to 22 mm) were printed, as shown in Figure 6.12a. It is found that the shape fidelity of the printed aortic valves increased with the valve size and the resolution mainly depended on the path speed and nozzle diameter. To study the biological characteristics of the valves, valvular stromal cells from the pig aorta were inoculated onto the surface of the printed structures, and the cells showed excellent viability (~100%) over 21 days. This study demonstrated the feasibility of using DIW to print aortic valves with various sizes and mechanical properties.

Duan et al. [23] developed an alginate/gelatin bioink and printed the aortic valve structure, as shown in Figure 6.12b. In this study, sodium chloride or phosphate buffer solution was used as the solvent during bioink preparation to

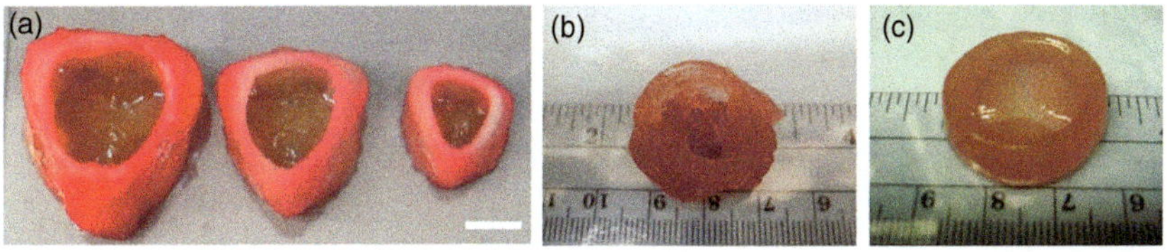

Figure 6.12 3D bioprinting of aortic valves using DIW. (a) 3D printed axisymmetric valve with diameters of 22, 17, and 12 mm. Source: Hockaday et al. [90]. Reproduced with permission of IOP Publishing, Ltd. (b) Engineered aortic valve from alginate/gelatin. Source: Duan et al. [23]. Reproduced with permission of John Wiley and Sons. (c) 3D printed valve conduit from HAMA/GelMA. Source: Duan et al. [54]. Reproduced with permission of Elsevier.

enhance the biocompatibility of the bioink. Stromal cells and smooth muscle cells were mixed with the bioink to print the aortic valve and the aortic sinus, respectively. Herein, a two-step cross-linking method was used. The rapidly thermal cross-linking of gelatin component in the bioink facilitated the printing process. After printing, the cellular construct was incubated in calcium chloride solution for 10 minutes for ionic cross-linking of alginate component. Thus, the cellular construct can maintain its shape during cell culturing. After 7-day incubation, cell viability was tested by living/dead assays, and the survival rate was nearly 82%.

Duan et al. [54] further developed a HAMA/GelMA bioink to print the clover catheter model. In this bioink, the concentration of GelMA was varied within 6–12% (w/v) to adjust the mechanical properties of the gelled structure. Human aortic valve mesenchymal cells were added to HAMA/GelMA to prepare the cellular bioink. The cell-laden clover catheter model was printed, as shown in Figure 6.12c. After 7-day incubation, the cells in the construct were found to be highly sustainable (~92%) and ECM proteins (collagen and glycosaminoglycan) produced by the cells were observed, which validated that this formula of bioink was promising to 3D print implantable aortic valves in the future.

6.4.2 Bone and Cartilage Tissue Printing

DIW-based cell printing technology has been widely used to print scaffolds in bone tissue engineering. For example, Yao et al. [115] printed the scaffolds from polycaprolactone–hydroxyapatite, which was able to support physiological loads. Wang et al. [116] printed the porous scaffold using poly (propylene glycol fumarate) and studied its degradation process within 224 days. They proved that the printed scaffold had long-term stability to be used for bone tissue engineering. Pati et al. [117] 3D printed the PCL-PLGA-TCP scaffolds with mesenchymal stromal cells derived from human subnasal mesenchymal tissue, as shown in Figure 6.13a. They found that the printed scaffold exhibited excellent bone induction and conduction properties.

The reconstruction of cartilage tissues has been popular recently [121]. Kundu et al. [118] printed the bioink composed of chondrocytes and alginate on the supporting PCL structures, as shown in Figure 6.13b, and chondrogenesis was observed

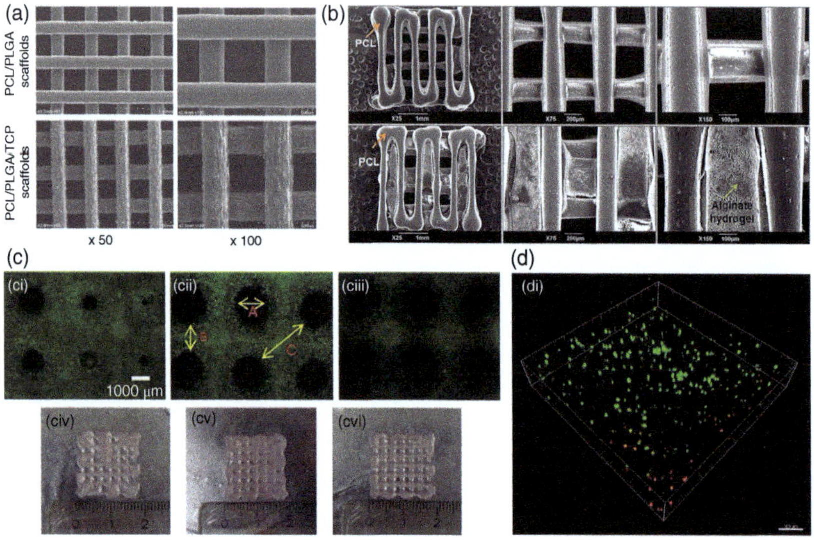

Figure 6.13 (a) 3D printed PCL-PLGA-TCP scaffolds with mesenchymal stromal cells for bone tissue engineering. Source: Pati et al. [117]. Reproduced with permission of Elsevier. (b) Porous 3D PCL scaffold and porous 3D PCL-alginate hybrid scaffold for cartilage tissue printing. Source: Kundu et al. [118]. Reproduced with permission of John Wiley and Sons. (c) Printed hCMPCs in alginate scaffolds with different concentrations as cardiac patches. Source: Gaetani et al. [119]. Reproduced with permission of Elsevier. (d) 3D stack of hepatocyte-like cells (HLCs) printed in hydrogel, recorded one-hour post-printing showing live cells in green and dead cells in red. Source: Faulknerjones et al. [120]. Reproduced with permission of IOP Publishing, Ltd.

in the printed structures after implantation into the human body. Lee et al. [122] used chondrocytes/PEG-PCL as the bioink to print biomimicry constructs such as ears. Markstedt et al. [88] developed a novel bioink which was a mixture of nanocellulose and alginate. The printed structures can support the growth of human septal cartilage cells. Recently, Pati et al. [123] developed decellularized extracellular matrix (dECM) of chondrocyte as the bioink to print 3D biological constructs using DIW. This bioink was able to support the chondrogenic differentiation and promote the formation of new cartilage tissues.

6.4.3 Cardiac Tissue Printing

Heart failure has attracted more and more attention, and 3D bioprinting of functional heart tissues has been envisioned as a promising solution [124]. Recently, DIW-based cell printing technology has been applied to the manufacturing of heart tissue *in vitro* Gaetani et al. [119] mixed human cardiac-derived cardiomyocyte progenitor (hCMPC) with alginate solution to prepare the bioink and printed cell-laden constructs using DIW, as shown in Figure 6.13c. The encapsulated hCMPC was found to be highly active and retained the potentials to cardiac lineage. Zhang et al. [125] used the mixture of alginate, GelMA, and endothelial cells as the cell-laden bioink to fabricate endothelialized myocardium. In their study, the encapsulated endothelial cells were firstly printed onto a 3D scaffold to form the endothelialized structure and vascular bed, on which cardiomyocytes were subsequently seeded. The encapsulated endothelial cells formed a confluent layer of endothelium outside of the microfilaments. And the seeded cardiomyocytes, which contracted spontaneously and synchronously, were fully aligned under the induction of the anisotropic scaffold. Maiullari et al. [126] developed the bioink by mixing human umbilical vein endothelial cells (HUVECs) and induced pluripotent cell-derived cardiomyocytes (iPSC-CMs) with a mixture of alginate and peg-fibrinogen (PF) solution. They printed 3D vascularized cardiac tissue. In the printed structure, the encapsulated iPSC-CMs were found to have a high orientation index.

6.4.4 Liver Tissue Printing

Liver failure, which is associated with multi-organ failure, has been proven as one of the major causes of death. Therefore, studies on liver tissue printing have been of great interest [67]. Due to the high sensitivity of liver tissues to drug toxicity, 3D printed liver tissues have great potentials in drug detection and screening. In addition, transplantation of engineered liver tissues will draw more concerns in the future. Faulkner-Jones et al. [120] used DIW-based 3D bioprinting to print human induced pluripotent stem cells (hiPSCs) into liver tissues for the first time. They systematically analyzed the effects of process parameters such as dispensing pressure and nozzle dimensions on the activity of stem cells (hiPSCs). The encapsulated hiPSCs were stimulated to differentiate into hepatocytes, as shown in Figure 6.13d and the printed structure was used in micro-liver organ engineering. Bertassoni et al. [127] developed the bioink with agarose, gelatin, GelMA, HepG2 cells, and fibroblasts. During printing, agarose was solidified rapidly by controlling the surrounding temperature to hold the structural integrity. Then, the printed structure was further cross-linked by cross-linking GelMA with UV radiation. Internal channels were formed by thermally removing the agarose component.

6.4.5 Lung Tissue Printing

3D printing of lung tissues has emerged recently. Hovath et al. [128] successfully printed a multi-layered air–blood barrier model to mimic the function of lungs *in vitro*. In their study, a thin layer of Matrigel was printed first, which was used as the basis to support the adhesion of cells. Then, another layer of Matrigel with epithelial cells was printed atop the previously deposited layers, as shown in Figure 6.14a. Finally, the model was incubated for several days. It was found that the cell viability of the epithelial cells and endothelial cells on day-5 reached 95% and 86%, respectively, and the compactness of the entire structure was excellent.

6.4.6 Neural Tissue Printing

The successful3D printing of the nervous system is viewed as an important step toward clinical applications of engineered tissues/organs. The human nervous system is usually divided into the central nervous system (CNS) and the peripheral

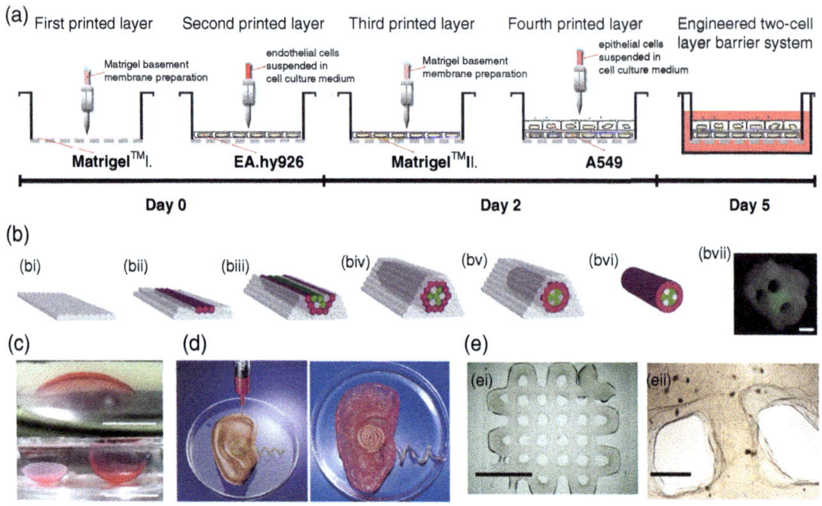

Figure 6.14 (a) Schematic of the timeline for bioprinting the two cell-layer barrier system. Source: Horváth et al. [128]. Reproduced with permission of Springer Nature. (b) Schematic of nerve graft fabrication. Source: Owens et al. [129]. Reproduced with permission of IOP Publishing, Ltd. (c) 3D printed alginate-gelatin cornea. Source: Zhang et al. [130]. Reproduced with permission of Springer Nature. (d) 3D printed bionic ear. Source: Mannoor et al. [131]. Reproduced with permission of American Chemical Society. (e) Alginate/gelatin plotted scaffold and plotted islets in the construct. Source: Marchioli et al. [132]. Reproduced with permission of IOP Publishing, Ltd.

nervous system (PNS). CNS mainly includes the nerves in the brain and spinal cord, and all the other nerves belong to PNS. To date, PNS has been printed by DIW. Owens et al. [129] used Schwann cells (SCs) and bone marrow stem cells (BMSCs) to fabricate nerve grafts that were implanted in mice. In their study, the cells were aspirated into capillary tubes with an inner diameter of 0.5 mm, and the cellular cylinders were subsequently extruded onto a 3D printed agarose scaffold to form an implantable nerve conduit, as shown in Figure 6.14b. The recovery of motor function and sensory function was evaluated 40 weeks after implantation, and the results proved the feasibility of using cell printing technology to make nerve grafts. Ning et al. [133] prepared the bioink by mixing Schwann cells with a mixture of hydrogel solution (consisting of alginic acid, fibrin, hyaluronic acid, and/or RGD peptide). They printed a scaffold as an alternative for autograft. The encapsulated Schwann cells in the constructs were found to have high viability, good proliferation, and cell protein expression ability.

6.4.7 Eye and Ear Printing

3D bioprinting has been utilized to print eye models, which can approximately simulate the optical properties of human eyes. For example, Isaacson et al. [134] used DIW to print a corneal structure similar to the native human corneal stroma. In their study, they mixed human corneal stromal cells with alginate and methacrylated Type I collagen as the bioinks. It was found that the encapsulated corneal keratocytes presented high cell viability after 1-day (>90%) and 7-day (83%) incubation. In addition, Zhang et al. [130] developed a hybrid 3D bioprinting system by combining digital light processing (DLP) and DIW. Using this system, they printed highly transparent and curved membranes with high water content, which had the geometrical characteristics of the human natural cornea, as shown in Figure 6.14c.

For ear printing, Mannoor et al. [131] printed biomimetic ear cartilage tissue *in vitro*, as shown in Figure 6.14d. The induction coil antenna was embedded in the ear cartilage tissue, which can monitor the inductively coupled signals from the cochlear electrode. These bionic ears printed in this study had a higher sense of hearing.

6.4.8 Pancreas Printing

Since pancreatic beta cells are difficult to survive outside the human body, only a handful of human stem cells have been attempted to differentiate into pancreatic beta cells, and the research on pancreatic tissue regeneration mainly used mice as examples. Recently, Marchioli et al. [132] mixed human and mouse pancreatic islet and rat insulinoma INS1E β cells with alginate or alginate-gelatin solution as the bioink, which was further printed on a two-layered scaffold to form a cellular construct, as shown in Figure 6.14e. This construct was subsequently implanted into diabetic mice and removed after seven days. It was found that although the alginate and the alginate-gelatin solution did not affect the cell bioactivity during printing, the cells still lost their functions seven days after printing.

6.4.9 Skin Tissue Printing

Studies on the regeneration of skin tissues started earlier, and a variety of bioprinting techniques have been used to fabricate transplantable skin tissues. Among these bioprinting techniques, DIW has become the mainstream cell printing technology for skin tissue printing. For example, Skardal et al. [135] separately mixed amniotic fluid-derived stem (AFS) cells and bone marrow-derived mesenchymal stem cells (MSCs) with fibrin-collagen mixture as the bioinks and *in situ* printed these bioinks on a skin wound with a size of $2\,\text{cm} \times 2\,\text{cm}$. In their study, two layers of fibrin/collagen with cells and three layers of thrombin without cells were printed alternately on the wound, as shown in Figure 6.15a. After printing, the degree of wound healing and re-epithelialization of the wound were evaluated on day-0, day-7, and day-14. It was found that printed skin tissue can accelerate the wound healing process. In addition, Cubo et al. [137] printed bi-layered human skin structures from the bioink composed of human plasma, primary human fibroblasts, and keratinocytes. After *in vitro* culturing and transplantation to immunodeficient mice, 3Dprinted skins presented a similar structure to human native skin.

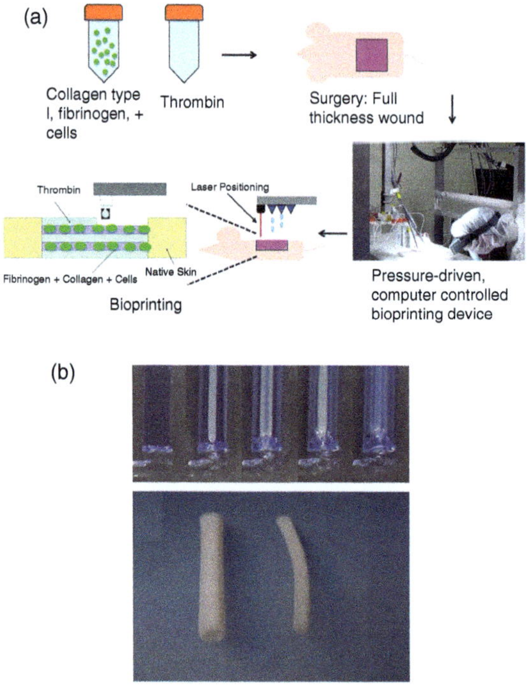

Figure 6.15 (a) A schematic describing the approach by which amniotic fluid-derived stem (AFS) cells are bioprinted in order to increase healing of a full-thickness skin wound. Source: Skardal et al. [135]. Reproduced with permission of John Wiley and Sons. (b) Layer-by-layer deposition of agarose cylinders and engineered pig SMC tubes of distinct diameters resulted after three days of post-printed fusion. Source: Norotte et al. [136]. Reproduced with permission of Elsevier.

6.4.10 Blood Vessel Printing

Blood vessels transport oxygen and nutrients for tissues/organs. The construction of tissue with vascular features has always been challenging in the fields of tissue engineering and cell printing. Gao et al. [138] developed a new coaxial nozzle-assisted DIW technique to print 3D structures with built-in microchannels. In their study, alginate and calcium chloride were co-extruded through the sheath channel and core channel in the coaxial nozzle, separately. Thus, calcium chloride can cross-link the surrounding alginate to form cylindrical filaments with hollow channels. By constructing these filaments layer by layer, 3D structures with built-in microchannels were manufactured. These microchannels can be used as blood vessels to deliver nutrients to the encapsulated cells. Norotte et al. [136] reported a new DIW approach for constructing vascular networks. In their study, multicellular spheroids or strands with controllable diameters ranging from 300 to 500 μm were printed with agarose rods (as shown in Figure 6.15b). After printing, the fusion of these spheroids or strands resulted in the formation of vessels with a small diameter.

6.5 Conclusions and Future Work

Recently, DIW has been well developed and widely utilized to fabricate complex 3D constructs with living cells. Due to its easy implementation and wide range of printable materials, it has become one of the mainstream technologies in 3D bioprinting, and it is promising for the fabrication of 3D functional tissues/organs. For rapid solidification-induced 3D bioprinting, future work may focus on two directions: (i) development of biocompatible materials with rapid solidification property and (ii) investigation on new rapid solidification strategies. For self-supporting material-assisted 3Dbioprinting, more yield-stress additive materials with excellent biocompatibility will be developed and mixed with hydrogel solutions to prepare hydrogel composites with self-supporting property in the future.

References

1 Lewis, J.A. (2006). Direct ink writing of 3D functional materials. *Advanced Functional Materials* 16 (17): 2193–2204.
2 Lewis, J.A. and Gratson, G.M. (2004). Direct writing in three dimensions. *Materials Today* 7 (7–8): 32–39.
3 Hao, L., Tang, D., Sun, T. et al. (2021). Direct ink writing of mineral materials: a review. *International Journal of Precision Engineering and Manufacturing-Green Technology* 8: 665–685.
4 Friedrich, L. and Begley, M. (2020). Corner accuracy in direct ink writing with support material. *Bioprinting* 19: e00086.
5 Davoodi, E., Sarıkhani, E., Montazerian, H. et al. (2020). Extrusion and microfluidic-based bioprinting to fabricate biomimetic tissues and organs. *Advanced Materials Technologies* 5 (8): 1901044.
6 Bertassoni, L.E., Cardoso, J.C., Manoharan, V. et al. (2014). Direct-write bio-printing of cell-laden methacrylated gelatin hydrogels. *Biofabrication* 6 (2): 024105.
7 Ehsan, S.M., Welch-Reardon, K.M., Waterman, M.L. et al. (2014). A three-dimensional in vitro model of tumor cell intravasation. *Integrative Biology (Camb)* 6 (6): 603–610.
8 Ozbolat, I.T. and Hospodiuk, M. (2016). Current advances and future perspectives in extrusion-based bioprinting. *Biomaterials* 76: 321–343.
9 Ning, L. and Chen, X. (2017). A brief review of extrusion-based tissue scaffold bio-printing. *Biotechnology Journal* 12 (8).
10 Duarte Campos, D.F., Blaeser, A., Korsten, A. et al. (2015). The stiffness and structure of three-dimensional printed hydrogels direct the differentiation of mesenchymal stromal cells toward adipogenic and osteogenic lineages. *Tissue Engineering Part A* 21 (3–4): 740.
11 Hellio, D. and Djabourov, M. (2006). Physically and chemically crosslinked gelatin gels. *Macromolecular Symposia* 241 (1): 23–27.
12 Cohen, D.L., Lipton, J.I., Bonassar, L.J. et al. (2010). Additive manufacturing for in situ repair of osteochondral defects. *Biofabrication* 2 (3): 035004.
13 Billiet, T., Gevaert, E., De, S.T. et al. (2014). The 3D printing of gelatin methacrylamide cell-laden tissue-engineered constructs with high cell viability. *Biomaterials* 35 (1): 49–62.
14 Chiu, Y.L., Chen, S.C., Su, C.J. et al. (2009). pH-triggered injectable hydrogels prepared from aqueous *N*-palmitoyl chitosan: in vitro characteristics and in vivo biocompatibility. *Biomaterials* 30 (28): 4877–4888.

15 Das, S., Pati, F., Choi, Y.J. et al. (2015). Bioprintable, cell-laden silk fibroin-gelatin hydrogel supporting multilineage differentiation of stem cells for fabrication of three-dimensional tissue constructs. *Acta Biomaterialia* 11 (1): 233–246.

16 Shao, L., Gao, Q., Xie, C. et al. (2020). Sacrificial microgel-laden bioink-enabled 3D bioprinting of mesoscale pore networks. *Bio-Design and Manufacturing* 3 (1): 30–39.

17 Landers, R., Pfister, A., Hübner, U. et al. (2002). Fabrication of soft tissue engineering scaffolds by means of rapid prototyping techniques. *Journal of Materials Science* 37 (15): 3107–3116.

18 Ling, Y., Rubin, J., Deng, Y. et al. (2007). A cell-laden microfluidic hydrogel. *Lab on A Chip* 7 (6): 756–762.

19 Das, S., Pati, F., Chameettachal, S. et al. (2013). Enhanced redifferentiation of chondrocytes on microperiodic silk/gelatin scaffolds: toward tailor-made tissue engineering. *Biomacromolecules* 14 (2): 311–321.

20 Zhang, W., Feng, C., Yang, G. et al. (2017). 3D-printed scaffolds with synergistic effect of hollow-pipe structure and bioactive ions for vascularized bone regeneration. *Biomaterials* 135: 85–95.

21 Zhu, J. and Marchant, R.E. (2011). Design properties of hydrogel tissue-engineering scaffolds. *Expert Review of Medical Devices* 8 (5): 607–626.

22 Heino, J., Huhtala, M., Kpyl, J. et al. (2009). Evolution of collagen-based adhesion systems. *International Journal of Biochemistry and Cell Biology* 41 (2): 341–348.

23 Duan, B., Hockaday, L.A., Kang, K.H. et al. (2013). 3D bioprinting of heterogeneous aortic valve conduits with alginate/gelatin hydrogels. *Journal of Biomedical Materials Research Part A* 101A (5): 1255–1264.

24 Xu, H., Casillas, J., and Xu, C. (2019). Effects of printing conditions on cell distribution within microspheres during inkjet-based bioprinting. *AIP Advances* 9 (9): 095055.

25 Xu, C., Chai, W., Huang, Y. et al. (2012). Scaffold-free inkjet printing of three-dimensional zigzag cellular tubes. *Biotechnology and Bioengineering* 109 (12): 3152–3160.

26 Zhang, M., Krishnamoorthy, S., Song, H. et al. (2017). Ligament flow during drop-on-demand inkjet printing of bioink containing living cells. *Journal of Applied Physics* 121 (12): 124904.

27 Behravesh, E., Jo, S., Zygourakis, K. et al. (2002). Synthesis of in situ cross-linkable macroporous biodegradable poly(propylene fumarate-*co*-ethylene glycol) hydrogels. *Biomacromolecules* 3 (2): 374–381.

28 Suggs, L.J. and Mikos, A.G. (1999). Development of poly(propylene fumarate-*co*-ethylene glycol) as an injectable carrier for endothelial cells. *Cell Transplantation* 8 (4): 345.

29 Fisher, J.P., Jo, S., Mikos, A.G. et al. (2004). Thermoreversible hydrogel scaffolds for articular cartilage engineering. *Journal of Biomedical Materials Research Part A* 71A (2): 268–274.

30 Peter, S.J., Lu, L., Kim, D.J. et al. (2000). Marrow stromal osteoblast function on a poly (propylene fumarate)/beta-tricalcium phosphate biodegradable orthopaedic composite. *Biomaterials* 21 (12): 1207–1213.

31 Behravesh, E., Zygourakis, K., and Mikos, A.G. (2003). Adhesion and migration of marrow-derived osteoblasts on injectable in situ crosslinkable poly(propylene fumarate-*co*-ethylene glycol)-based hydrogels with a covalently linked RGDS peptide. *Journal of Biomedical Materials Research Part A* 65A (2): 260–270.

32 Sungmin, H., Dalton et al. (2015). 3D printing of highly stretchable and tough hydrogels into complex, cellularized structures. *Advanced Materials* 27 (27): 4035–4040.

33 Xu, H., Zhang, Z., and Xu, C. (2019). Sedimentation study of bioink containing living cells. *Journal of Applied Physics* 125 (11): 114901.

34 Wu, D. and Xu, C. (2018). Predictive modeling of droplet formation processes in inkjet-based bioprinting. *Journal of Manufacturing Science and Engineering* 140 (10): 101007.

35 Xu, C., Zhang, M., Huang, Y. et al. (2014). Study of droplet formation process during drop-on-demand inkjetting of living cell-laden bioink[J]. *Langmuir* 30 (30): 9130–9138.

36 Bacelar, A.H., Silvacorreia, J., Oliveira, J.M. et al. (2016). Recent progress in gellan gum hydrogels provided by functionalization strategies. *Journal of Materials Chemistry B* 4 (37): 6164–6174.

37 Wuestenberg, T. (2014). Cellulose and cellulose derivatives in the food industry: fundamentals and applications[M]. *Institution of Electronic and Radio Engineers*: 161–182.

38 Oliveira, J.T., Martins, L., Picciochi, R. et al. (2010). Gellan gum: a new biomaterial for cartilage tissue engineering applications. *Journal of Biomedical Materials Research Part A* 93A (3): 852–863.

39 Melchels, F.P.W., Dhert, W.J.A., Hutmacher, D.W. et al. (2014). Development and characterisation of a new bioink for additive tissue manufacturing. *Journal of Materials Chemistry B* 2 (16): 2282–2289.

40 Lozano, R., Stevens, L., Thompson, B.C. et al. (2015). 3D printing of layered brain-like structures using peptide modified gellan gum substrates. *Biomaterials* 67: 264–273.

41 Alarcón, C.D.L.H., Pennadam, S., and Alexander, C. (2005). Stimuli responsive polymers for biomedical applications. *Cheminform* 34 (3): 276–285.

42 Heskins, M. and Guillet, J.E. (1968). Solution properties of poly(*N*-isopropylacrylamide). *Journal of Macromolecular Science: Part A Chemistry* 2 (8): 1441–1455.

43 Eeckman, F., Moes, A.J., and Amighi, K. (2004). Synthesis and characterization of thermosensitive copolymers for oral controlled drug delivery. *European Polymer Journal* 40 (4): 873–881.

44 Jones, M.S. Effect of pH on the lower critical solution temperatures of random copolymers of *N*-isopropylacrylamide and acrylic acid. *European Polymer Journal* 35 (5): 795–801.

45 Eeckman, F., Amighi, K., and Moës, A.J. (2001). Effect of some physiological and non-physiological compounds on the phase transition temperature of thermoresponsive polymers intended for oral controlled-drug delivery. *International Journal of Pharmaceutics* 222 (2): 259–270.
46 Huang, J. and Wu, X.Y. (1999). Effects of pH, salt, surfactant and composition on phase transition of poly (NIPAm/MAA) nanoparticles. *Journal of Polymer Science Part A Polymer Chemistry* 37 (14): 2667–2676.
47 Etsuo, K., Yong et al. (1993). Effects of surfactants on the phase transition of poly(*N*-isopropylacrylamide) gel. *Macromolecules* 26 (5): 1053–1059.
48 Malonne, H., Eeckman, F., Fontaine, D. et al. (2005). Preparation of poly(*N*-isopropylacrylamide) copolymers and preliminary assessment of their acute and subacute toxicity in mice. *European Journal of Pharmaceutics and Biopharmaceutics* 61 (3): 188–194.
49 Kikuchi, A. and Okano, T. (2005). Nanostructured designs of biomedical materials: applications of cell sheet engineering to functional regenerative tissues and organs. *Journal of Controlled Release* 101 (1–3): 69–84.
50 Shimizu, T., Yamato, M., Kikuchi, A. et al. (2001). Two-dimensional manipulation of cardiac myocyte sheets utilizing temperature-responsive culture dishes augments the pulsatile amplitude. *Tissue Engineering* 7 (2): 141–151.
51 Bromberg, L.E. and Ron, E.S. (1998). Temperature-responsive gels and thermogelling polymer matrices for protein and peptide delivery. *Advanced Drug Delivery Reviews* 31 (3): 197–221.
52 Vlierberghe, S.V., Dubruel, P., and Schacht, E. (2011). Biopolymer-based hydrogels as scaffolds for tissue engineering applications: a review. *Biomacromolecules* 12 (5): 1387–1408.
53 Collins, M.N. and Birkinshaw, C. (2013). Hyaluronic acid-based scaffolds for tissue engineering--a review[J]. *Carbohydrate Polymers* 92 (2): 1262–1279.
54 Duan, B., Kapetanovic, E., Hockaday, L.A. et al. (2014). Three-dimensional printed trileaflet valve conduits using biological hydrogels and human valve interstitial cells. *Acta Biomaterialia* 10 (5): 1836–1846.
55 Sun, G. and Mao, J.J. (2012). Engineering dextran-based scaffolds for drug delivery and tissue repair. *Nanomedicine* 7 (11): 1771–1784.
56 Pescosolido, L., Schuurman, W., Malda, J. et al. (2011). Hyaluronic acid and dextran-based semi-IPN hydrogels as biomaterials for bioprinting. *Biomacromolecules* 12 (5): 1831–1838.
57 Dash, M., Chiellini, F., Ottenbrite, R.M. et al. Chitosan—A versatile semi-synthetic polymer in biomedical applications. *Progress in Polymer Science* 36 (8): 981–1014.
58 Rabea, E.I., Badawy, E.T., Stevens, C.V. et al. (2003). Chitosan as antimicrobial agent: applications and mode of action. *Biomacromolecules* 4 (6): 1457–1465.
59 Molinaro, G., Leroux, J.C., Damas, J. et al. (2002). Biocompatibility of thermosensitive chitosan-based hydrogels: an in vivo experimental approach to injectable biomaterials. *Biomaterials* 23 (13): 2717–2722.
60 Ku, J., Seonwoo, H., Park, S. et al. (2020). Cell-laden thermosensitive chitosan hydrogel bioinks for 3D bioprinting applications. *Applied Sciences* 10 (7): 2455.

61 Suh, J. and Matthew, H. (2000). Application of chitosan-based polysaccharide biomaterials in cartilage tissue engineering: a review. *Biomaterials* 21 (24): 2589–2598.

62 Wu, Q., Therriault, D., and Heuzey, M.-C. (2018). Processing and properties of chitosan inks for 3D printing of hydrogel microstructures. *ACS Biomaterials Science and Engineering* 4 (7): 2643–2652.

63 Jin, Y., Liu, C., Chai, W. et al. (2017). Self-supporting nanoclay as internal scaffold material for direct printing of soft hydrogel composite structures in air. *Acs Applied Materials and Interfaces* 9 (20): 17457–17466.

64 Xavier, J.R., Thakur, T., Desai, P. et al. (2015). Bioactive nanoengineered hydrogels for bone tissue engineering: a growth-factor-free approach. *ACS Nano* 9 (3): 3109–3118.

65 Mealy, J.E., Chung, J.J., Jeong, H.H. et al. (2018). Injectable granular hydrogels with multifunctional properties for biomedical applications. *Advanced Materials* 30 (20): e1705912.

66 Xin, S., Wyman, O.M., and Alge, D.L. (2018). Assembly of PEG microgels into porous cell-instructive 3D scaffolds via thiol-Ene click chemistry. *Advanced Healthcare Materials* 7 (11): e1800160.

67 Hinton, T.J., Jallcrat, Q., Palchesko, R.N. et al. (2015). Three-dimensional printing of complex biological structures by freeform reversible embedding of suspended hydrogels. *Science Advances* 1 (9): e1500758.

68 Daly, A.C., Riley, L., Segura, T. et al. (2019). Hydrogel microparticles for biomedical applications. *Nature Reviews Materials* 5 (1): 20–43.

69 Highley, C.B., Song, K.H., Daly, A.C. et al. (2019). Jammed microgel inks for 3D printing applications. *Advanced Science (Weinheim)* 6 (1): 1801076.

70 Xin, S., Chimene, D., Garza, J.E. et al. (2019). Clickable PEG hydrogel microspheres as building blocks for 3D bioprinting. *Biomaterials Science* 7 (3): 1179–1187.

71 Murphy, S.V., Skardal, A., and Atala, A. (2013). Evaluation of hydrogels for bio-printing applications. *Journal of Biomedical Materials Research Part A* 101A (1): 272–284.

72 Malda, J., Visser, J., Melchels, F.P. et al. (2013). 25th anniversary article: engineering hydrogels for biofabrication. *Advanced Materials* 25 (36): 5011–5028.

73 Bhattacharjee, T., Zehnder, S.M., Rowe, K.G. et al. (2015). Writing in the granular gel medium[J]. *Science Advances* 1 (8): e1500655.

74 Ouyang, L., Highley, C.B., Sun, W. et al. (2017). A generalizable strategy for the 3D bioprinting of hydrogels from nonviscous photo-crosslinkable inks. *Advanced Materials* 29 (8): 1604983.

75 Lim, K.S., Klotz, B.J., Lindberg, G.C.J. et al. (2019). Visible light Cross-linking of gelatin hydrogels offers an enhanced cell microenvironment with improved light penetration depth. *Macromolecular Bioscience* 19 (6): 1900098.

76 Gao, G., Kim, B.S., Jang, J. et al. (2019). Recent strategies in extrusion-based three-dimensional cell printing toward organ biofabrication. *ACS Biomaterials Science & Engineering* 5 (3): 1150–1169.

77 Murphy, S.V. and Atala, A. (2014). 3D bioprinting of tissues and organs. *Nature Biotechnology* 32 (8): 773–785.

78 Daly, A.C., Critchley, S.E., Rencsok, E.M. et al. (2016). A comparison of different bioinks for 3D bioprinting of fibrocartilage and hyaline cartilage. *Biofabrication* 8 (4): 045002.

79 Park, J.Y., Choi, J.C., Shim, J.H. et al. (2014). A comparative study on collagen type I and hyaluronic acid dependent cell behavior for osteochondral tissue bioprinting. *Biofabrication* 6 (3): 035004.

80 Engler, A.J., Sen, S., Sweeney, H.L. et al. (2006). Matrix elasticity directs stem cell lineage specification. *Cell* 126 (4): 677–689.

81 Paxton, N., Smolan, W., Böck, T. et al. (2017). Proposal to assess printability of bioinks for extrusion-based bioprinting and evaluation of rheological properties governing bioprintability. *Biofabrication* 9 (4): 044107.

82 He, Y., Yang, F., Zhao, H. et al. (2016). Research on the printability of hydrogels in 3D bioprinting. *Scientific Reports* 6: 29977.

83 Mouser, V.H.M., Melchels, F.P.W., and Visser, J. (2016). Yield stress determines bioprintability of hydrogels based on gelatin-methacryloyl and gellan gum for cartilage bioprinting. *Biofabrication* 8: 35003.

84 Yuk, H. and Zhao, X. (2017). A new 3D printing strategy by harnessing deformation, instability, and fracture of viscoelastic inks. *Advanced Materials* 30: 1704028.

85 Jin, Y., Zhao, D., and Huang, Y. (2018). Study of extrudability and standoff distance effect during nanoclay-enabled direct printing. *Bio-Design and Manufacturing* 1 (2): 123–134.

86 Ouyang, L., Yao, R., Zhao, Y. et al. (2016). Effect of bioink properties on printability and cell viability for 3D bioplotting of embryonic stem cells. *Biofabrication* 8 (3): 035020.

87 Jia, J., Richards, D.J., Pollard, S. et al. (2014). Engineering alginate as bioink for bioprinting. *Acta Biomaterialia* 10 (10): 4323–4331.

88 Markstedt, K., Mantas, A., Tournier, I. et al. (2015). 3D bioprinting human chondrocytes with nanocellulose–alginate bioink for cartilage tissue engineering applications. *Biomacromolecules* 16 (5): 1489–1496.

89 Blaeser, A., Campos, D.D., Puster, U. et al. (2016). Controlling shear stress in 3D bioprinting is a key factor to balance printing resolution and stem cell integrity. *Advanced Healthcare Materials* 5 (3): 326–333.

90 Hockaday, L.A., Kang, K.H., Colangelo, N.W. et al. (2012). Rapid 3D printing of anatomically accurate and mechanically heterogeneous aortic valve hydrogel scaffolds. *Biofabrication* 4 (3): 035005.

91 Chung, J.H., Naficy, S., Yue, Z. et al. (2013). Bio-ink properties and printability for extrusion printing living cells. *Biomaterials Science* 1 (7): 763–773.

92 Yin, J., Yan, M., Wang, Y. et al. (2018). 3D bioprinting of low-concentration cell-laden gelatin methacrylate (GelMA) bioinks with a two-step cross-linking strategy. *ACS Applied Materials & Interfaces* 10 (8): 6849–6857.

93 Wüst, S., Godla, M.E., Müller, R., and Hofmann, S. (2014). Tunable hydrogel composite with two-step processing in combination with innovative hardware upgrade for cell-based three-dimensional bioprinting. *Acta Biomaterialia* 10 (2): 630–640.

94 Tabriz, A.G., Hermida, M.A., Leslie, N.R., and Shu, W. (2015). Three-dimensional bioprinting of complex cell laden alginate hydrogel structures. *Biofabrication* 7: 45012.

95 Kang, K.H., Hockaday, L.A., and Butcher, J.T. (2013). Quantitative optimization of solid freeform deposition of aqueous hydrogels. *Biofabrication* 5 (3): 035001.

96 Zhao, Y., Li, Y., Mao, S. et al. (2015). The influence of printing parameters on cell survival rate and printability in microextrusion-based 3D cell printing technology. *Biofabrication* 7 (4): 045002.

97 Jin, Y., Chai, W., and Huang, Y. (2017). Printability study of hydrogel solution extrusion in nanoclay yield-stress bath during printing-then-gelation biofabrication. *Materials Science and Engineering: C* 80: 313–325.

98 Duarte Campos, D.F., Blaeser, A., Weber, M. et al. (2013). Three-dimensional printing of stem cell-laden hydrogels submerged in a hydrophobic high-density fluid. *Biofabrication* 5 (1): 015003.

99 Smith, C.M., Stone, A.L., Parkhill, R.L. et al. (2004). Three-dimensional bioassembly tool for generating viable tissue-engineered constructs. *Tissue Engineering* 10 (9–10): 1566.

100 Snyder, J.E., Hamid, Q., Wang, C. et al. (2011). Bioprinting cell-laden matrigel for radioprotection study of liver by pro-drug conversion in a dual-tissue microfluidic chip. *Biofabrication* 3 (3): 034112.

101 Pfister, A., Landers, R., Laib, A. et al. (2004). Biofunctional rapid prototyping for tissue-engineering applications: 3D bioplotting versus 3D printing. 42 (3): 624–638.

102 Geng, L., Feng, W., Hutmacher, D.W. et al. (2005). Direct writing of chitosan scaffolds using a robotic system. *Rapid Prototyping Journal* 11 (2): 90–97.

103 Zhang, Y., Yu, Y., Akkouch, A. et al. (2015). In vitro study of directly bioprinted perfusable vasculature conduits. *Biomaterials Science* 3 (1): 134–143.

104 Lee, V., Singh, G., Trasatti, J.P. et al. (2014). Design and fabrication of human skin by three-dimensional bioprinting. *Tissue Engineering Part C Methods* 20 (6): 473–484.

105 Skardal, A., Zhang, J., Mccoard, L. et al. (2010). Photocrosslinkable hyaluronan-gelatin hydrogels for two-step bioprinting. *Tissue Engineering Part A* 16 (8): 2675–2685.

106 Wadnap, S., Krishnamoorthy, S., Zhang, Z. et al. (2019). Biofabrication of 3D cell-encapsulated tubular constructs using dynamic optical projection stereolithography. *Journal of Materials Science: Materials in Medicine* 30 (3): 36.

107 Krishnamoorthy, S., Wadnap, S., Noorani, B. et al. (2020). Investigation of gelatin methacrylate working curves in dynamic optical projection stereolithography of vascular-like constructs. *European Polymer Journal* 124: 109487.

108 Zhao, Y., Yao, R., Ouyang, L. et al. (2014). Three-dimensional printing of Hela cells for cervical tumor model in vitro. *Biofabrication* 6 (3): 035001.

109 Ghobril, C. and Grinstaff, M.W. (2015). The chemistry and engineering of polymeric hydrogel adhesives for wound closure: a tutorial. *Chemical Society Reviews* 44 (7): 1820–1835.

110 Wilson, S.A., Cross, L.M., Peak, C.W. et al. (2017). Shear-thinning and thermo reversible nanoengineered inks for 3D bioprinting. *ACS Applied Materials & Interfaces* 9 (50): 43449–43458.

111 Ahlfeld, T., Cidonio, G., Kilian, D. et al. (2017). Development of a clay based bioink for 3D cell printing for skeletal application. *Biofabrication* 9 (3): 034103.

112 Peak, C.W., Stein, J., Gold, K.A. et al. (2018). Nanoengineered colloidal inks for 3D bioprinting. *Langmuir* 34 (3): 917–925.

113 Han, K., Bergman, R., Matyal, R. et al. (2013). Three-dimensional echocardiography and en face views of the aortic valve: technical communication. *Journal of Cardiothoracic & Vascular Anesthesia* 27 (2): 376–380.

114 Misfeld, M. and Sievers, H.H. (2007). Heart valve macro- and microstructure. *Philosophical Transactions of The Royal Society B Biological Sciences* 362 (1484): 1421–1436.

115 Yao, Q., Wei, B., Guo, Y. et al. (2015). Design, construction and mechanical testing of digital 3D anatomical data-based PCL-HA bone tissue engineering scaffold. *Journal of Materials Science Materials in Medicine* 26 (1): 1–9.

116 Wang, M.O., Piard, C., Melchiorri, A.J. et al. (2015). Evaluating changes in structure and cytotoxicity during in vitro degradation of 3D printed scaffolds. *Tissue Engineering Part A* 21 (9–10): 1642.

117 Pati, F., Song, T.H., Rijal, G. et al. (2015). Ornamenting 3D printed scaffolds with cell-laid extracellular matrix for bone tissue regeneration. *Biomaterials* 37: 230–241.

118 Kundu, J., Shim, J.H., Jang, J. et al. (2015). An additive manufacturing-based PCL–alginate–chondrocyte bioprinted scaffold for cartilage tissue engineering. *Journal of Tissue Engineering and Regenerative Medicine* 9 (11): 1286.

119 Gaetani, R., Doevendans, P.A., Metz, C.H. et al. (2012). Cardiac tissue engineering using tissue printing technology and human cardiac progenitor cells. *Biomaterials* 33 (6): 1782–1790.

120 Faulknerjones, A., Fyfe, C., Cornelissen, D.J. et al. (2015). Bioprinting of human pluripotent stem cells and their directed differentiation into hepatocyte-like cells for the generation of mini-livers in 3D. *Biofabrication* 7 (4): 044102.

121 Tatman, P.D., Gerull, W., Sweeney-Easter, S. et al. (2015). Multiscale biofabrication of articular cartilage: bioinspired and biomimetic approaches. *Tissue Engineering Part B Reviews* 21 (6): 543.

122 Lee, J.S., Hong, J.M., Jung, J.W. et al. (2014). 3D printing of composite tissue with complex shape applied to ear regeneration. *Biofabrication* 6 (2): 024103.

123 Pati, F., Jang, J., Ha, D.H. et al. (2014). Printing three-dimensional tissue analogues with decellularized extracellular matrix bioink. *Nature Communications* 5: 3935.

124 Hirt, M.N., Hansen, A., and Eschenhagen, T. (2014). Cardiac tissue engineering: state of the art. *Circulation Research* 114 (2): 354–367.

125 Zhang, Y.S., Arneri, A., Bersini, S. et al. (2016). Bioprinting 3D microfibrous scaffolds for engineering endothelialized myocardium and heart-on-a-chip. *Biomaterials* 110: 45–59.

126 Maiullari, F., Costantini, M., Milan, M. et al. (2018). A multi-cellular 3D bioprinting approach for vascularized heart tissue engineering based on HUVECs and iPSC-derived cardiomyocytes. *Scientific Reports* 8 (1): 13532.

127 Bertassoni, L.E., Cecconi, M., Manoharan, V. et al. (2014). Hydrogel bioprinted microchannel networks for vascularization of tissue engineering constructs. *Lab on A Chip* 14 (13): 2202–2211.

128 Horváth, L., Umehara, Y., Jud, C. et al. (2015). Engineering an in vitro air–blood barrier by 3D bioprinting. *Scientific Reports* 5: 7974.

129 Owens, C.M., Marga, F., Forgacs, G. et al. (2013). Biofabrication and testing of a fully cellular nerve graft. *Biofabrication* 5 (4): 045007.

130 Zhang, B., Xue, Q., Hu, H.-Y. et al. (2019). Integrated 3D bioprinting-based geometry-control strategy for fabricating corneal substitutes. *Journal of Zhejiang University-SCIENCE B* 20 (12): 945–959.

131 Mannoor, M.S., Jiang, Z., James, T. et al. (2013). 3D printed bionic ears. *Nano Letters* 13 (6): 2634–2639.

132 Marchioli, G., Van, G.L., Van Krieken, P.P. et al. (2015). Fabrication of three-dimensional bioplotted hydrogel scaffolds for islets of Langerhans transplantation. *Biofabrication* 7 (2): 025009.

133 Ning, L., Sun, H., Lelong, T. et al. (2018). 3D bioprinting of scaffolds with living Schwann cells for potential nerve tissue engineering applications. *Biofabrication* 10 (3): 035014.

134 Isaacson, A., Swioklo, S., and Connon, C.J. (2018). 3D bioprinting of a corneal stroma equivalent. *Experimental Eye Research* 173: 188–193.

135 Skardal, A., Mack, D., Kapetanovic, E. et al. (2012). Bioprinted amniotic fluid-derived stem cells accelerate healing of large skin wounds. *Stem Cells Translational Medicine* 1 (11): 792–802.

136 Norotte, C., Marga, F.S., Niklason, L.E. et al. (2009). Scaffold-free vascular tissue engineering using bioprinting. *Biomaterials* 30 (30): 5910–5917.

137 Cubo, N., Garcia, M., Del Canizo, J.F. et al. (2017). 3D bioprinting of functional human skin: production and in vivo analysis. *Biofabrication* 9 (1): 015006.

138 Gao, Q., He, Y., Fu, J.Z. et al. (2015). Coaxial nozzle-assisted 3D bioprinting with built-in microchannels for nutrients delivery. *Biomaterials* 61: 203–215.

7

Liquid Support Bath–Assisted 3D Bioprinting

7.1 Introduction

Liquid support bath–assisted three-dimensional (3D) printing, an alternative extrusion-based 3D printing strategy, has emerged in the past 10 years to create 3D structures for various applications. Different from traditional direct ink writing, liquid support bath-assisted 3D printing is performed by printing 3D structures in a liquid support bath, which can securely hold the deposited filaments at a liquid state for a long time. Thus, the ink materials are not required to possess either rapid solidification or self-supporting property [1, 2]. In addition, the liquid support bath provides in situ supporting functionality during printing, making it unnecessary to build up any auxiliary supporting structures. As a result, this new extrusion 3D printing strategy has a wider range of printable materials, higher printing efficiency, and the capacity of creating scaled-up 3D structures with complex geometries. From a material standpoint, there are two key factors in liquid support bath–assisted 3D printing, including ink materials and liquid bath materials. Based on the materials being solidified, liquid support bath–assisted 3D printing can be further divided into two subcategories: embedded 3D printing (e-3DP) [3–6] and support bath–enabled 3D printing [7–10].

e-3DP technology is mainly used to fabricate bulk parts with embedded complex patterns. During printing, different liquid inks, including sacrificial inks [5, 6] and functional inks [4, 11], are deposited into complex two-dimensional (2D) or 3D patterns in a liquid support bath (Figure 7.1a). Then, the bath is solidified using different crosslinking mechanisms, while the 3D patterns remain liquid. Depending on the removal of liquid 3D patterns, the final products are either bulk structures with embedded hollow channels [5, 6] (3D liquid pattern removed as shown in Figure 7.1b) or bulk structures with encapsulated functional circuits [4, 11] (3D liquid pattern remained as shown in Figure 7.1c), which can be used as lab-on-a-chip, soft actuators, wearable sensors, soft robots, etc. The main challenge of e-3DP is the complexity of the bulk structures severely depending on the shape of the support bath container. As a result, it is difficult to fabricate complicated 3D structures using e-3DP. To overcome this challenge, an alternative strategy-support bath-enabled 3D printing has been proposed, in which various liquid build materials are printed into complex 3D structures in a support bath (as shown in Figure 7.1a). Different

Cell Assembly with 3D Bioprinting, First Edition. Yong He, Qing Gao, and Yifei Jin.

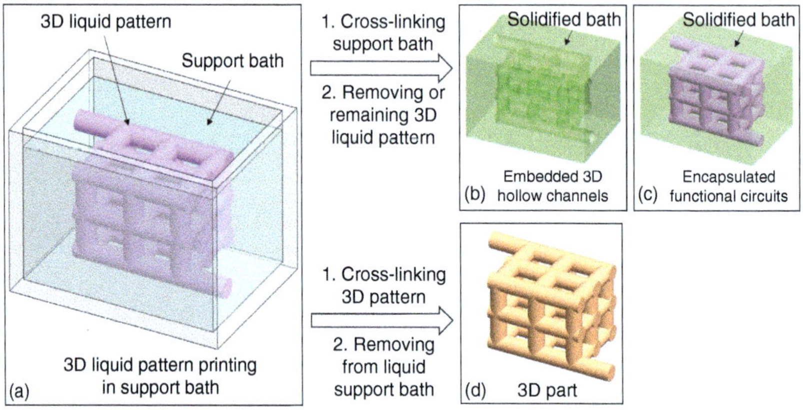

Figure 7.1 Liquid support bath-assisted 3D bioprinting techniques. (a) 3D liquid pattern printing in a support bath. Bulk structure with (b) embedded 3D hollow channels or (c) encapsulated functional circuits by crosslinking the support bath. (d) Complex 3D part by crosslinking the printed liquid structure. Source: Yifei Jin.

from e-3DP, liquid 3D structures instead of the support bath are solidified either during printing [7, 8] or after printing [9, 10]. Then, the solidified 3D structures are released from the support bath and used as functional 3D parts after removing residual support bath materials on surfaces (Figure 7.1d). Since this 3D printing strategy facilitates the fabrication of 3D parts from different build materials, such as hydrogels [8, 9, 12], composites [13–15], and elastomers [10, 16–18], it has been rapidly developed and demonstrates unimaginable potential for various applications [8–10, 15–17], especially for cell printing.

In this chapter, we will discuss the liquid support bath-assisted 3D printing technology in detail. In particular, liquid support bath materials will be introduced in Section 7.2, which are the core of liquid support bath-assisted 3D printing. Thus, the supporting mechanisms, bath material preparation methods, and design criteria of liquid support bath materials will be discussed sophisticatedly. Moreover, since the whole printing process is performed in a liquid support bath, the printing environment is liquid-in-liquid, which is quite different from the liquid-in-air environment in the traditional direct ink writing approaches, leading to different filament formation mechanisms. As a result, some technical issues during support bath-assisted 3D printing will be introduced in Section 7.3. After that, post-treatments of liquid support bath-assisted 3D printing technology will be reviewed in Section 7.4. Finally, some representative applications will be discussed in Section 7.5.

7.2 Liquid Support Bath Materials

Support bath materials are of great significance for both e-3DP and support bath-enabled 3D printing. To function as a support bath, liquid materials must have specific physical, chemical, and rheological properties. In this section, the supporting mechanisms of different categories of support bath materials will be introduced first. Then, some commonly used strategies to prepare liquid support bath materials will be summarized. Finally, the criteria for designing new support bath materials will be proposed.

7.2.1 Support Bath Materials Based on Different Supporting Mechanisms

7.2.1.1 Unrecoverable Matrix Materials

Unrecoverable matrix materials are usually gels and/or pastes that present a solid and/or a solid-like state. When a dispensing nozzle moves in an unrecoverable matrix bath, shear stress provided by the nozzle translation can create a crevasse behind the nozzle, allowing the deposition of ink materials. Since this crevasse is unrecoverable, additional fluid must be added to fill the crevasse, which is called filler. Thus, during printing, unrecoverable matrix material provides physical support to the deposited liquid pattern, while the filler functions to entrap the pattern in the matrix bath. Unrecoverable matrix materials are commonly used in e-3DP, in which both matrix materials and fillers are crosslinkable simultaneously after printing. However, the utilization of a filler makes it impossible to print 3D patterns in the unrecoverable matrix bath. As a result, only 2D patterns are reported to be printed by unrecoverable matrix bath–enabled e-3DP. The typical fabrication procedure in an unrecoverable matrix bath is illustrated in Figure 7.2.

A representative unrecoverable matrix material is Pluronic F127-DA, which is a Pluronic F127 copolymer modified with diacrylate. Wu et al. [5] proposed the omnidirectional printing strategy to successfully fabricate biomimetic microvascular networks as shown in Figure 7.2a. In their studies, F127-DA at two different concentrations was used as both unrecoverable matrix bath and filler, while the unmodified Pluronic F127 was used as sacrificial ink. In particular, higher concentration (25% (w/w)) F127-DA exhibited a high plateau shear elastic modulus G'_{matrix} to support printed patterns and suitable yield stress $\tau_{y,\text{matrix}}$ to accommodate the movement of the nozzle within the matrix [3–5], so it was selected as the matrix bath. In contrast, lower concentration (20% (w/w)) F127-DA had relatively low viscosity and good flowability, enabling it to effectively fill the crevasse during printing as shown in Figure 7.2b. The synthesized F127-DA was a photocurable hydrogel. Thus, after printing, both the matrix bath and the filler were crosslinked under ultraviolet (UV) radiation, as shown in Figure 7.2c, while the microvascular pattern from F127 was still at a liquid state. By decreasing the temperature to 4 °C, F127 presented excellent flowability and could be easily removed from the solidified F127-DA matrix (Figure 7.2d), resulting in the formation of hollow channels in the matrix as shown in Figure 7.2e,f.

7.2.1.2 Buoyant Support Fluids

Buoyancy refers to the force that an object receives in a fluid against gravity. The simple equation of buoyancy is $B = \rho_f gV$, where ρ_f is the density of the fluid, g is gravitational acceleration, and V is the volume of the object submerged in the fluid [19].

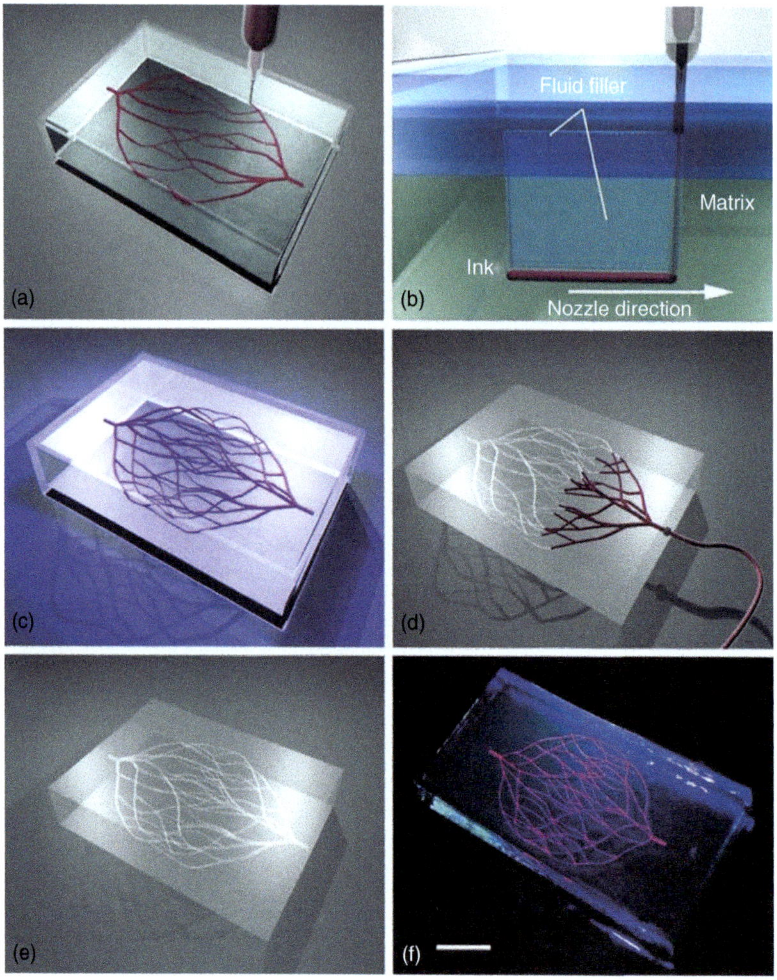

Figure 7.2 (a–e) Schematics of omnidirectional printing of 3D microvascular networks within a hydrogel reservoir. (a) Deposition of a sacrificial ink into a physical gel reservoir allows hierarchical, branching networks to be patterned. (b) Voids induced by nozzle translation are filled by a liquid that migrates from the fluid-capping layer. (c) Subsequent photopolymerization of the reservoir yields a chemically crosslinked, hydrogel matrix. (d, e) The ink is liquified and removed under a modest vacuum to expose the microvascular channels. (f) Fluorescent image of a 3D microvascular network fabricated via omnidirectional printing of a sacrificial ink (dyed red) within a photopolymerized Pluronic F127-diacrylate matrix (Scale bar: 10 mm). Source: Wu et al. [5]. Reproduced with permission of John Wiley and Sons.

During 3D bioprinting, liquid support bath materials and cell-laden bioinks usually possess similar density. Thus, the buoyancy provided by the liquid support bath is equivalent to the gravity of the deposited structure, leading to the structure suspending in the liquid support bath. Because the buoyant support fluids are usually uncrosslinkable materials with good flowability, they are potentially able to facilitate the fabrication of complex 3D structures using support bath–enabled 3D printing.

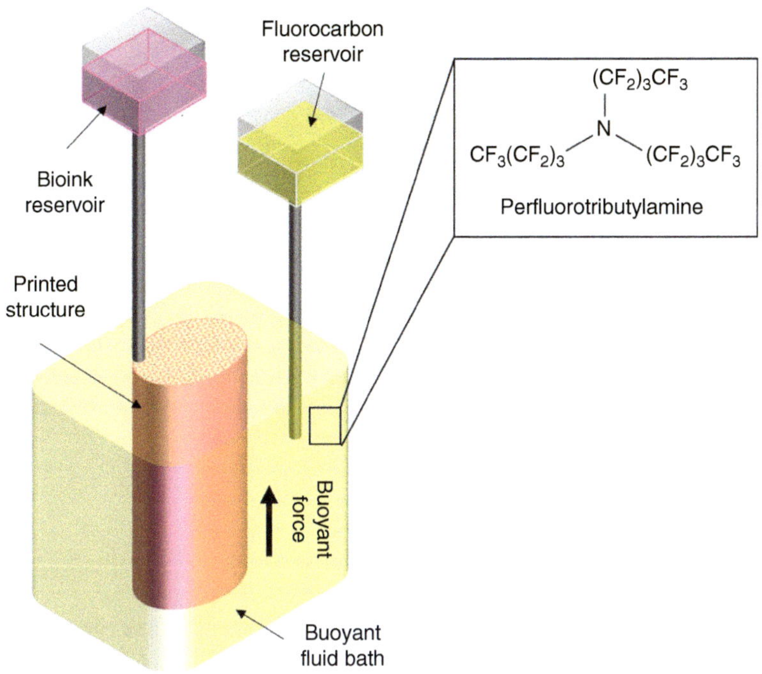

Figure 7.3 Schematic diagram of printing in a buoyant fluid bath. Source: Yifei Jin.

However, buoyancy cannot provide strong support during printing and stably holds the printed structure in situ. Both the location and the fidelity of the structure in the buoyant support bath can be easily affected and disrupted by the nozzle movement. As a result, this type of support bath material has been merely investigated. Only one literature was reported to use the buoyant fluid as support bath material for 3D bio-printing applications. Campos et al. [20] utilized perfluorotributylamine ($C_{12}F_{27}N$) to print the cell-laden agarose hydrogel as shown in Figure 7.3. They found that the high-density, hydrophobic $C_{12}F_{27}N$ enabled better control of the deposition of each droplet, and the encapsulated cells had a good survival rate and proliferation after printing.

7.2.1.3 Reversibly Self-Healing Hydrogels

Reversibly self-healing hydrogels are supramolecular hydrogels with guest–host recognition and nature matrix (e.g. gelatin, chitosan, hyaluronic acid [HA]). The guest–host recognition has non-covalent bonds for forming supramolecular assembly as well as covalent bonds with a natural matrix [21]. When a dispensing nozzle moves in a self-healing hydrogel bath, non-covalent bonds are disrupted by shear stress, leading to the decrease of the viscosity. This shear-thinning behavior allows the nozzle to move freely in the bath to deposit bioinks based on designed trajectory. The covalent bonds of the supramolecular hydrogels can facilitate the formation of stable networked structures that promote the re-construction of the

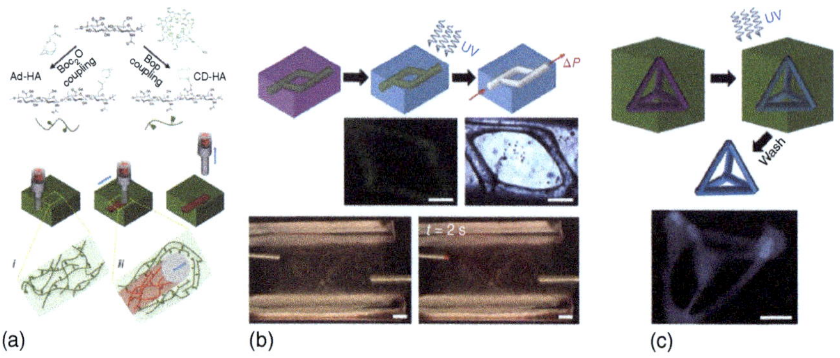

Figure 7.4 (a) Schematic diagram of self-healing hydrogels and self-healing hydrogel bath-enabled 3D printing approach. (b) 3D printing of hollow channels by UV crosslinking self-healing support bath. (c) Fabrication of 3D structures by UV crosslinking shear-thinning bioink material. Source: Highley et al. [22]. Reproduced with permission of John Wiley and Sons.

non-covalent bonds. Thus, reversibly self-healing hydrogels can rapidly switch from gel to sol states and fill the gaps behind the nozzle.

One representative self-healing hydrogel was investigated and proposed by Highley et al. [22]. In their study, they developed a supramolecular hydrogel assembly through guest–host complexes. HA was selected as the main component of the supramolecular hydrogel due to its amenability to chemical modification and excellent biocompatibility. HA was modified with either adamantane (Ad) or β-cyclodextrin (β-CD) (Ad-HA and CD-HA, as shown in Figure 7.4a), and supramolecular assemblies rapidly formed upon mixing of Ad-HA and CD-HA through intermolecular guest–host bonds (between Ad and β-CD moieties). Moreover, methacrylates (≈20% of repeat units) were introduced to further modify Ad-HA and CD-HA for enhancing the mechanical strength. Modified hydrogels can form stable covalent bonds under UV radiation. Using this guest–host hydrogel printing technology ("ghost writing"), they successfully printed encapsulated hollow channels and freeform 3D structures as shown in Figure 7.4b,c, respectively.

7.2.1.4 Yield-Stress Fluids

Among different support bath materials, yield-stress fluids are of great significance, which are usually composed of nano and/or microparticles with given inherent microstructures. Yield stress is a result of the threshold energy required to disrupt the inherent microstructures before the fluids flow. Thus, yield-stress support bath materials have two unique properties: (i) a solid-like state when the applied stress is lower than the yield stress and (ii) repeatable transition between liquid and solid-like states upon different stressed conditions. For the former, except buoyancy provided by solvents of yield-stress fluids, the microstructures of the fluids can provide additional supporting force to hold deposited 3D structures stably in situ, which makes yield-stress fluids to have excellent supporting functionality. For the latter, yield-stress fluids present a liquid state when the applied stress is higher than the yield stress and recover to a solid-like state when the applied stress is removed or decreased. Thus, during printing, the higher shear stress–induced liquefied yield-stress fluid around the dispensing nozzle can fill the crevasse behind the nozzle and entrap the deposited 3D structure in the support bath, making it unnecessary to use an additional filler to facilitate this encapsulation process [5]. In addition, by designing and/or selecting suitable solvents to prepare yield-stress fluids, the chemical stability of yield-stress property can be ensured, which may prevent the dysfunctionality of the supporting capability caused by some possible reactions with the guest–host hydrogel support bath [22]. As a result, various yield-stress support bath materials have been developed and widely used in support bath–enabled 3D printing.

Currently, several yield-stress fluids have been investigated for 3D bioprinting applications. For example, Bhattacharjee et al. [7], Hinton et al. [17], Jin et al. [12], and Zhang et al. [23] investigated a granular gel medium (Carbopol microgels) as yield-stress support bath material, which was composed of numerous sponge-like microgels, making the granular gel medium to have a unique jamming/unjamming transition property. When a nozzle moves in the granular gel medium, the localized shear stress around the nozzle tip is in the range of 1–200 Pa, leading to the unjamming status of microgels in microscopic view as well as the sol state of the bath in macroscopic view. Thus, the nozzle can be easily inserted and move rapidly within the bath. When the nozzle moves through that region, the localized unjamming microgels can switch to a jamming state. As a result, the fluidized gel rapidly solidifies in the wake to firmly hold the deposited material in place as shown in Figure 7.5a. Hinton et al. [8], Lee et al. [25], and Mirdamadi et al. [24] proposed gelatin microparticles as support bath material. In their studies, gelatin was gelled at a lower temperature and then stirred up into microscale particles. After that, these microparticles were collected together as a support bath. The whole printing and crosslinking were performed at room temperature or even lower temperatures to maintain the gel state of gelatin microparticles. Finally, the temperature of the support bath was increased to a higher temperature range (~37 °C) to melt the gelatin microparticles. Thus, the printed structure can be easily released from the support bath as shown in Figure 7.5b. Excluding these two yield-stress bath materials, Jin et al. [9, 26] proposed a versatile support bath material – Laponite nanoclay suspension, which is composed of numerous nanosilicates. When dispersed in an aqueous solvent, these nanosilicates can form a unique "house-of-cards" arrangement, leading to the generation of the yield-stress property. When the nozzle moves in the nanoclay suspension, the shear stress at the nozzle tip disrupts the "house-of-cards" arrangement of localized nanosilicates, making the suspension to behave as a liquid. In contrast, nanoclay suspension far away from the nozzle tip has no shear stress. Thus, the nanosilicates remain their "house-of-cards" arrangement, behaving as solid-like as shown in Figure 7.5c. Comparing with other yield-stress fluids, nanoclay suspension has excellent thermal stability, ionic insensitivity, UV transparency, and good biocompatibility. Therefore, it is a promising support bath material to 3D print complex constructs for various biomedical applications.

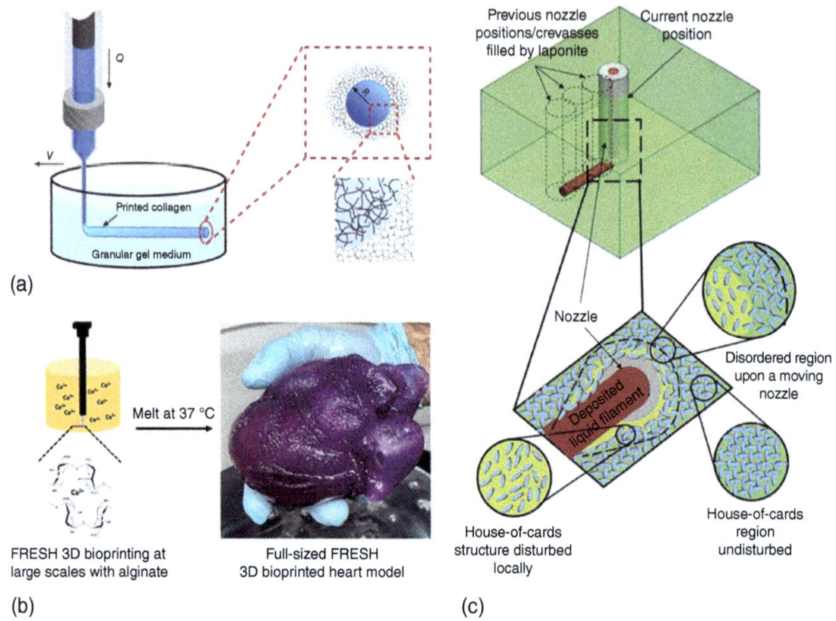

Figure 7.5 Schematics of printing mechanisms using (a) granular gel medium. Source: Zhang et al. [23]. Reproduced with permission of Elsevier. (b) Gelatin microparticles. Source: Mirdamadi et al. [24]. Reproduced with permission of American Chemical Society. (c) Laponite nanoclay suspension. Source: Jin et al. [9]. Reproduced with permission of American Chemical Society.

7.2.2 Preparation Methods

Various methods have been utilized to make liquid support bath materials based on different supporting mechanisms. In this section, some representative preparation methods are introduced.

7.2.2.1 Microparticle Aggregation

Preparation of microparticles is the first step to make granular gels or microparticle gels as support bath materials. There are three typical methods to generate microparticles: (i) physical sedimentation, (ii) mechanical stirring, and (iii) microfluidic system.

Physical sedimentation was introduced by O'Bryan et al. [27] to make granular gels. In their study, they first prepared acrylamide, ionizable comonomer, poly(ethylene glycol) diacrylate, and azobis (isobutyronitrile) ethanol solutions separately. Then, these solutions were mixed together with nitrogen for 30 minutes and placed in an oil bath at 60 °C. After 30 minutes, white precipitates were generated. Using vacuum filtration, these precipitates were collected. Finally, the precipitates were rinsed and soaked with ethanol and then dried in a vacuum oven at 50 °C, leading to the formation of microparticles with anion or cation.

Comparing with physical sedimentation, mechanical stirring is an easier method to make microgel particles. There are only two main steps in mechanical stirring, including crosslinking and stirring. In the former, different hydrogels are crosslinked

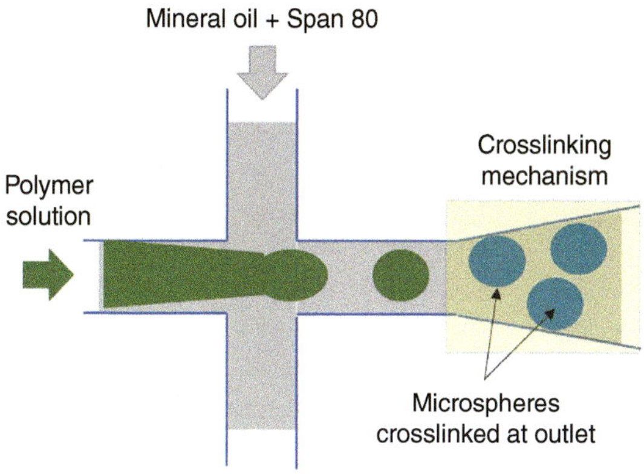

Figure 7.6 Formation of microgel particles using the microfluidic system. Source: Yifei Jin.

into solid bucks using various crosslinking mechanisms, while in the latter, these gelled hydrogel bucks are broken up into microparticles by mechanical forces. For example, Hinton et al. [8] mixed gelatin (Type A) and $CaCl_2$ solution in a mason jar and stored it at 4 °C for 12 hours and then added the $CaCl_2$ solution again for stirring by a commercial blender. After that, the resulted gelatin slurry was centrifuged for collection. After removing the supernatant on top, the remaining slurry was re-dispersed into the $CaCl_2$ solution and centrifuged again until there was no bubble in the supernatant. Thus, the collected gelatin slurry can be used as a support bath for printing purposes. Theoretically, mechanical stirring is able to make microgel particles from any crosslinkable materials, especially hydrogels with excellent biocompatibility. As a result, it is the most promising method to prepare support bath materials for 3D bioprinting applications.

The microfluidic system is an emerging method to prepare microgel particles as a liquid support bath. In the system, a hydrophilic hydrogel solution and a hydrophobic liquid are perfused into a microfluidic device from two independent channels. Due to the interfacial tension effect, the hydrophilic hydrogel solution can break up into microspheres when intersecting with the hydrophobic liquid. At the end of the outlet channel, a corresponding crosslinking mechanism is used to solidify the microspheres, which are collected as support bath material as shown in Figure 7.6. For example, Highley et al. [28] pumped mineral oil, Span 80, and NorHA/PEGDA/agarose solutions into a microfluidic system. At the end of the outlet channel, polymer droplets were formed and crosslinked by UV radiation.

7.2.2.2 Homogenous Suspensions with Micro/Nanostructures

Preparing homogenous suspensions with inherent micro and/or nanostructures is the most commonly used method in the preparation of a yield-stress support bath from commercial products. In this methodology, powders are dispersed in suitable media and mixed uniformly to form a homogenous and stable suspension as a support bath. Generally, the preparation step is easy to handle and the preparation efficiency is relatively high. For example, to make Carbopol granular microgels, Carbopol powder was first dissolved with water and then an appropriate amount of NaOH solution was added to adjust the pH value to around 7 [7, 12]. For the preparation of the Laponite nanoclay support bath, Jin et al. [13] dispersed Laponite powder in deionized water and continued mixing for 60 minutes to obtain homogenous suspensions. Since the formation of "house-of-cards" arrangements needed a given time period, the suspensions were usually stored in a dark, sealed container for 24 hours. After that, the suspension possessed stable rheological properties and was ready to function as a yield-stress support bath.

7.2.2.3 Chemical Synthesis

Except for the aforementioned two physical methods to prepare liquid support baths, some chemical approaches are also used to synthesize fluids with suitable rheological properties. One representative example is the preparation of self-healing hydrogel. Although HA has excellent biocompatibility and has been widely used for various biomedical applications, it is difficult to be directly applied in liquid support bath–assisted 3D bioprinting due to lack of certain rheological properties. As a result, Highley et al. [22] used Ad and β-CD to modify HA. The resultant Ad-HA and CD-HA can form guest–host bonds through supramolecular assemblies, thereby enhancing the mechanical strength of gel networks. In addition, to achieve photo crosslinking capacity, they further added methacrylic anhydride to HA so that the synthesized hydrogel can be cured by UV radiation. Thus, the prepared HA-based support bath material and bioink had different crosslinking mechanisms to facilitate both e-3DP and support bath–enabled 3D printing. Another example of the synthesized hydrogel support bath is Pluronic F127-DA [5]. Pure Pluronic F127 is thermal crosslinkable. By modifying with a functional group of diacrylate, F127-DA can be used as a UV-curable support bath material [29].

7.2.2.4 Other Methods

Other methods are used to prepare specific support bath materials for 3D bioprinting applications. For example, Skylar-Scott et al. [30] cultivated embryoid bodies (EBs) composed of high density–induced pluripotent stem cells (iPSCs) in extracellular matrices (ECMs) at 0–4 °C to form organoids. These organoids can further be used as a matrix material for e-3DP purpose. In their study, sacrificial ink was printed into a vascular networked pattern in the organoid matrix bath. After removing sacrificial ink, hollow vascular channels were formed in the matrix bath. Thus, nutrients and oxygen were transported to cells in the organoids via the hollow channels, enabling these organoids to grow up and differentiate at 37 °C.

7.2.3 Design Criteria for Ideal Liquid Support Bath Material

There are several basic requirements for designing ideal liquid support bath materials in the future, including rheological properties, physical/chemical stability, biocompatibility, hydrophilicity, and dissolvability.

7.2.3.1 Rheological Properties

First, the yield-stress property is one of the primary rheological properties of support bath materials to facilitate liquid support bath–assisted 3D printing. During printing, support bath materials must present solid-like behavior at low shear stresses and behave as a liquid at high shear stresses. This unique rheological property can be described by the Herschel–Bulkley model: $\sigma = \sigma_y + k\dot{\gamma}^n$, where σ is shear stress, σ_y is the yield stress of support bath material, $\dot{\gamma}$ is shear rate, and n and k are two dimensionless constants. When $n = 1$, the support bath is Bingham plastic fluid, while $n < 1$ corresponds to the Herschel–Bulkley fluid [9]. The ideal range of yield stress is summarized in Table 7.1. Second, support bath materials must have a sufficient shear elastic modulus (G′) higher than the shear loss moduli (G″) to hold the geometries of printed structures when the applied shear stress is below the yield point [4]. Third, to effectively fill the crevasses behind nozzle translation, support bath materials should be able to rapidly transit between liquid and solid-like states. Hydrostatic stress is one factor to induce this liquid–solid-like transition, which is defined as $\sigma_h = \rho gh$ (where ρ is the density of support bath material, g is the acceleration of gravity, and h is depth). When hydrostatic stress is larger than the yield stress σ_y, that is $\rho gh/\sigma_y > 1$, the support bath material can switch to a liquid state to fill the crevasses [7]. The thixotropic time scale, also defined as recovery time, is another critical rheological parameter to evaluate the functionality of support bath materials. Usually, a short thixotropic time scale indicates a rapid transition between liquid and solid-like. For example, the recovery time of Pluronic F127-DA is extremely long, making it necessary to use the filler solution to fill the crevasse during printing. In contrast, the recovery time of Laponite nanoclay is only 0.02 seconds. Thus, the liquified localized nanoclay suspension can easily fill the gaps behind the nozzle, effectively facilitating the embedded printing of complex 3D structures in the support bath.

7.2.3.2 Chemical Stability

During liquid support bath–assisted 3D printing, ink material is directly deposited in a support bath without crosslinking. If ink material has the potential to chemically react with support bath material, this chemical reaction may destroy the inherent molecular structures of the bath material, leading to the dysfunctionality as a support bath. In addition, crosslinking mechanisms should ideally not induce chemical reactions of support bath materials. For example, Carbopol granular microgel has good chemical stability during printing and has no reactions with most biocompatible materials, such as alginate [12], gelatin [7, 12], and polydimethylsiloxane (PDMS) [17]. However, the addition of Ca^{2+} and Al^{3+} as crosslinkers of alginate may bring damage to the jamming status of granular microgels. Thus, Carbopol loses its supporting function before the alginate structure is crosslinked and cannot be used for alginate printing purposes.

7.2.3.3 Physical Stability

Physical stability usually refers to the stability of support bath materials in different physical conditions. One typical example is the temperature change. For some thermal-sensitive support bath materials, such as gelatin microparticles, both the printing temperature and the crosslinking temperature must be controlled below the melt temperature of gelatin (~30 °C depending on concentration). Otherwise, gelatin microparticles will melt and lose the functionality as the support bath. In contrast, printing in a Pluronic F127 bath should be performed at higher temperature

Table 7.1 Summary of critical properties of commonly used liquid support bath materials.

Materials	Inherent structures	Yield stress	Recover time	Sensitivity	Affinity	Biocompatibility	References
Carbopol[a)]	Granular microgels (jamming/unjamming)	1–200 Pa (0.05–1 wt%)	~0.3 s	pH cations	Hydrophilic	Yes	[7]
Laponite nanoclay[b)]	Nanosilicates ("house-of cards" arrangement)	9–10 Pa (2–4 wt%)	~0.1 s	Thermal (>60 °C)	Hydrophilic	Yes	[9]
Modified HA[c)]	Polymer molecules	n/a	n/a	Chemical reaction	Hydrophilic	Yes	[22]
Pluronic F127-DA	Paste with high plateau shear elastic modulus	~60–900 Pa (23–30 wt%)	Long (filler required)	Thermal	Amphiphilic	Yes	[5]
Gelatin slurry	Microparticles (particle aggregates)	Minimum 314.63 Pa	fast	Thermal (>37 °C)	Hydrophilic	Yes	[8, 31]

a) Carbopol ETD 2020 [7].
b) Laponite EP [9, 26].
c) HA modified with either adamantane (Ad) or β-cyclodextrin (β-CD) [22].

since Pluronic F127 is completely dissolved at a temperature of 4 °C. As a result, ideal liquid support bath materials must have excellent physical stability to maintain their rheological properties in different physical conditions.

7.2.3.4 Biocompatibility

For cell printing applications, biocompatibility is one of the basic requirements of liquid support bath materials. When using as matrix baths for e-3DP, support bath materials not only are nontoxic but also provide an excellent ECM environment for cell growth and proliferation. For support bath–enabled 3D printing, since printed living tissue constructs are finally removed from the support bath and cultured in the culture medium, support bath materials are only required to be cytocompatible. The biocompatibility of commonly used liquid support bath materials is summarized in Table 7.1.

7.2.3.5 Hydrophilicity and Hydrophobicity

Different from traditional direct ink writing, liquid support bath–assisted 3D printing is performed in a liquid-in-liquid environment. Thus, different hydrophilicity/hydrophobicity between ink materials and support bath materials may lead to interfacial tension, which can induce breakup of continuous filaments into droplets, damaging the integration of printed structures. In cell printing cases, cell-laden bioinks are aqueous solvent-based, and most commonly used support bath materials also use water as a solvent. As a result, the effect of the aqueous affinity is negligible. However, for 3D printing of biomedical devices, some hydrophobic ink materials may be utilized, such as PDMS. In these cases, new support bath materials need to be designed to minimize the effects of interfacial tension as reported in Refs. [10, 16].

7.2.3.6 Others

Dissolvability is the ability of support bath materials to completely lose their supporting function, which is of great significance during post-treatments. Gelatin microparticles are a typical dissolvable support bath material. After the printed structure is crosslinked in the gelatin microparticle bath, the bath temperature increases to physiological temperature. Thus, gelatin microparticles melt and dissolve. Crosslinked structures can be easily removed from the low-viscosity gelation solution bath. In addition, the particle size is also a critical factor when designing a granular gel–based support bath material. Generally, the particle size affects the surface morphology of the deposited filaments and finally impacts the resolution of printed 3D structures. As a result, it is necessary to make the particle size as small as possible. However, the size of microparticles in granular gels should be controlled above 1 μm to eliminate colloidal scale diffusion [7]. Therefore, when designing new granular gel–based support bath materials, both resolution and diffusion must be taken into consideration simultaneously. Finally, the transparency of liquid support bath materials is also important, which can facilitate in situ imaging of cell growth and proliferation in 3D parts fabricated by e-3DP and/or online observation of the printing process during support bath–enabled 3D printing.

7.3 Scientific Issues During Liquid Support Bath-Assisted 3D Printing

Liquid support bath–assisted 3D printing is featured by printing liquid ink materials in a liquid support bath. Thus, the printing process is in a complicated liquid-in-liquid environment and quite different from traditional direct ink writing approaches. In this section, some interesting topics during liquid support bath–assisted 3D printing are introduced.

7.3.1 Effects of Operating Conditions on Filament Formation in Support Bath

Herein, operating conditions mainly include process parameters, such as nozzle diameter, dispensing pressure, and path speed as well as ink material properties. One mathematical model has been created by Jin et al. [26] to numerically investigate the filament formation in a yield-stress support bath. The schematic is illustrated in Figure 7.7a. In their study, the velocity of ink material at the exit of the dispensing nozzle (V_{out}) can be expressed as $V_{out} = \frac{(P_0 - \sigma) + \rho g L}{8\eta_0 L}R^2$, which is a function of dispensing pressure (P_0), ink material properties (density [ρ] and zero-shear-rate viscosity [η_0]), dispensing nozzle geometries (radius [R] and length [L]), as well as stress (σ) from the support bath. From this mathematical model, it is found that more materials can be extruded in the support bath when increasing the dispensing pressure and the nozzle diameter, thus resulting in the increase of the filament diameter. In contrast, the increase of ink material viscosity brings more resistance to the extrusion process, which reduces the velocity of ink materials and further decreases the filament diameter. The relationship between V_{out} and path speed (V_{path}) also affects the filament diameter. In particular, when $V_{out} > V_{path}$, more ink materials are deposited in place at a given time, leading to the formation of swelling filaments with the filament diameter being larger than the nozzle diameter. When $V_{out} < V_{path}$, the dragging effect by nozzle movement can stretch the deposited filament, resulting in the formation of stretched filaments with the filament diameter being smaller than the nozzle diameter. It is only when $V_{out} \approx V_{path}$ that equivalent-diameter filaments with a diameter close to the nozzle diameter can be formed, as shown in Figure 7.7b [26].

7.3.2 Effects of Support Bath Materials on Filament Morphology

During liquid support bath–assisted 3D printing, support bath materials affect the morphology of printed filaments significantly from three aspects: (i) filaments with irregular shapes resulting from rheological properties of support bath materials, (ii) diffusion-induced filament/structure geometrical change, and (iii) interfacial tension–induced filament deformation.

7.3.2.1 Rheological Properties of Support Bath Materials

When printing liquid filaments in a liquid support bath, support bath material properties determine the morphology of filaments directly. However, there are only a

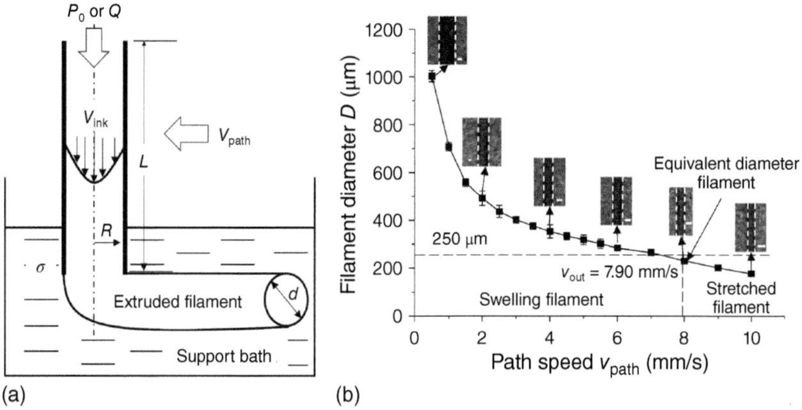

Figure 7.7 (a) Schematic diagram of filament printing in a liquid support bath. Source: Yifei Jin. (b) Filament diameter and category as a function of path speed. Source: Jin et al. [26]. Reproduced with permission of Elsevier.

handful of papers investigating the function and effects of support bath materials. Muth et al. [4] found that for support bath-enabled 3D printing, the ink and bath materials must meet several requirements: (i) the inks possess a shear elastic modulus and yield stress which are approximately an order of magnitude larger than those of the support bath materials, (ii) the support bath materials must possess a sufficiently high elastic modulus to support the printed structures and an appropriate yield stress, which allows the nozzle to move freely in the bath, and (iii) the ink and support bath materials should be chemically compatible. Hinton et al. [17] investigated the filament morphology in a Carbopol yield-stress bath during PDMS printing and found that both Carbopol 940 and ETD 2020 can produce smooth, cylindrical filaments, while Carbopol Ultrez 30 can only produce filaments with a rough surface as shown in Figure 7.8a. Jin et al. [26] systematically investigated the filament morphologies by printing the alginate solutions in the nanoclay suspensions with different rheological properties. It is found that by varying the combinations of material properties (both bioink materials and support bath materials) as well as operating conditions, seven filament types (Figure 7.8b) are possibly to be formed during liquid support bath-assisted 3D printing, including three types of well-defined filaments (swelling filament, equivalent diameter filament, stretched filament) and four types of irregular filaments (rough surface filament, overdeposited filament, compressed filament, discontinuous filament). Thus, in liquid support bath-assisted 3D printing, the rheological properties of support bath materials must match with the rheological properties of the ink materials to facilitate the formation of well-defined filaments with controllable diameters.

7.3.2.2 Diffusion of Ink Materials into Surrounding Support Bath

When printing miscible bioink materials in a support bath, the concentration difference between the bioink and the support bath leads to the existence of osmotic pressure, which makes ink materials to gradually diffuse from the deposited filaments

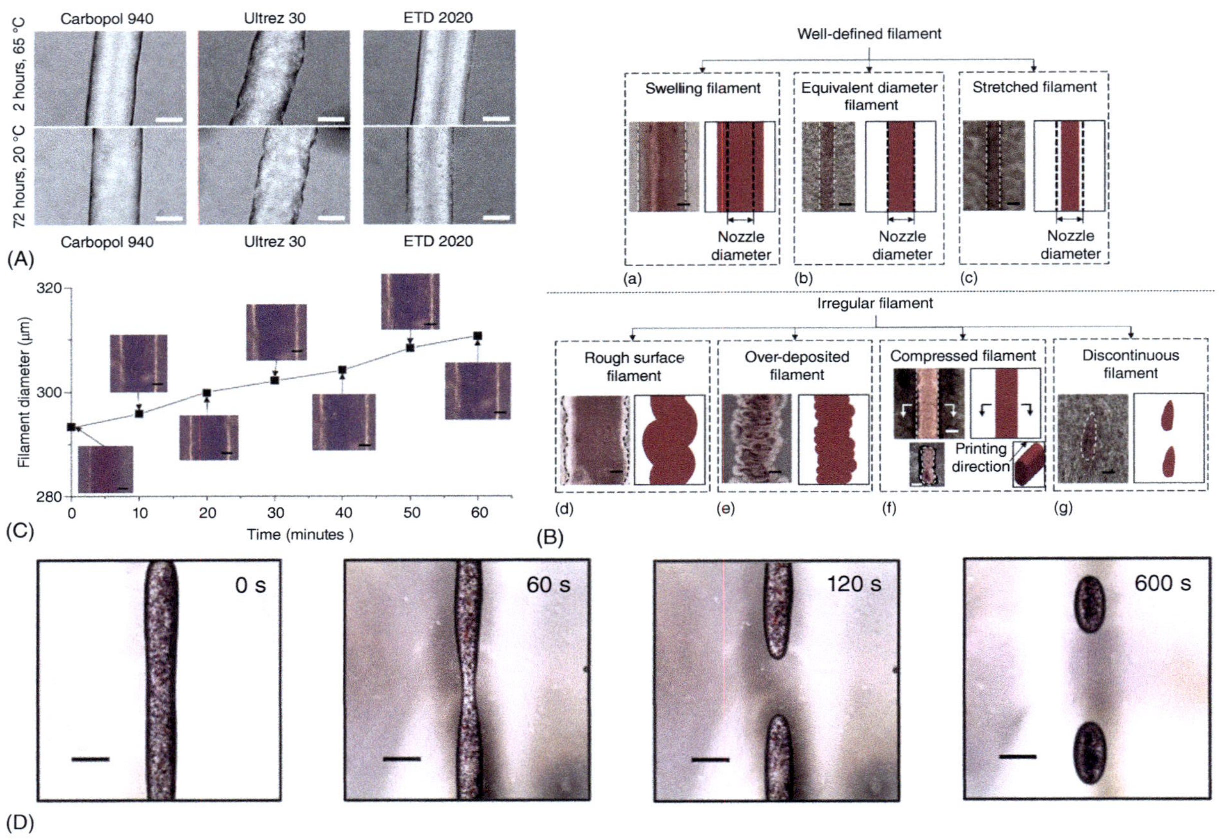

Figure 7.8 (a) PDMS filament printing in different Carbopol support baths. Source: Hinton et al. [17]. Reproduced with permission of American Chemical Society. (b) Representative images and schematics of seven types of filaments being printed in support baths with different rheological properties. Source: Jin et al. [26]. Reproduced with permission of Elsevier. (c) Filament diameter increase due to diffusion. Source: Jin et al. [9]. Reproduced with permission of American Chemical Society. (d) Hydrophobic PDMS filament breaking up into segments in hydrophilic nanoclay suspension due to interface tension effect. Source: Jin et al. [10]. Reproduced with permission of American Chemical Society.

into the surrounding support bath. Thus, the size of the filament may increase over time [32]. For example, Jin et al. [9] measured the alginate filament diameters in the nanoclay suspension at different time intervals and found that without crosslinking, the diameter of alginate filaments increased 5.8% in 60 minutes due to alginate diffusion as shown in Figure 7.8c. Thus, crosslinking filaments immediately after printing can effectively reduce the diameter change.

For theoretical investigation, Fick's second law of diffusion, $\frac{\partial C}{\partial t} = D_c \frac{\partial^2 C}{\partial x^2}$, can be used to quantitively predict the diffusion as a function of submerging time, where $C = C_{(x,t)}$ is the concentration of ink material at length x from the filament surface at time t and D_c is the effective diffusion coefficient. The unidirectional diffusion can be expressed as: $C_{(x,t)} = C_0 \left[1 - \mathrm{erf}\left(\frac{x}{2\sqrt{D_c t}}\right)\right]$, where erf(·) is the error function. Jin et al. [12] used this model to successfully explain the geometrical changes of alginate/gelatin tubes being printed in the Carbopol support bath. As a result, during liquid support bath–assisted 3D printing, controlling crosslinking time is one of the most convenient ways to improve the fidelity of printed structures.

7.3.2.3 Interfacial Tension–Induced Filament Deformation

When a hydrophilic ink material is printed in a hydrophobic support bath or vice versa, the density differences between the ink and support bath material lead to the generation of interfacial tension at the boundary of these two liquids, which may deform the printed liquid filament and even cause the breakup of continuous filaments into droplets [10, 16]. For example, Jin et al. [10] printed a hydrophobic PDMS filament in a hydrophilic nanoclay suspension and found that the filament with a diameter of ~400 μm broke up into segments in only 60 seconds as shown in Figure 7.8d. This phenomenon can be quantitively predicted by the Plateau–Rayleigh instability. During printing, the interfacial tension (γ) between the ink material and the support bath material must be balanced by the yield stress (σ_y) of the support bath material. Thus, there is a critical filament diameter (l_c), which is described as $l_c = \frac{\gamma}{\sigma_y}$. When the printed filament has a diameter larger than this critical filament diameter, it can overcome the interfacial tension–induced filament deformation [33, 34]. Otherwise, it will break up into droplets. As a result, for hydrophilic (hydrophobic) ink material printing in a hydrophobic (hydrophilic) support bath, two strategies can be used to maintain the integration of the printed structures: increasing the filament size and crosslinking the filament before deformation occurs.

7.3.3 Effects of Nozzle Movement on the Printed Structure

When a dispensing nozzle moves in a yield-stress support bath, the localized bath material around the nozzle tip is disrupted by the movement, which may affect the location and shape of the deposited structures [35, 36]. The disrupted area is defined as the affected thixotropic region. Bhattacharjee et al. [7] and Jin et al. [9] used a high-speed camera to record the movement of support bath materials around the nozzle tip separately and analyzed by particle image velocimetry as shown in

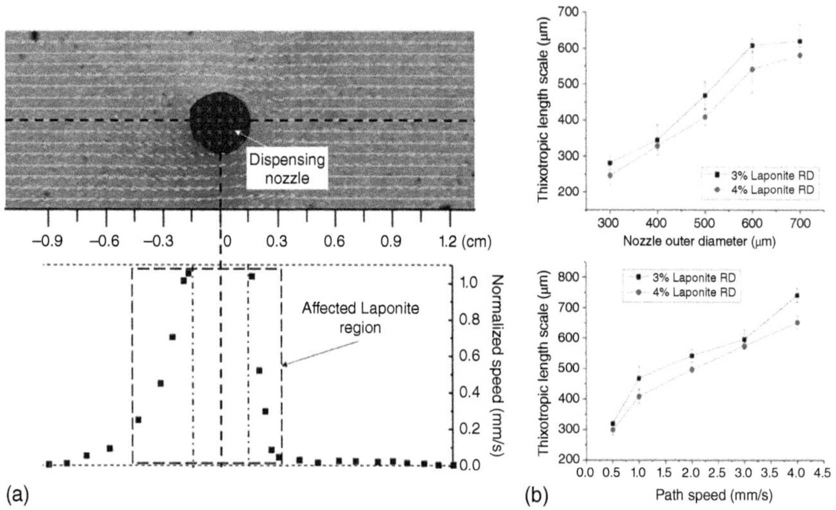

Figure 7.9 (a) Dispensing nozzle moving in a support bath of yield-stress fluids. (b) Thixotropic length scale as a function of the nozzle outer diameter (top) and path speed (bottom). Source: Jin et al. [9]. Reproduced with permission of American Chemical Society.

Figure 7.9a. They found that the thixotropic length scale was close to the diameter of the dispensing nozzle and also increased linearly with path speed as shown in Figure 7.9b. As a result, during liquid support bath-assisted printing, controlling path speed and using microscale nozzle with smaller outer diameter are two potential solutions to reduce the effects of nozzle movement on a printed structure.

7.3.4 Path Design in Liquid Support Bath-Assisted 3D Printing

In traditional direct ink writing approaches, 3D models are sliced into numerous layers before printing, and the printing trajectory is generated based on the 2D pattern on each sliced layer. Thus, during printing, the dispensing nozzle moves along the horizontal directions to finish the printing of the entire layer and then moves upward to print the adjacent layer. This path design method is defined as a conventional layer-by-layer strategy. However, during liquid support bath-assisted 3D printing, the support bath material can hold the deposited materials in place at any location in the support bath. As a result, the printing path design is completely different from the conventional layer-by-layer strategy [37, 38]. For example, Jin et al. [9] proposed a localized "layer-by-layer" printing method to design a trajectory. In this new method, it is no longer necessary to consider force balance or the supporting capacity of the previously deposited structures. Instead, the printing efficiency is primary. The steps for printing a vascular network structure are illustrated in Figure 7.10. This method improves the printing efficiency by maximizing the continuous deposition of ink materials and minimizing the non-depositing travel time among various deposition locations, thereby reducing the total print time from 75 minutes, using the conventional layer-by-layer strategy, to only 35 minutes, using the localized "layer-by-layer"

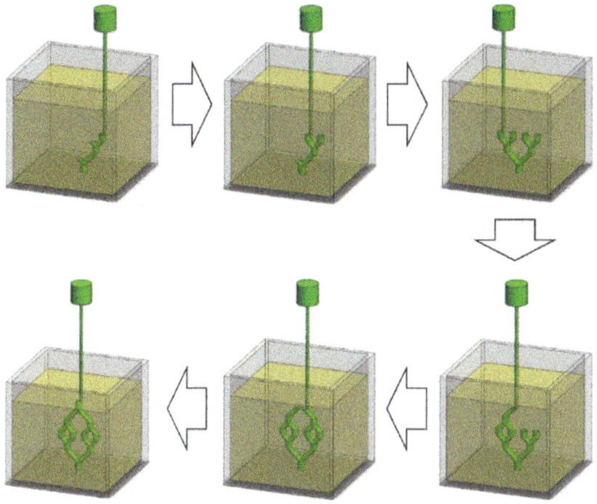

Figure 7.10 Localized "layer-by-layer" path design strategy. Source: Jin et al. [9]. Reproduced with permission of American Chemical Society.

printing method. In the future, more trajectory design strategies will be proposed to further optimize the printing efficiency.

7.4 Post-treatments for Liquid Support Bath–Assisted 3D Printing

Comparing with traditional direct ink writing, post-treatments for liquid support bath–assisted 3D printing are more tedious and time-consuming. In particular, e-3DP and support bath–enabled 3D printing have different requirements of post-treatments.

7.4.1 Post-treatments in e-3DP

In e-3DP, post-treatments consist of two main steps: (i) crosslinking the matrix bath material and (ii) removing the sacrificial inks, if necessary. During the crosslinking step, various crosslinking mechanisms are used to solidify matrix bath materials and entrap printed liquid patterns inside. The crosslinking mechanisms are introduced in detail in Chapters 2 and 6. Usually, UV radiation and temperature change are the two most commonly used crosslinking methods in e-3DP. If liquid patterns are made from functional inks, such as conductive inks and cell aggregates, they are unnecessary to be removed from the solidified matrix bath. However, if the purpose of embedded printing is to form entrapped hollow channels, the sacrificial inks must be removed from the matrix. The details of post-treatments in e-3DP are summarized in Table 7.2. Pluronic F127 is one of the most popular sacrificial inks, which behaves like paste at the room temperature and presents excellent flowability at a lower temperature (e.g. 4 °C). Thus, during post-treatments, the environment temperature is decreased to ~4 °C. Then, the Pluronic F127 solution can be easily siphoned from the solidified matrix.

Table 7.2 Post-treatments for liquid support bath–assisted 3D printing.

	Matrix materials and inks	Matrix curing methods	Removal of inks	Ink functionality	References
e-3DP	M: Pluronic-DA I: Pluronic[a]	UV radiation with photoinitiator	Yes	Sacrificial ink ($<4\,^{\circ}$C: sol; $>22\,^{\circ}$C: gel)	[5]
	M: GelMA I: Pluronic[a]		Yes		[11]
	M: OBBs[b]/ECM I: Gelatin	Temperature change	Yes		[30]
	M: PDMS I: Fumed silica/EMIM-ES[c]	Cured by crosslinking agent	No	Conductive ink	[6]
	M: PDMS I: Carbon conductive grease[d]		No		[4]

	Bath materials and inks	Ink curing methods	Bath material removing methods	References
Support bath-enabled 3D printing	B: Laponite I: SU-8/alginate/gelatin	SU-8 cured by UV radiation Alginate crosslinked by $CaCl_2$ Gelatin gelled by temperature change	Mechanical cleaning Chemical agent (HCl)	[9]
	B: Gelatin microparticle I: Acidified collagen/fibronectin	Collagen crosslinked by pH change Fibronectin cured by thrombin	Dissolvable support bath ($37\,^{\circ}$C)	[8, 25]
	B: Carbopol I: PDMS	Cured by crosslinking agent	Mechanical cleaning	[17]
	B: Fumed silica I: PDMS			[10]
	B: Self-healing HA[e] I: Photopolymerized HA[e]	UV radiation	Dissolvable support bath (β-CD solution)	[22]
	B: Gelatin and OAlg[f] I: Gel-CDH[g]	Crosslinked between aldehyde in OAlg and amino in Gel-CDH	Dissolvable support bath (temperature change)	[31]

a) Pluronic F127.
b) Organ building blocks (OBBs).
c) 1-Ethyl-3-methylimidazolium ethyl sulfate.
d) Carbon particles in silicone oil.
e) HA modified with adamantane (Ad) or β-cyclodextrin (β-CD).
f) Oxidized alginate.
g) Carbohydrazide-modified gelatin.

7.4.2 Post-treatments in Support Bath–Enabled 3D Printing

Post-treatments in support bath–enabled 3D printing (summarized in Table 7.2) also consist of two steps: crosslinking the printed 3D structures and removing the residual support bath materials. How to completely remove the residual support bath materials from solidified 3D structures is the main challenge during post-treatments. Based on the dissolvability, liquid support bath materials can be categorized into dissolvable and non-dissolvable. For non-dissolvable bath materials, such as granular gel medium and Laponite EP suspension, the main method to remove residual yield-stress materials is mechanical cleaning, in which mechanical force is applied to wipe and/or rinse residual bath materials from surfaces [9, 12, 26]. This may destroy the integration of printed biomedical structures commonly made from hydrogels, such as collagen [8, 25], that have poor mechanical stiffness. In addition, in 3D organ printing, printed organs/tissues usually have numerous microscale channels and/or end-capped chambers, which makes it even more challenging to completely remove residual bath materials via mechanical cleaning. For dissolvable bath materials, either physical methods, such as temperature change [22, 25, 39], or chemical agents, such as acids and alkali [9], are used to completely dissolve yield-stress bath materials to low-viscous liquids, which can be easily removed. Gelatin microparticles are a representative dissolvable support bath material [8, 25]. After printed structures are solidified in the gelatin microparticle bath, the bath temperature increases to physiological temperature, at which gelatin microparticles melt into a gelation solution with low viscosity and are rinsed from the printed structures. Jin et al. [9] used hydrochloric acid to dissolve Laponite RD suspension. When 5.0% (v/v) hydrochloric acid was perfused into the Laponite RD bath, the "house-of-cards" arrangement in the suspension was destroyed. Thus, Laponite RD lost the yield-stress property and behaved as a low-viscosity liquid. However, the chemical agents may damage the structures' functionality, for instance by killing cells in engineered organs/tissues. As a result, how to mildly remove the residual support bath materials after printing is a major technical challenge that may constrain the wide application of the support bath–enabled 3D printing technology for various biomedical applications.

7.5 Representative Biomedical Applications

Due to a wide range of ink selections and high printing efficiency, liquid support bath–assisted 3D printing has been widely used in different fields, such as tissue engineering, lab-on-a-chip, soft robotics, and wearable electronic devices. In this section, some representative achievements of liquid support bath–assisted 3D printing are introduced.

7.5.1 Organ Printing

Organ printing is considered as the promising technical solution to overcome the organ shortage around the world. In organ printing, cells are mixed with hydrogel matrices as bioinks that are deposited into organ-like constructs. During incubation, the adjacent cellular layers undergo tissue fusion and finally complete the transformation into functional organs/tissues by self-assembly [40–42]. Liquid support bath-assisted 3D bioprinting has been widely used to print organoids in recent years. In particular, e-3DP technology benefits engineered organ fabrication by creating vascular network systems in the organoids, which can simulate the transportation channels of human blood vessels. Support bath-enabled 3D printing is able to print organoids with complex geometries and multiple living cells.

For example, Bhattacharjee et al. [7] successfully printed a hierarchically branched vascular network within the granular gel medium. This network with multi-scale hollow channels can be used to mimic blood vessel systems in the human body for transporting nutrients and oxygen to and expelling metabolic waste from organs and tissues. In addition, cell-laden vascular networked constructs were printed in nanoclay suspension as shown in Figure 7.11a [9]. After printing, the constructs were crosslinked in situ and released from the bath. After 3-day incubation, cell viability and metabolic activity were excellent, which validates the effectiveness of support bath-enabled 3D printing to create living tissue constructs. Furthermore, Ding et al.

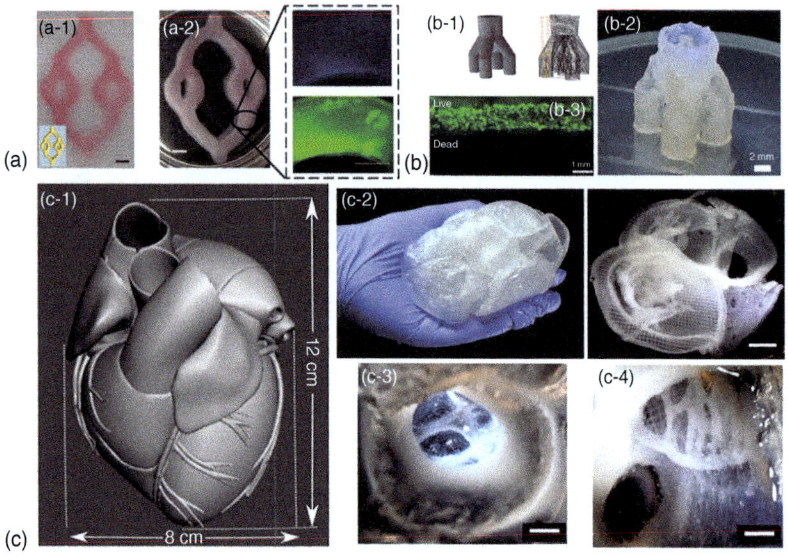

Figure 7.11 Representative applications in organ printing. (a) (a-1) Vascular network printed in Laponite EP support bath and (a-2) crosslinked gelatin/alginate construct with living 3T3 fibroblasts (Scale bar: 3 mm). Source: Jin et al. [9]. Reproduced with permission of American Chemical Society. (b) (b-1) The CAD model of the multi-tubular structure. (b-2) Printed tubular structure from alginate/gelatin bioink. (b-3) The viability of NHDFs after printing and post-treatment. Source: Ding et al. [43]. Reproduced with permission of Multidisciplinary Digital Publishing Institute. (c) (c-1) The CAD model of a full-scale human heart. (c-2) 3D printed heart model from alginate and its cross section to show the internal structure (Scale bar: 1.5 cm). (c-3) Aortic valves in the heart model (Scale bar: 1 cm) and (c-4) 3D printed trabeculae carnae (Scale bar: 1 cm). Source: Mirdamadi et al. [24]. Reproduced with permission of American Chemical Society.

[43] printed the blend of alginate and gelatin with normal human dermal fibroblast (NHDF) cells in the Laponite EP nanoclay to form the multi-tubular hydrogel structure as shown in Figure 7.11b. For organ printing applications, Mirdamadi et al. [24] printed a full-scale human heart model from alginate ink. The entire printing process was performed in the gelatin microparticle bath as shown in Figure 7.11c. The printed cardiac model presented the mechanical properties similar to those of a real adult heart, making it an ideal training tool for surgical purposes. Moreover, Skylar-Scott et al. [30] used e-3DP to create perfusable channels in a high cell density OBBs/ECM bath. With the help of these channels, nutrients and oxygens were able to be delivered to the cardiac tissue, enhancing the survival rate and differentiation ability of cells.

In the future, e-3DP and support bath–enabled 3D printing may be combined as a revolutionary 3D bioprinting technique in which support bath–enabled 3D printing is used to print macroscopic organoids, while e-3DP is simultaneously utilized to create vascular systems in the printed organoids. Therefore, this new 3D bioprinting technology may have the potential to fabricate full-scaled human organs with a sophisticated multi-scale blood vessel system to promote the development of clinical transplant studies.

7.5.2 Lab-on-a-Chip

Lab-on-a-chip is a technology that cultures cells on a microfluidic device to mimic partial or completed functions of tissues and organs. It is of great significance to facilitate the understanding of the mechanisms of some diseases in vitro and provide a rapid solution for drug screening [44]. Comparing with traditional tissue-culture methods, lab-on-a-chip offers a complicated microfluidic system and logic control unit, which make culturing cells more efficient and reduce the cost of experimental consumables and drugs.

Liquid support bath–assisted 3D printing, in particular e-3DP, has been used to fabricate microchips with different geometries. For instance, Kolesky et al. [11] applied e-3DP approach to print embedded vascularized heterogeneous tissue constructs with different cell types (Figure. 7.12a-1). In their study, GelMA was selected as a photocurable ECM, which also functioned as a support bath. Pluronic F-127 was served as the sacrificial ink to print vascularized channels (Figure 7.12a-2), while 15% (w/v) GelMA were mixed with 10T1/2 mouse fibroblasts and human neonatal dermal fibroblasts (HNDFs), respectively, to prepare two cellular inks, which were printed into vascularized heterogeneous tissues in the matrix bath. After printing, UV radiation was used to crosslink both the matrix bath and the cellular tissues. Finally, Pluronic F-127 was removed from the channels, and human umbilical vein endothelial cells (HUVEC) were perfused along with culture media through the channels to (i) form endothelial layers at the wall and (ii) transfer nutrients to and remove metabolites from co-cultured heterogeneous tissues as shown in Figure 7.12a-3. The cell-viability results revealed that this 3D bioprinting approach can print heterocellular microchips with excellent cytocompatibility (Figure 7.12a-4). In addition, Noor et al. [45] printed a cardiac patch with bionic vessel networks using e-3DP (Figure 7.12b-1). In their studies, they proposed a

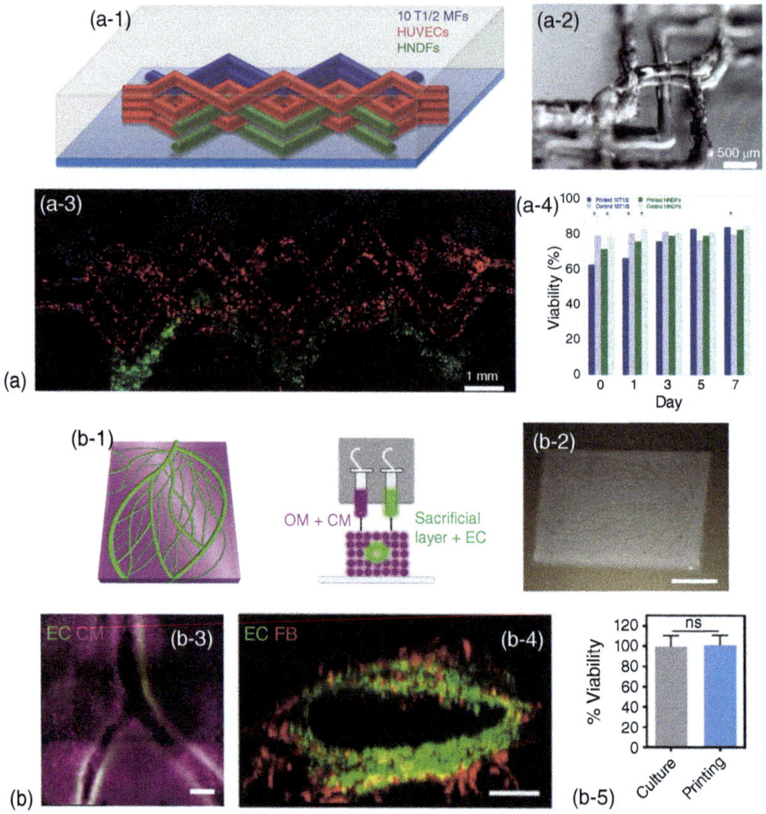

Figure 7.12 Representative lab-on-a-chip applications. (a) Heterogeneous cellular tissue printed in the GelMA matrix bath via e-3DP. (a-1) Schematic view of a heterogeneous engineered tissue construct. (a-2) Image showing the spanning and out-of-plane nature of the 3D printed construct. (a-3) Composite image of the 3D printed tissue construct acquired using three fluorescent channels. (a-4) Cell-viability assay results. Source: Kolesky et al. [11]. Reproduced with permission of John Wiley and Sons. (b) Cardiac patch with an embedded vessel system (Scale bars: 1 cm for (b-2), 500 μm for (b-3), and 100 μm for (b-4)). (b-1) A 3D model of the cardiac patch and the schematic of the printing concept. (b-2) A printed vascularized cardiac patch. (b-3) A printed iPSCs-derived cardiac patch. (b-4) Cross-sections of a single lumen. (b-5) Cell-viability after printing. Source: Noor et al. [45]. Reproduced with permission of John Wiley and Sons.

temperature-sensitive hydrogel as the matrix bath, which was directly extracted from human or pig omental tissues. This hydrogel presented a relatively loose gel state at room temperature but behaved as a rigid bulk when the temperature increased to 37 °C. Gelatin was selected as the sacrificial ink to print vascular networks in the gel matrix, which can be easily removed at 37 °C (Figure 7.12b-2). This cardiac patch was used as a tissue-on-a-chip device to co-culture endothelial cells (ECs) and iPSCs-derived cardiomyocytes (CMs) as shown in Figure 7.12b-3,b-4. It was found that the viability of the after-printed cells was almost the same as the cells cultured in the normal cell medium (Figure 7.12b-5).

7.5.3 Other Bio-Related Applications

Except for cell printing and lab-on-a-chip applications, liquid support bath–assisted 3D printing is also applied to make some biomimetic soft robots and/or biomedical wearable devices. Soft robotics made from highly compliant materials are often bionics of biological tissues or organs. Comparing with rigid robotics, soft robotics are more adaptable to the surrounding environment and provide a safe environment when working with humans [18, 46, 47]. At present, e-3DP has been proven to fabricate autonomous robots. For example, Truby et al. [6] used e-3DP to fabricate a multilayered soft somatosensitive actuator (SSA), which consisted of a curvature sensor, an inflation sensor, and a contact sensor. The schematic diagrams of the SSA and 3D printing process are illustrated in Figure 7.13a-1,a-2, respectively. To further demonstrate the functionality of the actuator, they integrated three SSAs to fabricate a soft robotic gripper. This gripper can successfully grip a ball without auxiliary, and the resistance change of each sensor was recorded as a function of time during the working period as shown in Figure 7.13a-3.

Wearable devices, such as sensors, are made from stretchable electronic materials that are able to attach to clothing or bodies. Visual signals (e.g. electrical signals) transformed by the strain of the sensor materials under stretching are used to monitor different types of movements of organisms, such as breathing, speech, and finger motion. The most popular sensor material is carbon conductive grease (carbon particles suspending in silicone oil) when using e-3DP to fabricate such wearable sensors. For example, Muth et al. [4] printed carbon conductive grease into a conductive wire array inside the elastomeric matrix (Figure 7.13b-1), which served as strain sensors. They also printed multilayered pressure sensors using e-3DP, which can distinguish non-stretched (Figure 7.13b-2) and stretched (Figure 7.13b-3) conditions. In addition, Wei et al. [48] printed a wearable glove via e-3DP. In their study, the formula of carbon-conductive grease ink was improved by adding a given ratio of Ecoflex0030 (a soft platinum cure silicone) and printed into embedded circuits. In addition, an empty bag over the embedded sensors was created by removing the sacrificial ink (Pluronic F127) that left space for fingertips (as shown in Figure 7.13c).

Although the ink materials in the aforementioned studies are acellular functional inks, cell-laden bioinks have the potential to be used in these bio-related devices in the future. For example, some animal or human cells can produce therapeutic effects through injection or transplantation [49, 50], such as T cells of immunotherapy for cancer. Thus, soft drug carriers will be printed with embedded acellular functional inks and given cells, which can release therapy cells or/and drugs at the lesion to achieve efficient target treatments.

7.6 Conclusions and Future Directions

Liquid support bath–assisted 3D printing technology, a new material extrusion 3D printing strategy, has presented the unimaginable potential to fabricate either bulk structures with embedded complex 2D and/or 3D patterns or freeform 3D structures

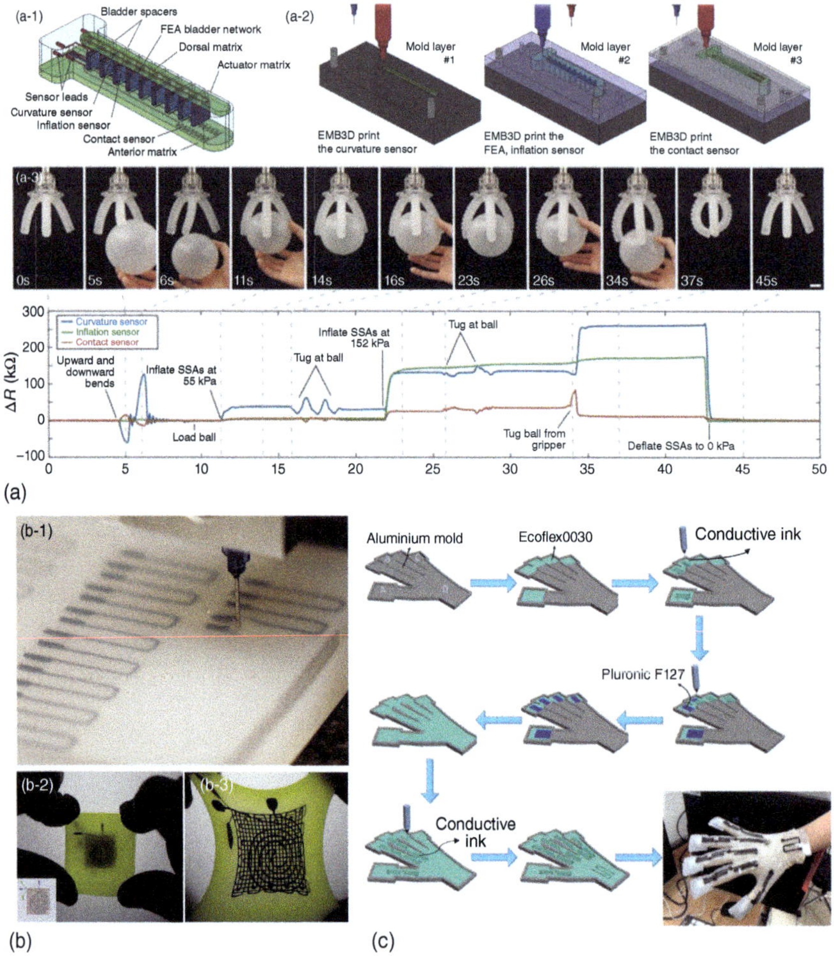

Figure 7.13 Representative bio-related applications via liquid support bath-assisted 3D printing. (a) Embedded 3D printing of a soft somatosensitive actuator (SSA) (Scale bar: 20 mm). (a-1) Schematic of SSA. (a-2) SSA printing process. (a-3) Images of an interaction sequence between a ball and a soft robotic gripper comprised of SSAs and ΔR of each sensor as a function of time during the interaction sequence. Source: Truby et al. [6]. Reproduced with permission of John Wiley and Sons. (b) Embedded 3D printing of a conductive wire array as strain sensors (b-1) and multi-layered pressure sensors under (b-2) non-stretched and (b-3) stretched conditions. Source: Muth et al. [4]. Reproduced with permission of John Wiley and Sons. (c) Fabrication process of wearable glove using e-3DP. Source: Wei et al. [48]. Reproduced with permission of John Wiley and Sons.

with complicated geometries. Due to the wide range of ink material selection, high printing efficacy, and in situ supporting functionality during printing, it is one of the most promising 3D printing techniques to biofabricate full-scaled human organs and tissues in the future. For liquid support bath-assisted 3D printing, more investigations will be performed on developing new versatile and dissolvable support bath materials, unveiling filament formation mechanisms in the liquid matrix and/or support bath and improving fabrication efficiency by optimizing the printing trajectory.

References

1 Ngo, T.D., Kashani, A., Imbalzano, G. et al. (2018). Additive manufacturing (3D printing): a review of materials, methods, applications and challenges. *Composites Part B: Engineering* 143: 172–196.

2 McCormack, A., Highley, C.B., Leslie, N.R. et al. (2020). 3D printing in suspension baths: keeping the promises of bioprinting afloat. *Trends in Biotechnology* 38 (6): 584–593.

3 Grosskopf, A.K., Truby, R.L., Kim, H. et al. (2018). Viscoplastic matrix materials for embedded 3D printing. *ACS Applied Materials & Interfaces* 10 (27): 23353–23361.

4 Muth, J.T., Vogt, D.M., Truby, R.L. et al. (2014). Embedded 3D printing of strain sensors within highly stretchable elastomers. *Advanced Materials* 26 (36): 6307–6312.

5 Wu, W., DeConinck, A., and Lewis, J.A. (2011). Omnidirectional printing of 3D microvascular networks. *Advanced Materials* 23 (24): H178–H183.

6 Truby, R.L., Wehner, M., Grosskopf, A.K. et al. (2018). Soft somatosensitive actuators via embedded 3D printing. *Advanced Materials* 30 (15): 1706383.

7 Bhattacharjee, T., Zehnder, S.M., Rowe, K.G. et al. (2015). Writing in the granular gel medium. *Science Advances* 1 (8): e1500655.

8 Hinton, T.J., Quentin, J., Palchesko, R.N. et al. (2015). Three-dimensional printing of complex biological structures by freeform reversible embedding of suspended hydrogels. *Science Advances* 1 (9): e1500758.

9 Jin, Y., Compaan, A., Chai, W. et al. (2017). Functional nanoclay suspension for printing-then-solidification of liquid materials. *ACS Applied Materials & Interfaces* 9 (23): 20057–20066.

10 Jin, Y., Song, K., Gellermann, N. et al. (2019). Printing of hydrophobic materials in fumed silica nanoparticle suspension. *ACS Applied Materials & Interfaces* 11 (32): 29207–29217.

11 Kolesky, D.B., Truby, R.L., Gladman, A.S. et al. (2014). 3D bioprinting of vascularized, heterogeneous cell-laden tissue constructs. *Advanced Materials* 26 (19): 3124–3130.

12 Jin, Y., Compaan, A., Bhattacharjee, T. et al. (2016). Granular gel support-enabled extrusion of three-dimensional alginate and cellular structures. *Biofabrication* 8 (2): 025016.

13 Jin, Y., Liu, C., Chai, W. et al. (2017). Self-supporting nanoclay as internal scaffold material for direct printing of soft hydrogel composite structures in air. *ACS Applied Materials & Interfaces* 9 (20): 17456–17465.

14 Parandoush, P. and Lin, D. (2017). A review on additive manufacturing of polymer-fiber composites. *Composite Structures* 182: 36–53.

15 Li, J., Monaghan, T., Nguyen, T.T. et al. (2017). Multifunctional metal matrix composites with embedded printed electrical materials fabricated by ultrasonic additive manufacturing. *Composites Part B: Engineering* 113: 342–354.

16 O'Bryan, C.S., Bhattacharjee, T., Hart, S. et al. (2017). Self-assembled micro-organogels for 3D printing silicone structures. *Science Advances* 3 (5): e1602800.

17 Hinton, T.J., Hudson, A., Pusch, K. et al. (2016). 3D printing PDMS elastomer in a hydrophilic support bath via freeform reversible embedding. *ACS Biomaterials Science & Engineering* 2 (10): 1781–1786.

18 Wehner, M., Truby, R.L., Fitzgerald, D.J. et al. (2016). An integrated design and fabrication strategy for entirely soft, autonomous robots. *Nature* 536 (7617): 451–455.

19 Liu, J.-L., Feng, X., and Wang, G. (2007). Buoyant force and sinking conditions of a hydrophobic thin rod floating on water. *Physical Review E* 76 (6): 066103.

20 Campos, D.F.D., Blaeser, A., Weber, M. et al. (2012). Three-dimensional printing of stem cell-laden hydrogels submerged in a hydrophobic high-density fluid. *Biofabrication* 5 (1): 015003.

21 Wang, Z., An, G., Zhu, Y. et al. (2019). 3D-printable self-healing and mechanically reinforced hydrogels with host–guest non-covalent interactions integrated into covalently linked networks. *Materials Horizons* 6 (4): 733–742.

22 Highley, C.B., Rodell, C.B., and Burdick, J.A. (2015). Direct 3D printing of shear-thinning hydrogels into self-healing hydrogels. *Advanced Materials* 27 (34): 5075–5079.

23 Zhang, Y., Ellison, S.T., Duraivel, S. et al. (2021). 3D printed collagen structures at low concentrations supported by jammed microgels. *Bioprinting* 21: e00121.

24 Mirdamadi, E., Tashman, J.W., Shiwarski, D.J. et al. (2020). FRESH 3D bioprinting a full-size model of the human heart. *ACS Biomaterials Science & Engineering* 6 (11): 6453–6459.

25 Lee, A., Hudson, A.R., Shiwarski, D.J. et al. (2019). 3D bioprinting of collagen to rebuild components of the human heart. *Science* 365 (6452): 482–487.

26 Jin, Y., Chai, W., and Huang, Y. (2017). Printability study of hydrogel solution extrusion in nanoclay yield-stress bath during printing-then-gelation biofabrication. *Materials Science and Engineering: C* 80: 313–325.

27 O'Bryan, C.S., Kabb, C.P., Sumerlin, B.S. et al. (2019). Jammed polyelectrolyte microgels for 3D cell culture applications: rheological behavior with added salts. *ACS Applied Bio Materials* 2 (4): 1509–1517.

28 Highley, C.B., Song, K.H., Daly, A.C. et al. (2019). Jammed microgel inks for 3D printing applications. *Advanced Science* 6 (1): 1801076.

29 Nasef, M.M. and Hegazy, E.-S.A. (2004). Preparation and applications of ion exchange membranes by radiation-induced graft copolymerization of polar monomers onto non-polar films. *Progress in Polymer Science* 29 (6): 499–561.

30 Skylar-Scott, M.A., Uzel, S.G.M., Nam, L.L. et al. (2019). Biomanufacturing of organ-specific tissues with high cellular density and embedded vascular channels. *Science Advances* 5.9: eaaw2459.

31 Heo, D.N., Alioglu, M.A., Wu, Y. et al. (2020). 3D bioprinting of carbohydrazide-modified gelatin into microparticle-suspended oxidized alginate for the fabrication of complex-shaped tissue constructs. *ACS Applied Materials & Interfaces* 12 (18): 20295–20306.

32 Lee, J.M. and Yeong, W.Y. (2015). A preliminary model of time-pressure dispensing system for bioprinting based on printing and material parameters: this paper reports a method to predict and control the width of hydrogel filament for bioprinting applications. *Virtual and Physical Prototyping* 10 (1): 3–8.

33 O'Bryan, C.S., Bhattacharjee, T., Niemi, S.R. et al. (2017). Three-dimensional printing with sacrificial materials for soft matter manufacturing. *MRS Bulletin* 42 (8): 571–577.

34 Pairam, E., Le, H., Fernández-Nieves, A. et al. (2014). Stability of toroidal droplets inside yield stress materials. *Physical Review E* 90 (2): 021002.

35 LeBlanc, K.J., Niemi, S.R., Bennett, A.I. et al. (2016). Stability of high speed 3D printing in liquid-like solids. *ACS Biomaterials Science & Engineering* 2 (10): 1796–1799.

36 Hockaday, L.A., Kang, K.H., Colangelo, N.W. et al. (2012). Rapid 3D printing of anatomically accurate and mechanically heterogeneous aortic valve hydrogel scaffolds. *Biofabrication* 4 (3): 035005.

37 Khalil, S. and Sun, W. (2007). Biopolymer deposition for freeform fabrication of hydrogel tissue constructs. *Materials Science and Engineering: C* 27 (3): 469–478.

38 Gong, J., Schuurmans, C.C.L., van Genderen, A.M. et al. (2020). Complexation-induced resolution enhancement of 3D-printed hydrogel constructs. *Nature Communications* 11 (1): 1–14.

39 Tcharkhtchi, A., Faivre, S., Roy, L.E. et al. (1996). Mechanical properties of thermosets. *Journal of Materials Science* 31 (10): 2687–2692.

40 Mironov, V., Boland, T., Trusk, T. et al. (2003). Organ printing: computer-aided jet-based 3D tissue engineering. *Trends in Biotechnology* 21 (4): 157–161.

41 Chia, H.N. and Wu, B.M. (2015). Recent advances in 3D printing of biomaterials. *Journal of Biological Engineering* 9 (1): 1–14.

42 Mironov, V., Visconti, R.P., Kasyanov, V. et al. (2009). Organ printing: tissue spheroids as building blocks. *Biomaterials* 30 (12): 2164–2174.

43 Ding, H. and Chang, R.C. (2018). Printability study of bioprinted tubular structures using liquid hydrogel precursors in a support bath. *Applied Sciences* 8.3: 403.

44 Huh, D., Hamilton, G.A., and Ingber, D.E. (2011). From 3D cell culture to organs-on-chips. *Trends in Cell Biology* 21 (12): 745–754.

45 Noor, N., Shapira, A., Edri, R. et al. (2019). 3D printing of personalized thick and perfusable cardiac patches and hearts. *Advanced Science* 6.11: 1900344.

46 Trivedi, D., Rahn, C.D., Kier, W.M. et al. (2008). Soft robotics: biological inspiration, state of the art, and future research. *Applied Bionics and Biomechanics* 5 (3): 99–117.

47 Rus, D. and Tolley, M.T. (2015). Design, fabrication and control of soft robots. *Nature* 521 (7553): 467–475.

48 Wei, P., Yang, X., Cao, Z. et al. (2019). Flexible and stretchable electronic skin with high durability and shock resistance via embedded 3D printing technology for human activity monitoring and personal healthcare. *Advanced Materials Technologies* 4.9: 1900315.

49 Kim, S.U. and De Vellis, J. (2009). Stem cell-based cell therapy in neurological diseases: a review. *Journal of Neuroscience Research* 87 (10): 2183–2200.

50 June, C.H., O'Connor, R.S., Kawalekar, O.U. et al. (2018). CAR T cell immunotherapy for human cancer. *Science* 359 (6382): 1361–1365.

8

Bioprinting Approaches of Hydrogel Microgel

8.1 Introduction

Microgel has become a new researching and application unit in the field of 3D bioprinting. Different from hydrogel bulk, microgel can provide a suitable growing microenvironment for the encapsulated cells and realize enough material exchanging because of its inherent high water content and small size. Thus, it's significant to explore more novel methods to establish microgel.

The basic principle to generate a microgel/microdroplet is to divide integrity of hydrogel prepolymer solution or a bulk hydrogel into smaller parts to microscale size. In terms of hydrogel in solid state before forming microgel, a common method is used to crush the bulk hydrogel into smaller fragments. In terms of prepolymer solution in liquid state, the fundamental principle is to break up the mechanical equilibrium of the liquid system. From the energy standpoint, at the initial state, the competitive relationship between surface energy and gravitational potential energy makes the liquid system maintain a stable static state. To generate a smaller droplet, external energy is required to enter into the liquid system to produce disturbance and it to break into smaller size. From the force standpoint, for example, the droplet is affected by gravity, atmosphere pressure, liquid pressure, and surface tension. And, to break its equilibrium, external force is required to make the droplet size much smaller.

According to these principles, a variety of forces are applied to generate microgels. In this chapter, five categories of approaches are summarized, namely auxiliary dripping, diphase emulsion, space constraint, photolithography, and bulk crushing.

8.2 Auxiliary Dripping

The conventional 2D printing process gave inspiration to 3D printing. At the earlier developing stage of 3D printing, traditional 2D printers were adapted or modified to form 3D structures. Microdroplet is the printing unit in these processes. Thus, introducing the droplet dripping method into the fabrication of microgels is considered. As we know, liquid can directly drip down from the output of a tube or nozzle with the effect of gravity. However, the size of the droplet formed in this

Cell Assembly with 3D Bioprinting, First Edition. Yong He, Qing Gao, and Yifei Jin.

way is always much bigger than microdroplets/microgels. Thus, auxiliary force or energy is needed to break up the basic mechanical equilibrium of the liquid system to generate microscale droplets. According to the source of force/energy, auxiliary dripping can be divided into three kinds: inkjet printing, laser-assisted printing, and electrohydrodynamic (EHD) printing.

8.2.1 Inkjet Printing

8.2.1.1 Piezoelectric Inkjet

Piezoelectric printer/nozzle is commonly used in the process of microdroplet generation. Following the code of piezoelectricity, as soon as an electric pulse signal is conducted to the piezoelectric nozzle, the electric energy transforms into mechanical energy and produces a pulse pressure to the liquid ink inside the nozzle, so that a microdroplet is produced.

In 2015, inspired by piezoelectric printing and ionic gelation method, Gao et al. (Figure 8.1) described the fabrication process of calcium alginate microgels owning varied shapes with a piezoelectric printhead (inner diameter: 80 μm) controlled by a periodic voltage waveform consisting of bipolar trapezoid pluses [1]. The sodium alginate prepolymer solution was held in a reservoir and maintain a suitable liquid level for the microdroplet generation with a negative pressure control system and a lifting platform. Thc positivc pulse stage could form a pressure wave on the printing head following piezoelectricity to generate sodium alginate microdroplets. The negative pulse stage was applied to weaken the residual pressure wave in the ink remained in the printhead to recover the fluidic equilibrium. An imaging system was added to visualize the microdroplet formation process. The dripping sodium alginate microdroplets were received and cross-linked by calcium chloride solution. Because of the liquid collision and gradual cross-linking in the receiving process, calcium alginate microparticles with different shapes, namely irregular, tail-shaped, spherical, elliptical, and satellitic outline were formed by modifying the dripping velocity of the microdroplet and concentration of the calcium chloride solution (Figure 8.2). By increasing these two parameters, the received calcium chloride microgels tend to transform from nonspherical microparticles to microspheroids. Weber number ($We = qm^2d/r$), which is equal to the ratio between the inertia force (facilitates the microdroplet deforming) and the restoring force (prevents the microdroplet deforming), was used to describe the collision process of a microdroplet during receiving (q: microdroplet density, m: impact velocity, d: microdroplet diameter, r: the surface tension of the microdroplet). With the increase of velocity and microdroplet diameter, the Weber number increased and the microparticles were deformed and compressed to the elliptical shape. For lower velocity, because of the slower transit of the microdroplet across the calcium chloride solution surface, calcium alginate microgels became tail-shaped. When the dripping velocity was further decreased, the microdroplet could not rush into the cross-linking solution and floated on the solution surface forming an irregular microgel. In terms of the concentration of cross-linking solution, its effect on the microgel shape was similar.

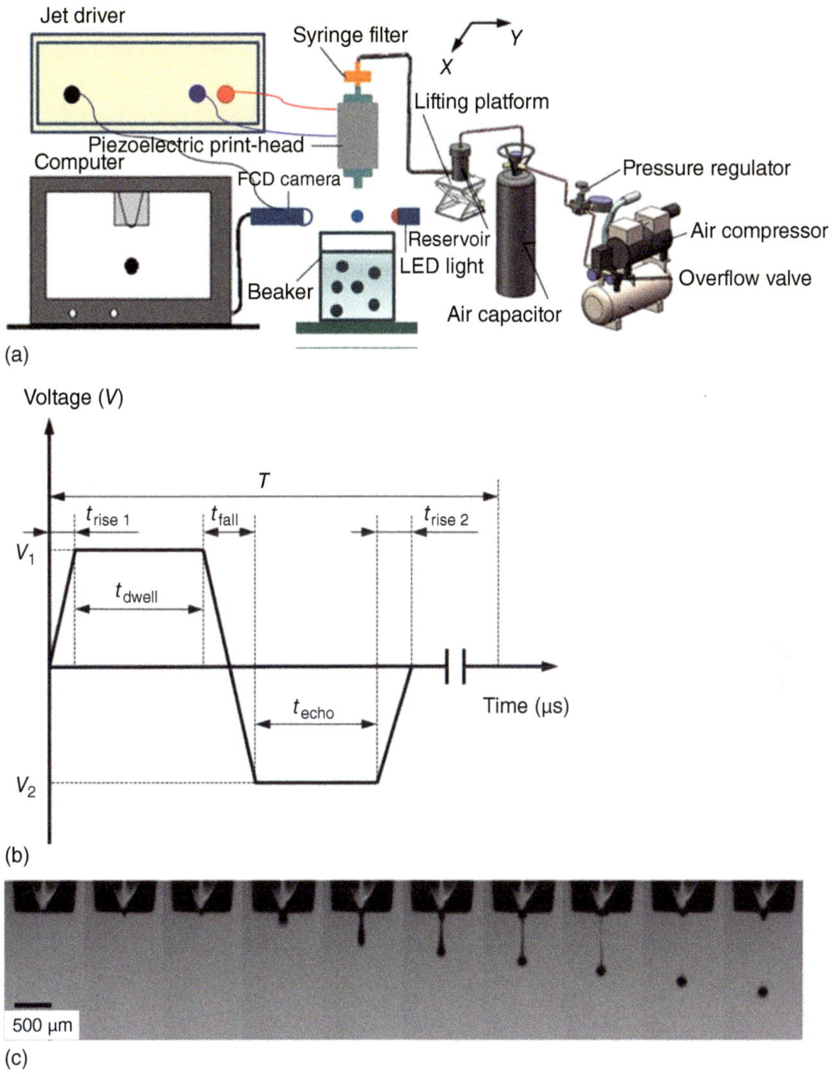

Figure 8.1 (a) Schematic diagram showing the experimental setup. (b) Bipolar voltage pulse used in the experiment. (c) A typical microdroplet formation sequence captured at different times. The times are 0, 40, 80, 120, 160, 200, 240, 280, 320, and 360 μs. Source: Gao et al. [1].

Higher concentration could increase the cross-linking speed and tend to make microgel formed as a spheroid.

Furthermore, Xu et al. also published an article to study the cell-laden sodium alginate microdroplet formation process in piezoelectric inkjet printing [2] (Figure 8.3). Because of the addition of live cells in the bioink, the rheological properties of the printing solution were modified. Thus, the dripping process of the microdroplets could change. In their work, the authors carried out a series

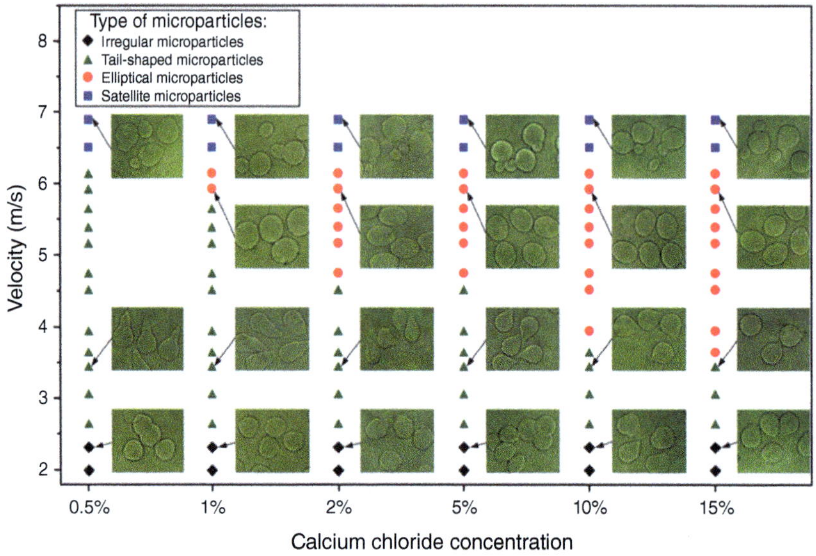

Figure 8.2 Formed alginate microparticles with a combination of the microdroplet velocity and the $CaCl_2$ concentration. Source: Gao et al. [1].

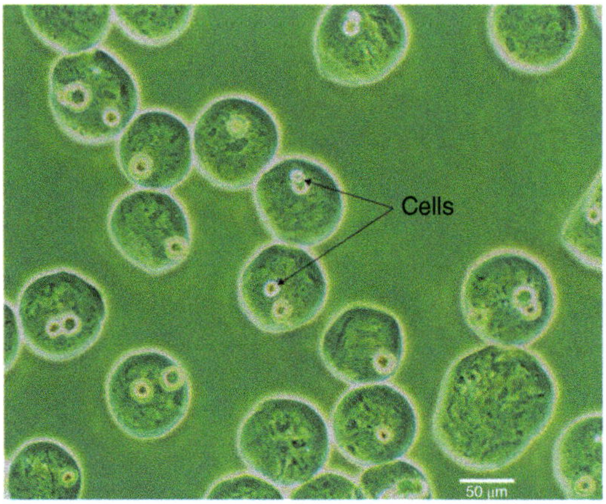

Figure 8.3 Alginate microspheres embedded with living cells. Source: Xu et al. [2].

of experiments to explore the effect of cell density (without cells, 1×10^6, 5×10^6, 1×10^7 cells/ml) on the microdroplet dripping state. Moreover, polystyrene beads, which are much harder than cells, were also applied to replace cells to further comprehend the influence of the physical properties of the encapsulated particles on the microdroplet formation. In conclusion, the research results demonstrated that with the increase in cell density in the bioink, the dripping microdroplet size and velocity would decrease. Moreover, the phenomenon of satellite droplet was limited and the breakup duration increased. In terms of the mixed particles, compared to the polystyrene bead suspension, the bioink with cells displayed a less ejected liquid volume, lower dripping velocity, and longer breakup duration because it owned a higher loss modulus G'' according to the rheological examination.

8.2.1.2 Thermal Bubble Inkjet

The thermal inkjet printing method is another commonly used drop-on-demand approach. Clara López-Iglesias et al. published a contribution in 2018 to generate salbutamol sulfate-loaded alginate aerogel microgels with thermal bubble inkjet printhead followed by supercritical drying [3] (Figure 8.4). Sodium alginate solutions were poured into the cartridges (HP51626A) of a thermal inkjet printer (Hewlett Packard Deskjet 340; Palo Alto, CA, USA). The printer was further modified so that the height between the printhead and receiving part could be tuned. The dripping microdroplets were received by calcium chloride solution bath with magnetically stirring. The appropriate concentrations of sodium alginate and calcium chloride solution were roughly determined by the microgel generation experiment by the prilling method. Different from the prilling method, the viscosity of the alginate solution is more significant for the microdroplet formation process in the thermal inkjet method. When the bioink has higher viscosity, the fluid jet tends to be continuous rather than independent droplets. Thus, the alginate solution concentration needed to be controlled inside a certain range. In the *XY* plane, the printhead could move according to the designed routine program with the original motion module to avoid the droplet integrating. The diameters of the undried alginate microgels were tested below 100 μm. In the further drying step to form aerogel microgels, ethanol was used to replace the solvent inside the cross-linked microgels. The method of sequential exchange of the microgel solvent with aqueous solutions of increasing ethanol concentrations was thought to be better than supercritical drying because it could preserve the initial structure of the wet microgels. After that, free-flowing powder was obtained which owned excellent

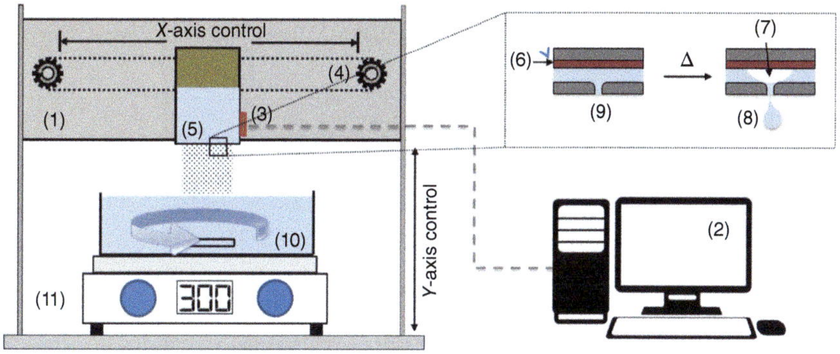

Figure 8.4 The setup used for the preparation of the alginate gel microspheres. The inkjet printer (1) is connected to a computer (2) that allows the control of the printing operation by the heating element (3) of the cartridge and controls the x-axis movement of the printer belt (4). The printer is located in metallic support with moving parts allowing its *y*-axis control (i.e. the printhead-to-gelation bath distance). The alginate solution is located in the printhead (5) containing several printing chambers. An electrical signal is applied to a microthermal heater (6) located at each printing chamber, and the sudden increase in the temperature generates a vapor bubble (7) that ejects the alginate solution as a picodroplet (8) through the nozzle (9). The droplets are jetted to a gelation bath (10) constantly. Source: López-Iglesias et al. [3].

and homogenous textural properties falling in the porous range from the scanning electron microscopy (SEM) morphology images.

8.2.2 Laser-Assisted Printing

In Zhang's previous work (2016), the impingement types and printing quality in laser-assisted printing of sodium alginate were analyzed in detail by building a microarray on the substrate, which was the representative ink for soft structural bioprinting [4] (Figure 8.5). The laser-assisted printer was implemented as matrix-assisted pulsed-laser evaporation direct-write (MAPLE DW) with alginate solution, which was composed of an optical beam delivery system, an argon fluoride (ArF) excimer laser, a ribbon consisting of an optically transparent quartz disk coated with a thin film of ink on its bottom side, and a poly-L-lysine coated glass slide as the receiving substrate. Motion units were assembled in this printing system to tune the depositing position and the printhead height. In the analysis of microdroplet diameters, the diameter data of the microdroplets with regular shapes were directly measured by ImageJ software. In terms of the ones with irregular shapes, the equivalent diameters were calculated with their volumes found by assuming a constant contact angle. The experimental results demonstrated that three printing parameters, namely the concentration of the sodium alginate, the intensity of the laser fluence, and the printhead height, had an obvious influence on the formation of the microdroplets. Meanwhile, during the dripping process, three printing states were also defined as droplet-impingement printing, jet-impingement printing with multiple breakups, and jet-impingement printing with a single breakup. When alginate with lower concentration was applied, microdroplets tended to be independent on the substrate. With the increase of the concentration, the bioink viscosity got higher and the formation of secondary droplets was suppressed and the

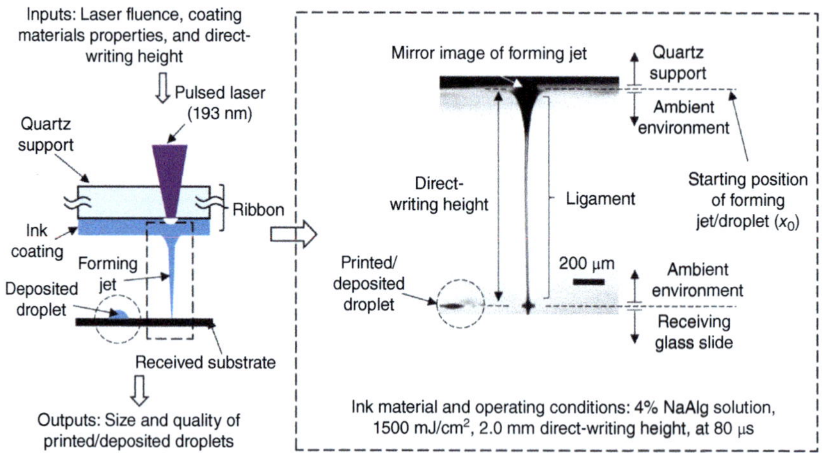

Figure 8.5 Laser printing schematic with an inset illustrating a printing process taken at 80 μs during the printing of 4% alginate solution at a 1500 mJ/cm laser fluence. Source: Zhang et al. [4].

diameter of the microdroplet got lower. Furthermore, higher laser fluence would optimize the droplet formation of alginate with a lower concentration, so that the diameter could be increased. In terms of printhead height, a higher height would change the printing state from jet-impingement printing with a single breakup to jet-impingement printing with multiple breakups or droplet-impingement printing because the jet could not touch the substrate before the jet crack. However, in a certain height range, the microdroplet diameter would not change obviously. These results established an excellent basement and could be a significant guide for further fabrication of microgels research by the laser-assisted method.

8.2.3 Electrohydrodynamic Printing

Xie et al. combined the EHD principle with the bioprinting of cell-laden GelMA microgels with low cost, cell damage, and high production [5] (Figure 8.6). Compared to the sodium alginate solution, GelMA solution owning better biocompatibility is always with low viscosity. A high voltage electric field was formed between a metal nozzle and a metal ring. GelMA bioink mixed with bone marrow stem cells (BMSCs) was fed by a gas pressure controlling system. The microdroplets were generated with the attraction of electrostatic force and cross-linked in silicone oil by 405 nm wavelength light.

To guide the further high throughput fabrication, the formation states and the microsphere diameter distribution were analyzed in detail (Figure 8.7). With the increase of the applied voltage, four printing states occurred, namely microdripping, Taylor jet, oscillating jet, and shaking spindle. Among them, microdripping and Taylor jet were the optimal states for GelMA microgels bioprinting. In the microdripping state, stable independent microdroplets were generated, because the diameters of the microgels tended to be larger (480~2200 μm). In the Taylor jet state, the surface tension of the single microdroplet cannot tolerate the increasing voltage and break up into smaller microdroplets (90~230 μm), which is known as the Coulomb fission phenomenon.

Furthermore, the growing state of the encapsulated cells was tested. The results showed that the encapsulated BMSCs kept high viability and maintained excellent spreading ability inside the 3D microenvironment (Figure 8.8). Due to the application of lower external forces to separate the droplets, cell damage during printing could be negligible, which often occurs in piezoelectric or thermal inkjet bioprinting. Moreover, the molecular controlled releasing was also tested (Figure 8.9). Because of the dense GelMA pore network, the encapsulated FITC-dextran was released gradually rather than explosively. This method showed great potential applications in cell therapy, drug delivery, etc.

Based on the research about pure GelMA microgels bioprinting with EHD, Xie et al. further published another contribution to bioprinting by using more complex hierarchical microgels, namely microfiber-laden minispheroids [6] (Figure 8.10). This approach settled the challenge of printing microfibers and microspheres simultaneously to effectively mimick the complicated situations in vivo. Apart from the EHD principle, another fluidic phenomenon, the "rope coiling effect" was applied

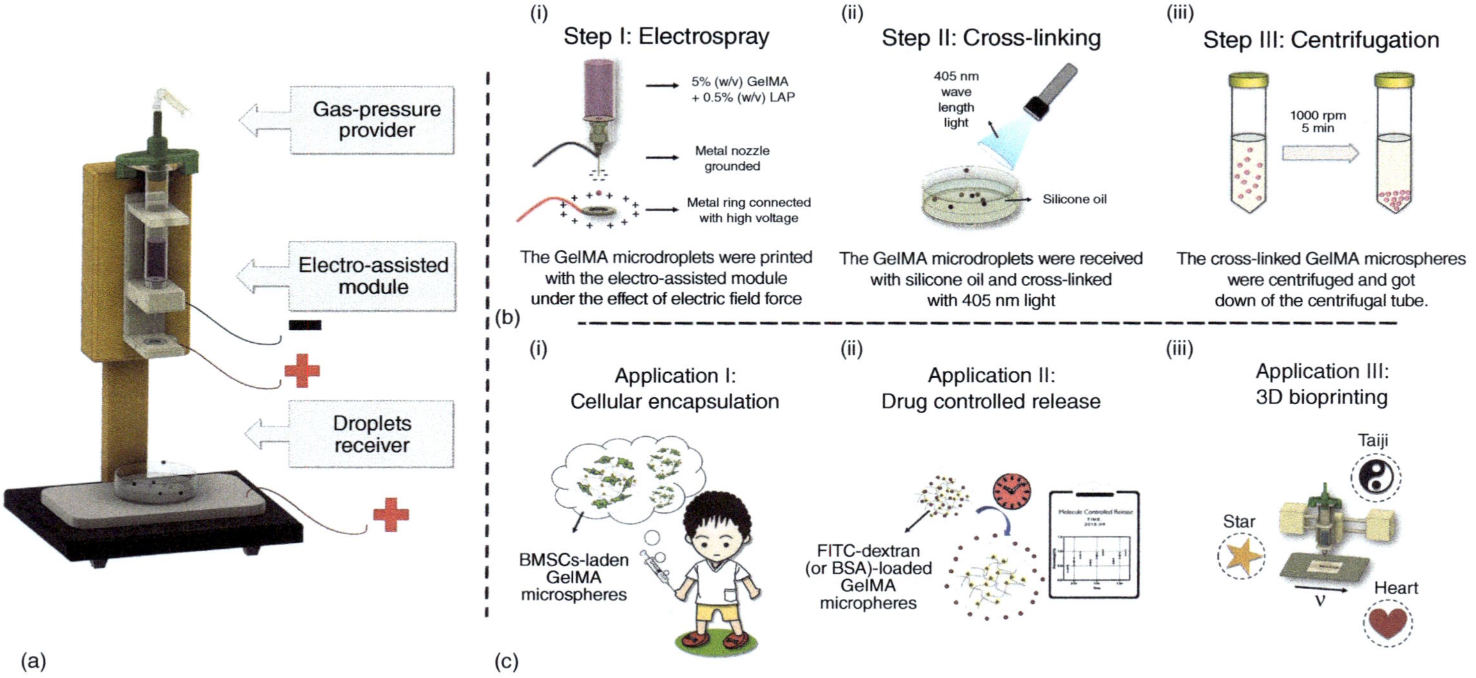

Figure 8.6 Sketch of the electro-assisted printing device, fabricating process, and application examination: (a) Diagram of the electro-assisted printing device for the fabrication of GelMA microdroplets. (b) The fabricating process of GelMA microspheres. (i) Step I: electrospray. (ii) Step II: cross-linking. (iii) Step III: centrifugation. (c) Biomedical applications carried out in this paper. (i) Application I: cellular encapsulation. (ii) Application II: drug controlled release. (iii) Application III: 3D bioprinting. Source: Xie et al. [5].

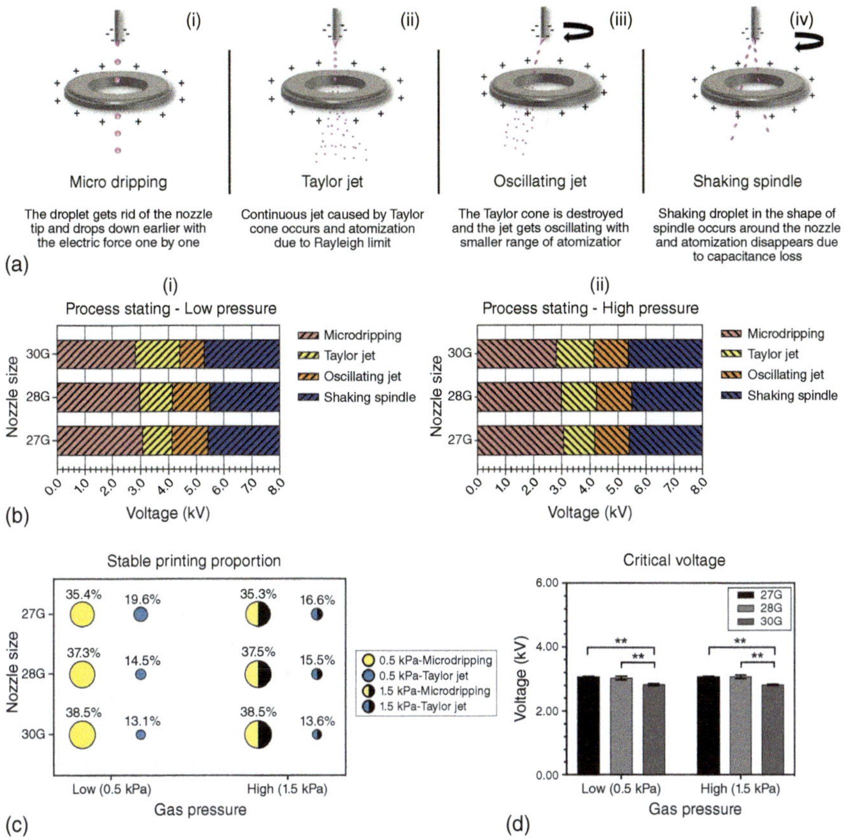

Figure 8.7 The analysis of different printing states. (a) The diagrammatic sketches of the four different printing states: (i) Microdripping. (ii) Taylor jet. (iii) Oscillating jet. (iv) Shaking spindle. (b) The voltage ranges of different printing states under different nozzle sizes and gas pressures. (i) Low pressure (0.5 kPa). (ii) High pressure (1.5 kPa). (c) Bubble chart of the controllable proportions of two stable fabrication methods: microdripping and Taylor jet in the range of 0–8 kV. (The data were indicated by the diameters of the bubbles.) (d) The critical voltage under different nozzle sizes and gas pressures. The nozzle size had an obvious effect on the critical voltage while the effect of gas pressure on it could be negligible. Source: Xie et al. [5].

in this strategy. Two fluids flowed from the inner nozzle and outer nozzle of the coaxial bioprinting nozzle, respectively. Pure GelMA solution, which was designed to become the spheroid part, flowed out from the outer nozzle. The GelMA solution containing sodium alginate (increasing the viscosity), which was designed to become the microfiber part, flowed out from the inner nozzle. With the attraction of the electric field force, the total size of the hierarchical microgels could be strictly designed. The diameter of the fiber could be controlled by modifying the flow rate ratio of the inner/outer fluid. In this work, the printing process was observed and analyzed in detail. The size distribution of the fibers (minimal diameter: 180 μm) and the spheroids (minimal diameter: 1800 μm) were measured and summarized by ImageJ software. Similarly, the dripping droplets were also received

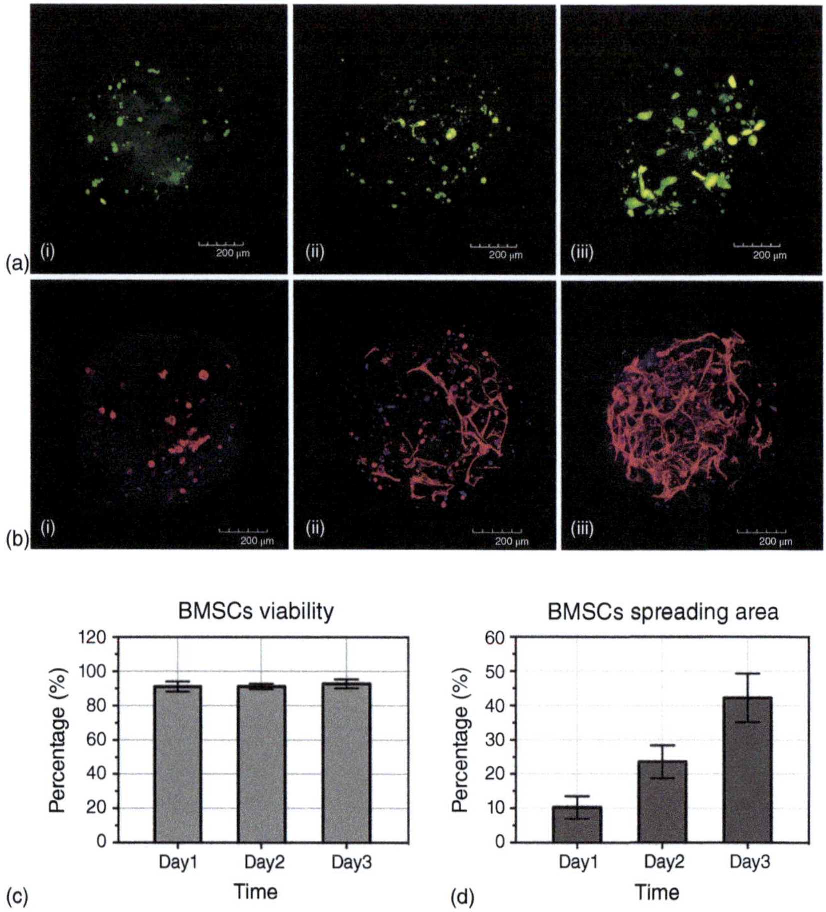

Figure 8.8 Cellular encapsulation. (a) Confocal images of LIVE/DEAD staining of the BMSCs GelMA microspheres. (i) one day, (ii) two days, and (iii) three days. (b) Confocal images of F-actin/nucleus markers from BMSCs encapsulation GelMA microspheres. (i) one day, (ii) two days, and (iii) three days. (c) Viability of BMSCs encapsulated in GelMA microspheres after one to three days of culturing (above 90%). (d) The spreading area proportion of BMSCs encapsulated in GelMA microspheres after one to three days of culturing. It was approved that BMSCs maintained the capability of spreading after undergoing the electro-assisted printing process. Source: Xie et al. [5].

by silicone oil and cross-linked by 405 nm wavelength light. The SEM morphology of the microfiber-laden minispheroids showed that the additional sodium alginate increased the pore size of the GelMA network, which could potentially provide more growing space for the encapsulated cells. To take tumor tissue with the vascular network as an example, the authors encapsulated breast tumor cells (MDA-MB-231s) and human umbilical vein endothelial cells (HUVECs) in the spheroids and fibers, respectively, to determine the biocompatibility of this coculturing system and the interaction of the two kinds of cells. The confocal fluorescent images showed that the encapsulated cells kept high viability after 13-day culturing and spreading

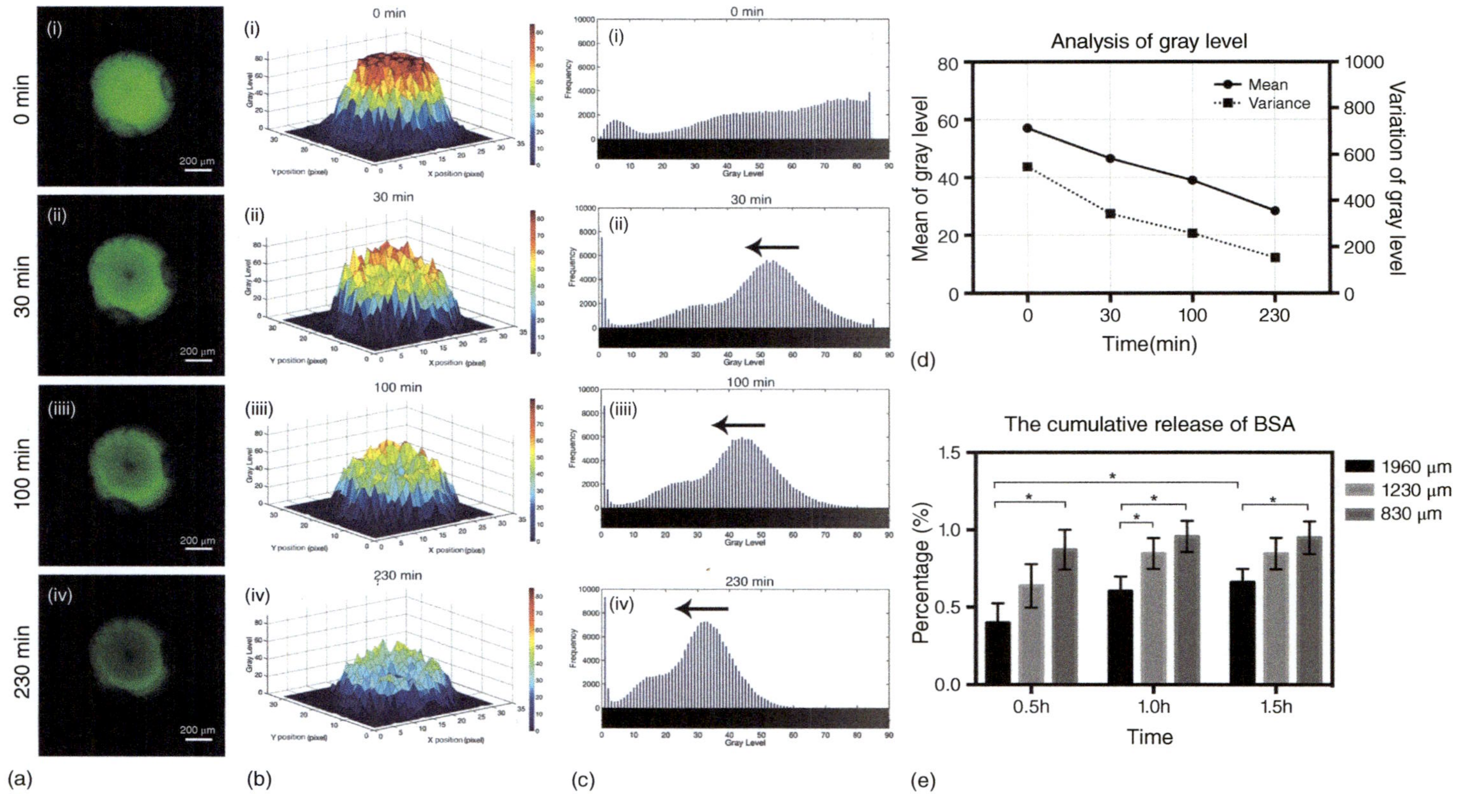

Figure 8.9 Drug-controlled release. (a) The confocal fluorescent microscopy images of the FITC-dextran release. The fluorescence intensity got lower with time. (i) 0 minute, (ii) 30 minutes, (iii) 100 minutes, and (iv) 230 minutes. (b) The 3D heat maps of the gray levels of the grayscale images transformed from the confocal fluorescent microscopy images. (i) 0 minute, (ii) 30 minutes, (iii) 100 minutes, and (iv) 230 minutes. (c) The distribution histograms of the grayscale images transformed from the confocal fluorescent microscopy images. At first, the gray levels mainly concentrated around higher values in 0 minute after immersing. Then, the peak of gray levels got moved in the negative direction to lower ones. (i) 0 minute, (ii) 30 minutes, (iii) 100 minutes, and (iv) 230 minutes. (d) The gray levels mean and variance in 0, 30, 100, and 230 minutes. Both the means and the variances of gray levels got decreased, which indicated that gray levels start concentrating around lower ones with time and the FITC-dextran release in-outside from the GelMA microspheres happened remarkably. (e) The cumulative release of BSA of 1960, 1230, and 830 μm GelMA microspheres in 0.5, 1.0, and 1.5 hours immersing in DPBS. The release rate got higher when the microspheres in smaller sizes were used. Source: Xie et al. [5].

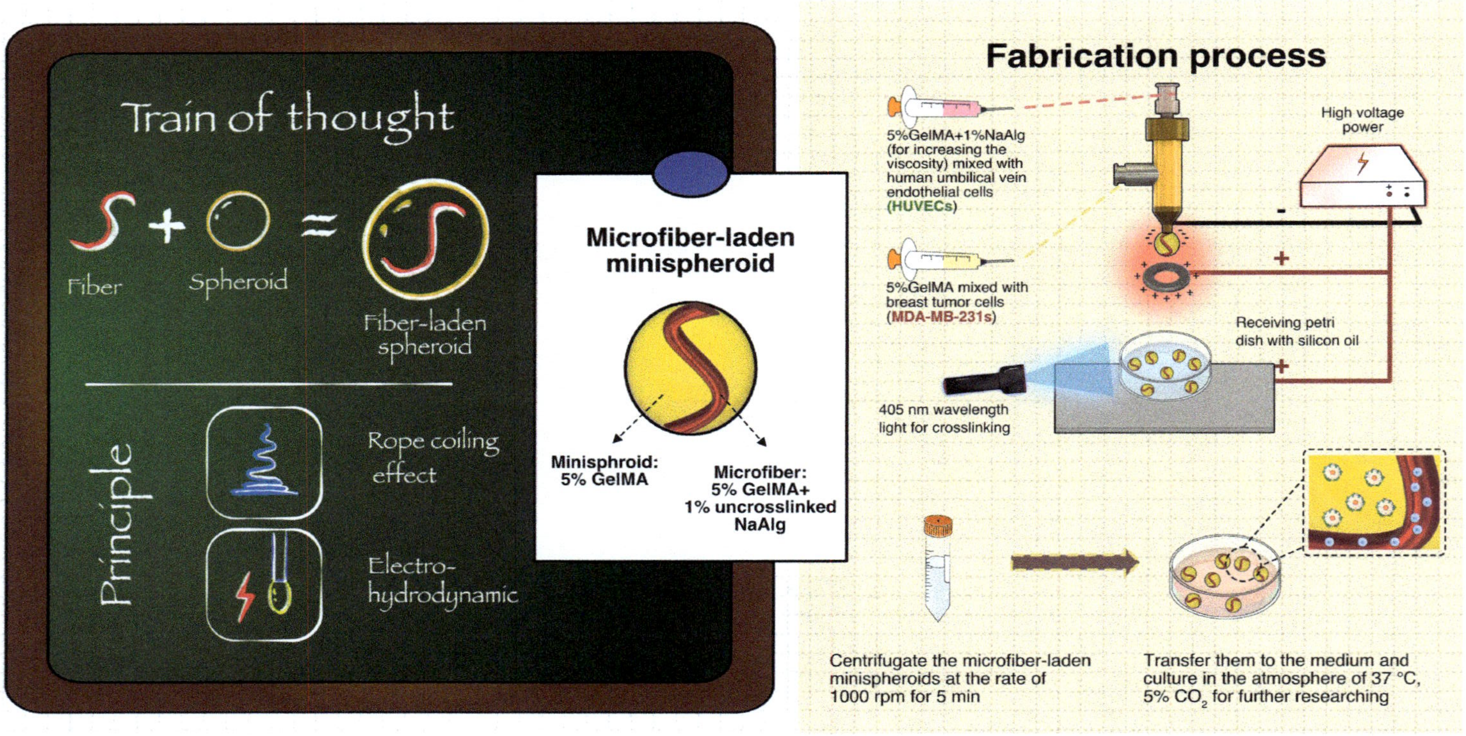

(a) (b)

Figure 8.10 The sketch of the train of thought and fabrication process. (a) Train of thought of this work. (i) The basic principle applied in this paper. The fiber and spheroid structures in vitro have been widely used in biomedical applications. It strikes us that we could fabricate the two structures simultaneously and form a brand-new "microfiber-laden minispheroid" structure with the "rope coiling effect" and "electro-hydrodynamic" principles. (ii) The sketch of the "microfiber-laden minispheroid." (b) The sketch of the fabrication process. (i) Basic fabrication system. (ii) Centrifugal process. (iii) Coculturing. Source: Xie et al. [6].

capacity. Importantly, the F-actin of HUVECs tended to elongate toward the cocultured tumor cells, in contrast to the HUVECs cultured alone. As a combination of two common printing structural units, namely microspheres and microfibers, the hierarchical microgels could rebuild the actual hierarchical tissues composed of spheroid and fiber such as tumors, embryos, and glomeruli in vitro and be a potential system for 3D cell coculturing research in the future.

Besides, another heterogeneous microgel, namely versatile spiral microarchitectures inside the microspheroids, was introduced in the study of Zhao et al. [7] (Figure 8.11). A special Y-type PDMS microchip was prepared to converge two fluidic flows (alginate and alginate with cells) in the laminar state. Toward the outlet of the metal nozzle inserted in the microchip, another nozzle supply airflow was added to rotate the suspended droplets so that the two fluidic flows get twisted with each other and form spiral shapes. To decrease the total size of the spheroids, EHD was applied as an auxiliary force to attract the droplets. The dripping microdroplets were received by a calcium chloride bath placed on the 3D bioprinter, which could change the receiving spot automatically according to the routine program.

By further modifying the flow rate ratio of the two fluidic flows, more fascinating hydrogel structures such as the spherical helix, rose, saddle, Tai chi, and single-cell line can be easily printed (Figure 8.12).

With these morphologies, the authors encapsulated L929 cells in the microgels with different microarchitecture morphologies to verify the 3D cell organization

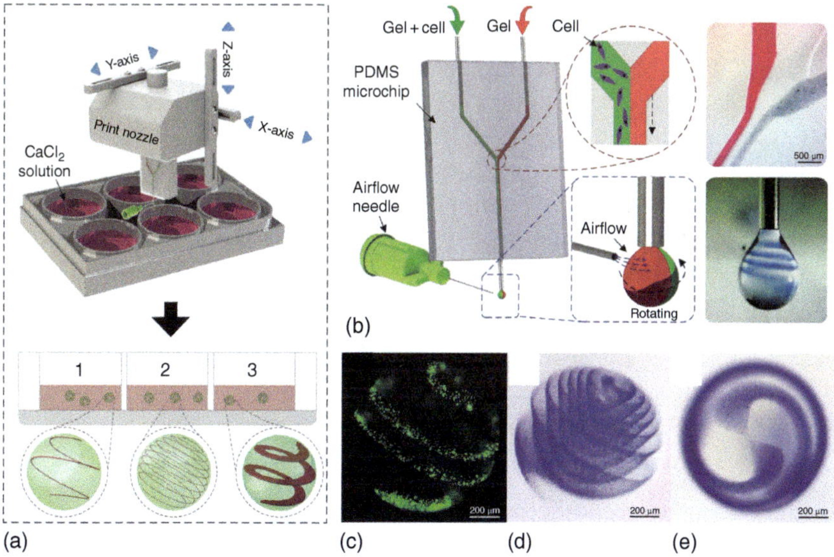

Figure 8.11 3D spiral-based droplet formation from airflow-driven rotation. (a) Schematic diagram of the bioprinting process, the six-well plate is introduced to receive different batches of microspheroids. (b) Schematic representation depicting the process of fabrication of 3D spiral-based cell-laden spheroids. Sodium alginate solutions were proportionally extruded out from the polydimethylsiloxane (PDMS) microchip and rotated by the adjustable airflow. (c) Spherical microspheroids. (d) Rose-like microspheroids. (e) Tai chi-like microspheroids (top view). Source: Zhao et al. [7].

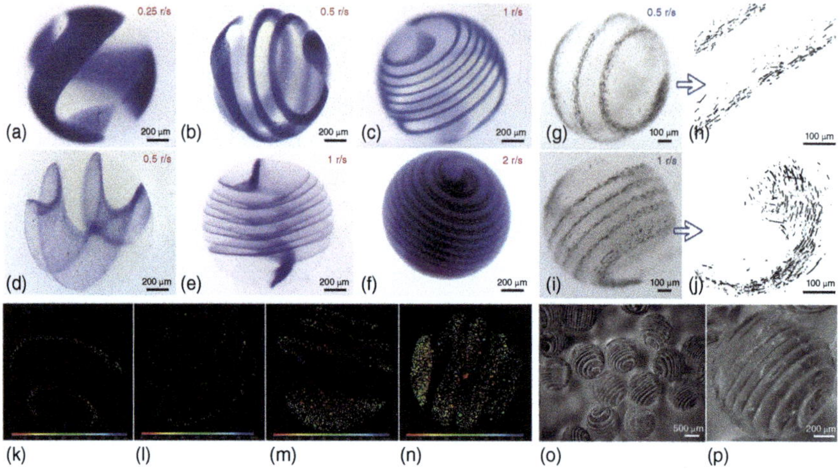

Figure 8.12 (a–c) Spheroids with spiral-compartmental geometries with spiral component (marked with blue dye) distributed on the spheres, the rotation speeds were 0.25; 0.5; 1r/s; (d–f) with spiral component distributed inside the spheres, at the rotation speeds of 0.5; 1; 2r s − 1; (g–j) the spiral component is made of carbon fiber, indicated the flow direction of the spiral structure at the middle and top part of the sphere; (k–n) confocal fluorescence microscope images of the spiral-compartmental spheres with depth information, the composition ratio ($Q_{\text{spiral}}/Q_{\text{nonsprial}}$) were 1 : 20; 1 : 6; 1 : 4; and 1 : 2; (o, p) with spiral texture outside the spheres using glycerol as sacrificial material. Source: Zhao et al. [7].

and proliferation. And HUVECs and MSCs were cocultured in the microgels to establish a complex microenvironment with osteogenic medium added with vascular endothelial growth factor (VEGF) for 10 days. The results showed high CD31 expression inside the outer part and obvious osteogenic functionalization by alkaline phosphatase (ALP) and Alizarin Red S (ARS) staining (Figure 8.13).

The theoretical model during printing was also built in detail and the printing parameters were systematically analyzed (Figures 8.14 and 8.15). In terms of the position of the airflow needle, the relevant position could be defined as (x_1, y_1, z). To simplify the evaluation parameters, the authors defined an angle (θ) to represent this relationship. The calculation was as follows:

$$\theta_g = \sin^{-1}\frac{|y_1 - y_0|}{\sqrt{(x_1 - x_0)^2 + (y_1 - y_0)^2}}$$

$$(x_0, y_0) = (0, 0)$$

According to the practical experience and basic geometric relationship, x_1 was restricted as below:

$$x_1 = \pm\frac{D_{\text{inlet}} + d_{\text{drop}}}{2}$$

where D_{inlet} was the nozzle tip diameter, d_{drop} was the average diameter of the droplet.

It was because disturbance would occur when $|x_1| < \frac{d_{\text{drop}}}{2}$ and airflow force would be shortened. Therefore, in actual bioprinting, the authors only tuned y_1.

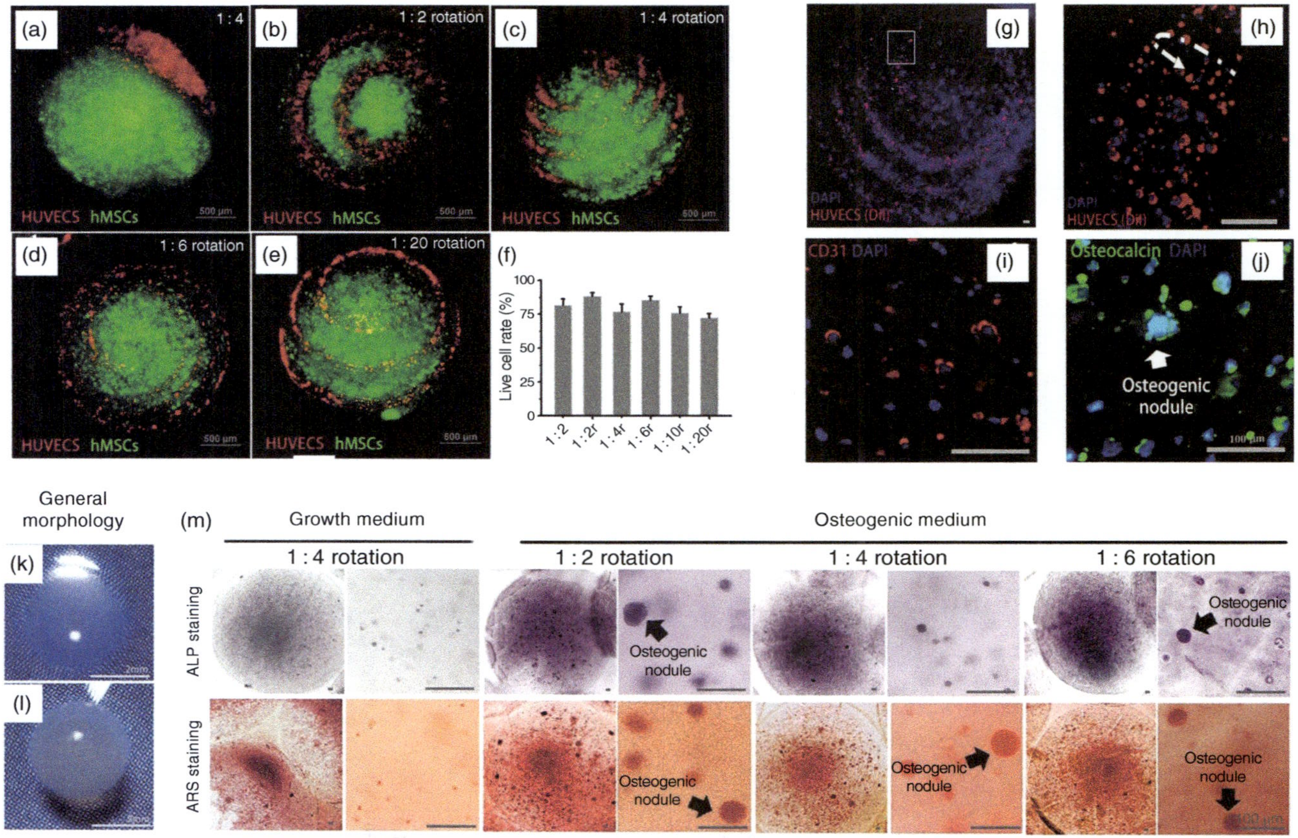

Figure 8.13 Organoids construction through the coculture of HUVECS and HMSCs in spiral-based microspheroids. (a) Fluorescence microscope images of traditional compartmentalized microspheroid, the composition ratio (Q_{HUVECs}/Q_{HMSCs}) was 1 : 4. (b–e) HUVECs cells were spiral distributed on the surface of the spheroids, with composition ratio (Q_{HUVECs}/Q_{HMSCs}) = 1 : 2; 1 : 4; 1 : 6; and 1 : 20. (f) Cell viability of all cells in microspheroids after 10 days of in vitro culture, quantified by Trypan blue staining. (g, h) Vascular geometrical feature of HUVECs cells (marked with dye DiI) inside the spheroid after 10 days of culture (added with VEGF); HUVECs cells were finally arranged as a circular cavity along the spiral path in panel (h). (i, j) Immunostaining of two types of cells (CD31 for HUVECs, and osteocalcin for HMSCs) after 10 days of culture (osteogenic medium added with VEGF). (k, l) Morphological difference between spheroids cultured in growth medium on day 10 (k) and osteogenic medium (l). (m) Obvious osteogenic nodules positively stained with ALP and ARS were found in microspheroids under the microscope on day 10. Source: Zhao et al. [7].

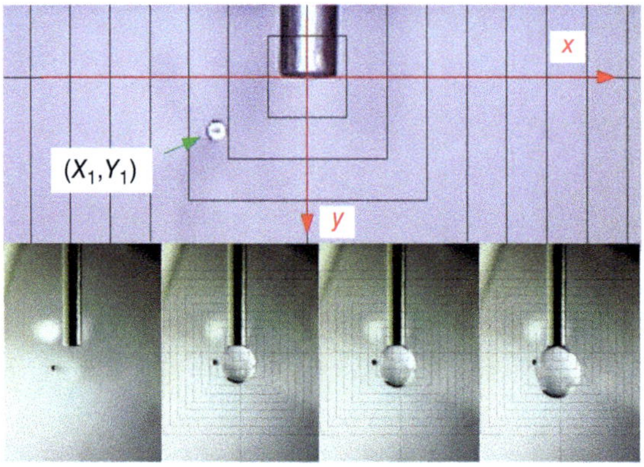

Figure 8.14 Principle of visual positioning of the airflow needle. Source: Zhao et al. [7].

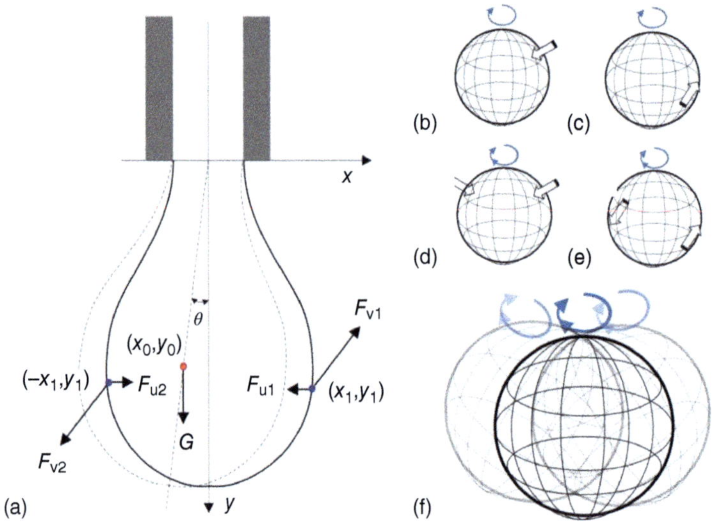

Figure 8.15 Mathematic analysis of the spinning process of microspheroids. (a) Mechanical model of the droplet. (b–e) Four types of airflow. (f) When using two airflow needles, oscillation of droplet occurs. Source: Zhao et al. [7].

Another parameter z was defined as below:

$$z = 1.2x_1 = \pm 0.6(D_{\text{inlet}} + d_{\text{drop}}) > \frac{D_{\text{inlet}}}{2}$$

Because the positions and airflows of two nozzles could not be absolutely symmetrical, the force exerted by the droplet could not be uniform, which caused droplet oscillation. The calculation could be expressed as below:

When $F_{u1} + F_{u2} \neq 0$, the moment of inertia (I) of the droplet was:

$$I\frac{\mathrm{d}\omega_d}{\mathrm{d}t} = |y_1(F_{u1} + F_{u2})|$$

where ω_d was the rotational angular velocity of the droplet and assumed that the two nozzles were strictly adjusted to the symmetrical position. This would produce an initial velocity that led the droplet to reach the dotted position. With the gravity effect and airflow force, the moment of inertia I of the droplet could be expressed as the differential equation as below:

$$-G\sqrt{x_0^2 + y_0^2}\sin\theta + |F_{u1} + F_{u2}| = I\frac{\mathrm{d}^2\theta}{\mathrm{d}t}$$

which was a second-order system, and would cause oscillation. On the contract, the mechanical model of the droplet was expressed as below when a single nozzle was used:

$$G\sin\theta = F_u \cdot y_1$$

$$\theta = \sin^{-1}\frac{F_u \cdot y_1}{G}$$

where F_u was the normal component of airflow force. The angle θ could be very small as long as F_u was small enough. It was easy to control by modifying the position (x_1, y_1) and the nozzle size.

In terms of incompressible fluids, the N–S equation could be expressed as below:

$$\begin{cases} f_x - \frac{1}{\rho}\frac{\partial p}{\partial x} + \nu\nabla^2 v_x = \frac{dv_x}{dt} = \frac{\partial v_x}{\partial t} + v_x\frac{\partial v_x}{\partial x} + v_y\frac{\partial v_x}{\partial y} + v_z\frac{\partial v_x}{\partial z} \\ f_y - \frac{1}{\rho}\frac{\partial p}{\partial y} + \nu\nabla^2 v_y = \frac{dv_y}{dt} = \frac{\partial v_y}{\partial t} + v_x\frac{\partial v_y}{\partial x} + v_y\frac{\partial v_y}{\partial y} + v_z\frac{\partial v_y}{\partial z} \\ f_x - \frac{1}{\rho}\frac{\partial p}{\partial z} + \nu\nabla^2 v_z = \frac{dv_z}{dt} = \frac{\partial v_z}{\partial t} + v_x\frac{\partial v_z}{\partial x} + v_y\frac{\partial v_z}{\partial y} + v_z\frac{\partial v_z}{\partial z} \end{cases}$$

According to the labeled size of the microfluidic Y-type chip, the Bernoulli equation at the inlet orifices could be expressed as below:

$$p_1 + \rho gh = \frac{1}{2}\rho_1 w_3^2 + \rho_1\frac{\partial w_{11}}{\partial t}L_{11} + \rho_1\frac{\partial w_2}{\partial t}L_2 + \rho_1\frac{\partial w_3}{\partial t}L_3 + \Delta p_{11} + \Delta p_2 + \Delta p_3 + \Delta p_{\text{inlet}}$$

$$p_2 + \rho gh = \frac{1}{2}\rho_2 w_3^2 + \rho_2\frac{\partial w_{12}}{\partial t}L_{12} + \rho_2\frac{\partial w_2}{\partial t}L_2 + \rho_2\frac{\partial w_3}{\partial t}L_3 + \Delta p_{12} + \Delta p_2 + \Delta p_3 + \Delta p_{\text{inlet}}$$

where p_1 was the pressure at the orifice on the lift channel and p_2 was the pressure on the right side. h was the vertical height of the microchip, Δp_{11} was the pressure loss at L_{11}, Δp_2 was the pressure loss at L_2, etc.

8.3 Diphase Emulsion

8.3.1 Nonaqueous Liquid Stirring

Caldwell et al. used nonaqueous stirring method to generate clickable microgels containing excess dibenzo cyclooctyne (DBCO) or azide groups for the establishment of microporous scaffolds [8]. PEO-DBCO (or PEG-N_3), azide functionalized fluorophore (AlexaFluor-594-N_3), and azide functionalized adhesive peptide

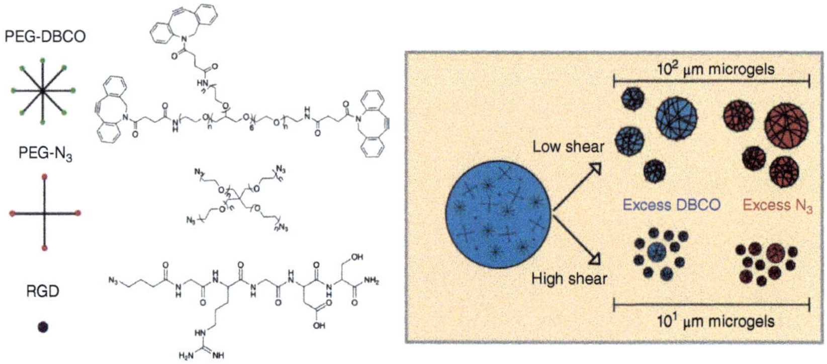

Figure 8.16 Inverse suspension polymerization of microgel building blocks. Scheme for microgel formation. 8 arm 20 kDa PEG-DBCO, 4 arm 10 kDa PEG−N3, and azide labeled RGD were combined in PBS and dispersed in hexane. Solutions were exposed to either continuous vortexing or sonication, while polymerizations were performed off-stoichiometry to yield microgels with excess DBCO or N3 surface functionalities. Source: Caldwell et al. [8].

(N_3-GRGDS) were mixed to become the aqueous prepolymer solution (Figure 8.16). After that, the prepared solution was rapidly poured into the organic solvent phase containing hexane, Span-80, and Tween-20. High and low vortexing speeds were applied, respectively, to the mixture of two phases to control microgel size. 10^2 μm microgels could be generated with low vortexing speed, with average diameters of 120 and 130 μm for microgels owing excess DBCO and azide groups, respectively. 10^1 μm microgels could be generated with high vortexing speed, with average diameters of 16 and 15 μm for microgels owing excess DBCO and azide groups, respectively. The microgels were washed with hexanes (3×), isopropanol (3×), and transferred to PBS. Moreover, the mechanical properties of the PEG-based microgels were estimated by testing the moduli of the bulk PEG hydrogels at identical cross-linking densities. The results showed that the compressive modulus of the microgels with excess DBCO and azide was 15 and 8.7 kPa.

Franco et al. also proposed and optimized a novel emulsion-based strategy with dual-photoinitiator for rapid generation of cell-laden PEG (poly(ethylene glycol))-based microgels [9]. Briefly, 20 μl hydrogel prepolymer solution was poured into a glass test tube (13 mm by 75 mm) loaded 1 ml of mineral oil and was vortexed at the maximum rate for three seconds (Figure 8.17). After that, the metal halide lamp was used to cross-link the hydrogel microdroplets. To remove the mineral oil, 500 μl of phosphate-buffered saline (PBS, pH 7.4) was added into the mixture and centrifuged at the speed of 300 g for two minutes. Removing the oil and the washing steps were repeated. It is worth mentioning that this work came up with a dual-photoinitiator method, in which two photoinitiators were applied simultaneously in the microgel fabrication process, namely 2,2-dimethoxy-2-phenyl acetophenone (I651) and eosin Y. In the previous works, the two photoinitiators were commonly used in the PEG cross-linking process. However, they have their natural advantages and disadvantages. Considering the suitable concentration

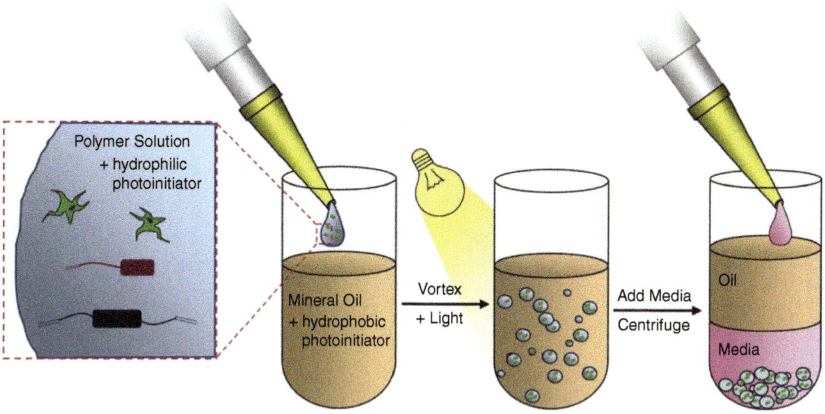

Figure 8.17 Diagram of vortex-induced emulsion encapsulation method. A polymer precursor solution containing cells (green), adhesive ligands (red), and degradable polymer (blue) is combined with mineral oil in a glass test tube. The solution is vortexed briefly during exposure to a cross-linking light source. A dual-initiator system with a hydrophilic photoinitiator in the aqueous phase and a hydrophobic initiator in the oil phase was found to be ideal for consistent particle formation. Cross-linked microspheres were isolated from the oil by the addition of buffer or media solution followed by centrifugation and aspiration of the oil layer. (For interpretation of the references to colors in this figure legend, the reader is referred to the web version of this paper.) Source: Franco et al. [9].

range of the photoinitiator for cell growth, in terms of I651, it is a kind of hydrophobic photoinitiator and its solubility in water is extremely low while high in oil so that the yield of cross-linked microgels could be extremely low. In terms of eosin Y, it is a kind of hydrophilic photoinitiator that could be dissolved in the water easily. It could make the emulsified microdroplets cross-linked better. However, the generated microgels were not of standard spheroid shape and could not be totally cross-linked, especially at the surface. Thus, the dual-photoinitiator method was an excellent optimization for the PEG microgels generation. For one thing, eosin Y was added to the hydrogel prepolymer solution to improve the total cross-linking density. For another, I165 was added into the mineral oil to rapidly cross-link the surface of the dispersed hydrogel microdroplets to make the microgels display a more standard spheroid outline. For further cell culturing with the PEG microgels, other modified types such as PEG-RGDS for cell attachment and PEG-PQ-PEG for responding to the enzyme treatment were also introduced to this strategy. The cell experiment results of both surface seeding and encapsulating culturing showed that the PEG-based microgels fabricated by this emulsion-based method owned high feasibility and great potential to be used in more biomedical applications.

8.3.2 Air-Assisted Atomization

The air-assisted atomization principle is known as a rapid way to spray a hydrogel prepolymer solution to fabricate microgels. To solve the problem of the nonhomogeneous cell distribution and growing for a long duration on lots of biomedical

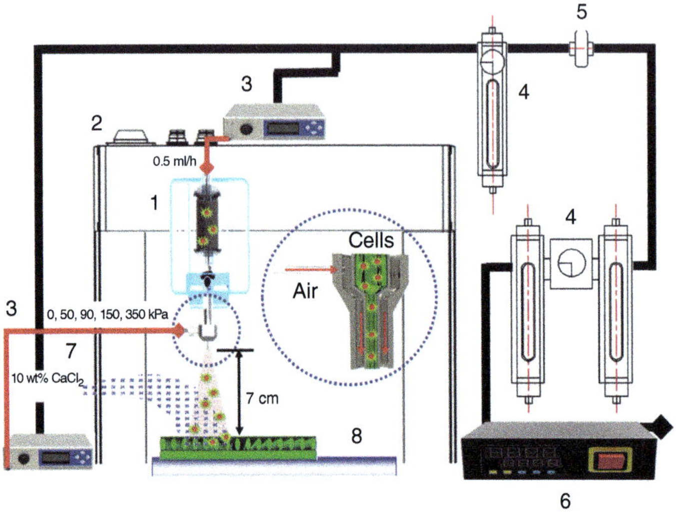

Figure 8.18 Scheme of the cell-spraying device: (1) a cell-laden alginate spraying system using a core–shell nozzle, (2) three-axis robot system, (3) a controller system for dispensing air, (4–6) filtering systems for air, (7) an aerosol system for CaCl2 solution, and (8) a working stage. Source: Ahn and Kim [10].

scaffolds, Ahn and Kim introduced the air-assisted atomization process to combine with the preprinted structures [10]. To achieve the aim, an aerodynamic force was applied to break the combined influence of viscosity and surface tension of the bioink (Figure 8.18). Cells were first mixed into sodium alginate solution and loaded in a syringe pump. The bioink would flow out from the inner nozzle of the coaxial nozzle. Compressed air would flow out from the outer nozzle at a high flow rate. When the aqueous phase was confronted with the gaseous phase, the high-speed gas flow would break up the aqueous bioink into microdroplets. To tentatively cross-link them, calcium chloride solution aerosol was generated by a humidifier (Tess-7400; Paju, South Korea). As a demonstration, the authors displayed the feasibility of establishing a hybrid cell-laden scaffold by spraying cell-laden alginate microgels on the surface of a preprinted PCL mesh scaffold. Briefly, the three-axial motion system was connected to the melt-depositing system to print a PCL layer-by-layer scaffold. After that, an air-assisted atomization approach was used to spray the alginate microgels onto the scaffold surface. The pneumatic pressure was set as 260 ± 5 kPa and the nozzle speed was set as 10 mm/s. After cross-linking, the hybrid scaffolds were cultured in vitro. The cell experiment results demonstrated that the cells could attach and proliferate on the PCL surface well. This strategy would serve as an efficient tool to form various cell-laden hybrid scaffolds for more applications in the future.

8.3.3 Microfluidic Technology

Choi et al. introduced a one-step strategy to fabricate cell-laden hydrogel microgels utilizing double emulsion droplets with a sacrificial ultra-thin oil shell [11].

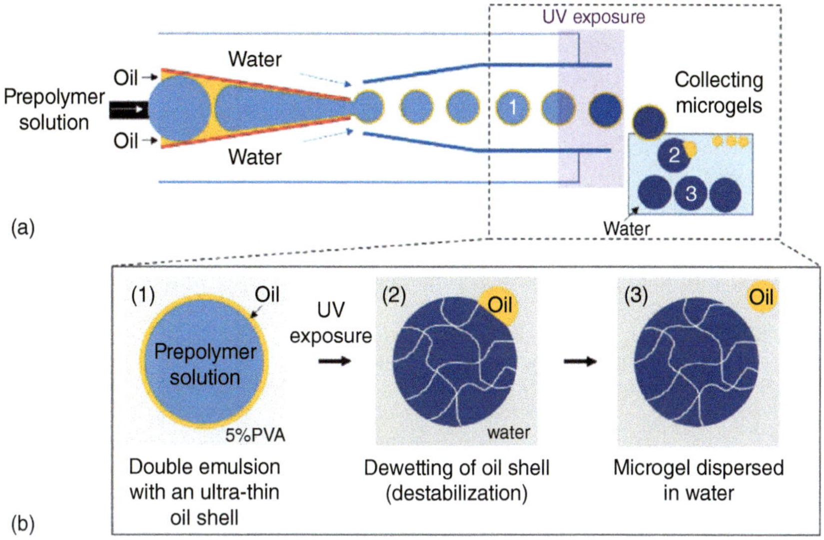

Figure 8.19 One-step production of microgels by the use of double emulsion drops with an ultra-thin oil layer. (a) Schematic illustration showing a glass capillary microfluidic device for the preparation of double emulsion drops. The innermost drops are solidified to form microgels upon UV exposure. (b) Schematic illustration showing the detailed procedure to form solid microgels from double emulsion drops through a dewetting process of the oil layer. Source: Choi et al. [11].

A microfluidic device was built with several glass capillaries (Figure 8.19). Two tapered circular capillaries, namely the injection part and cross-linking part, were inserted into a square capillary, which constituted the main fluid road. To make the circular injection capillary hydrophobic, *n*-octadecyltrichlorosilane was applied to treat its inner wall. Moreover, another tiny circular capillary was inserted into the inlet of the device to provide an aqueous hydrogel prepolymer solution. Oil flowed in the injection circular capillary and another aqueous phase flowed in the square capillary so that the hydrogel microdroplets were generated with a thin oil shell. After that, in the cross-linking part, the microdroplets were treated with UV irradiation to form cross-linked microgels. All of the cross-linked microgels with oil shells were collected with a water bath. The surrounding thin oil shell spontaneously dewetted upon cross-linking of the innermost microdroplets so that the oil phase was separated from the microgels directly. In this work, mineral oil containing 0.5% Span-80 surfactants or fluorinated oil (HFE 7500) containing 0.5% Krytox-PEG-Krytox surfactants was selected as the oil phase and was labeled by oil-soluble staining (Neil Red). 5% (w/v) PVA solution was the outermost continuous aqueous phase. The receiving water bath without surfactant was used to remove the oil shell from the microgel surface so that the emulsion could get unstable. PEGDA prepolymer solution was selected to test the microgel fabrication process and summarize the size distribution. For further cell-laden microgel, GelMA was chosen as the basement and printed into GelMA microgels encapsulating myelogenous leukemia cells (K562), Madin–Darby canine kidney (MDCK) epithelial cells, and NIH/3T3 fibroblast cells, respectively. All of them showed high viability after undergoing the fabricating process and the culturing in the 3D microenvironment. MDCK cell cluster inside GelMA microgels gradually got larger with time and NIT/3T3 fibroblast cells showed obvious spreading in the 3D space, which demonstrated the potential of this method to build cell-laden microgels.

Apart from mono-component microgels, microfluidic technology can also be applied to print complex microgels. For a long time, first, conventional microfluidic chips with two T-junctions were limited for their device preciseness and wettability, although it can produce monodisperse double emulsions. Second,

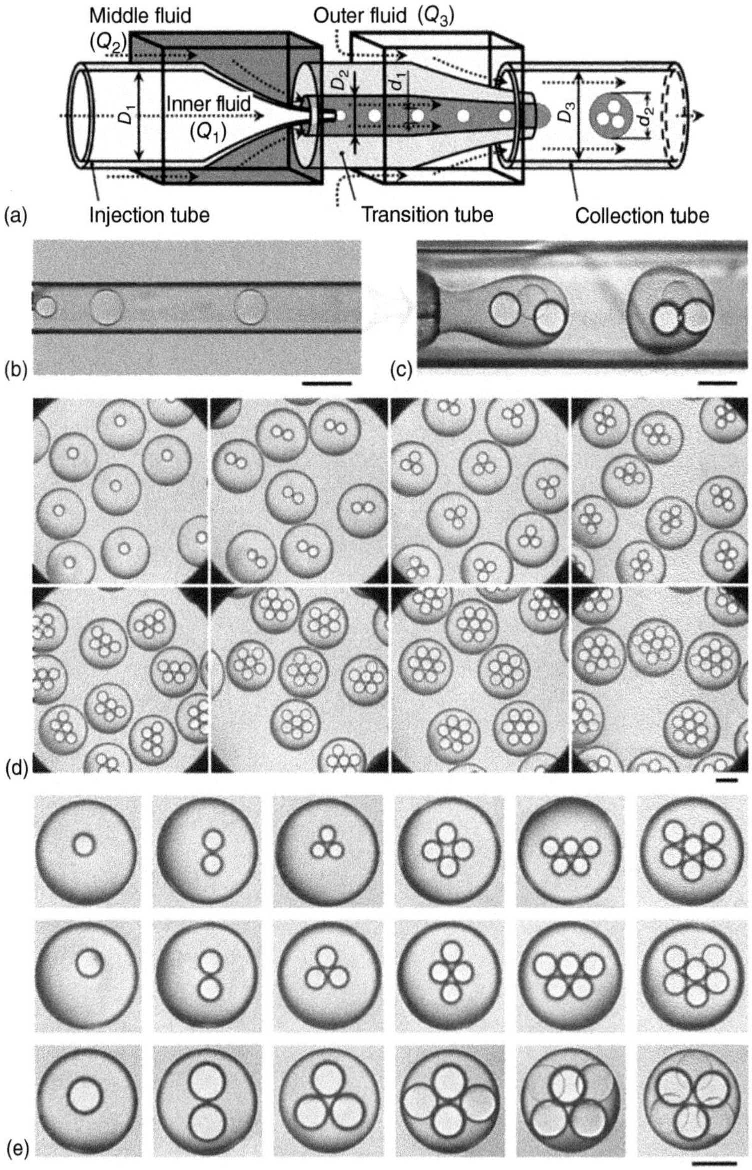

Figure 8.20 Capillary microfluidic device and the formation of precisely controlled monodisperse double emulsions. (a) Schematic diagram of the device geometry. The outer fluid must be immiscible with the middle fluid and the middle fluid must be immiscible with the inner fluid. (b) and (c) High-speed optical micrographs of the first (b) and second (c) emulsification stages. (d) Optical micrographs of monodisperse double emulsions containing a controlled number of monodisperse single emulsions. (e) Optical micrographs of monodisperse double emulsions showing a controlled increase of the diameter of the inner droplets while the number is constant. All double emulsions were made in the same device and with the same fluids. The flow rates of the inner, middle, and outer fluids in (b) and (c) are $Q_1 = 350$, $Q_2 = 2000$, and $Q_3 = 5000\,\mu l/h$, respectively. In (d), the flow rates of middle and outer fluids are fixed at $Q_2 = 2000$ and $Q_3 = 5000\,\mu l/h$, and those of the inner fluid are $Q_1 = 20, 55, 70, 85, 150, 200, 225$, and $240\,\mu l/h$, from the left top to the right bottom. In (e), the variation ranges for the flow rates of the inner, middle, and outer fluids are $Q_1 \approx 20-600$, $Q_2 \approx 1600-5000$, and $Q_3 \approx 2000-8000\,\mu l/h$ (in each case, Q_3 is larger than Q_2). All scale bars are 200 mm. Source: Chu et al. [12].

coaxial flow-focusing geometries solve the wettability but the fluid types that can be precisely controlled are limited. Considering these problems, in 2007, Chu et al. published a contribution to propose a novel microfluidic system to realize controllable monodisperse multiple emulsions [12]. The microfluidic chips consisted of several levels (Figure 8.20). The first emulsification step was realized with a coaxial and co-flow geometry. It was composed of the injection tube (a cylindrical capillary with a tapered outlet) inserted into the transition tube (a secondary cylindrical capillary with an inner diameter D_2). Both cylindrical capillary tubes were assembled coaxially within a larger square capillary. The other outlet of the transition tube was also tapered and inserted into the third, coaxially assembled cylindrical capillary tube (the collection tube of inner diameter D_3). With this novel microfluidic system, monodisperse multiple emulsions were generated in two emulsification steps. Innermost microdroplets were emulsified at the first level by the coaxial flow of the middle fluid. The single emulsion was subsequently emulsified at the second level through the coaxial flow of the outermost fluid, which was injected in the outer stream through the square capillary. These two levels guaranteed the formation of highly monodisperse microdroplets. By controlling the flow rates of the different fluids, the size of the innermost and outermost microdroplets and the number of the innermost droplets could be easily achieved. Furthermore, the authors extended this approach to establish more complex emulsions, triple emulsions, which were consisting water-in-oil-in-water-in-oil (W/O/W/O) microdroplets. To realize this structure, another level was added to form the third emulsification level, which led to the triple emulsions. The emulsion images showed that the formation process was still stable though large deformation would occur at the beginning of the last level. Similarly, the size and encapsulating numbers could be easily controlled by modifying the flow rate parameters. At last, as a demonstration, a thermosensitive hydrogel, poly(*N*-isopropylacrylamide), was introduced to this novel microfluidic device and to form a triple emulsion, which was potential to be applied in the molecular controlled releasing. The polymerization reaction of the hydrogel was performed in a specific layer. A W/O/W/O triple emulsion was formed and the monomer, cross-linker, and initiator was added into the outermost aqueous layer.

Moreover, the accelerator that could diffuse into the outermost aqueous shell and promote the polymerization process was added into the oil phase. As a result, a microencapsulation owning thermosensitive hydrogel shell encapsulated an oil microdroplet containing several water microdroplets. By heating the multiple emulsions, the thermosensitive hydrogel shell was broken and provided spontaneous, pulsed release of the innermost aqueous microdroplets into the continuous oil phase, which was figuratively described as Trojan-horse-like behavior.

8.4 Lithography Technology

8.4.1 Replica Mold

Kim et al. proposed a novel replica molding technology, named degassed micromolding lithography (DML), which realized the efficient loading of the hydrogel precursor into the gas-permeable micromold without defects and extremely homogeneous anisotropic microgels synthesis with high resolution [13]. Developing from the existing molding fabrication process, the mold in this work was degassed in a vacuum chamber (Figure 8.21). First, the master mold was built by hardening SU-8 structures on a silicon wafer with photolithography technology. Then, PDMS liquid was poured on the master mold with a glass cover and degassed in a vacuum chamber at

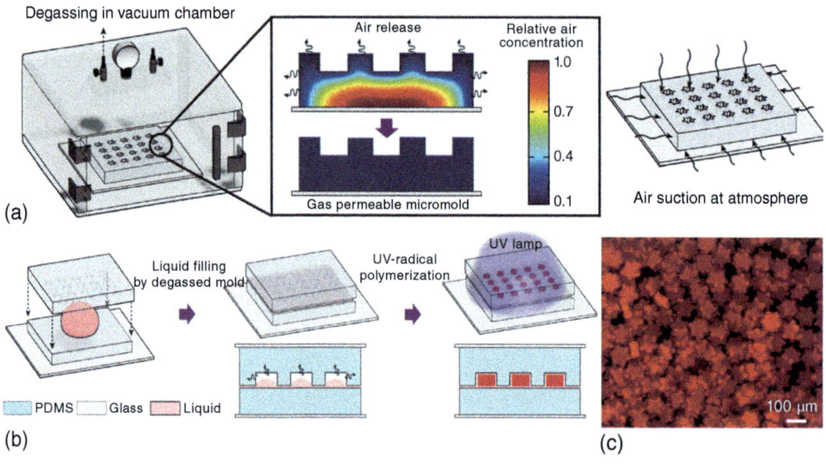

Figure 8.21 Schematic representation of degassed micromolding lithography. (a) Degassing of a gas-soluble PDMS mold template in a vacuum chamber. The air concentration in the mold template decreases to a new equilibrium concentration after degassing. When taken out from the chamber into the atmosphere, the degassed mold template can act as an air-absorbing pump. (b) Microparticle fabrication in a degassed micromold. The precursor is sandwiched between the cover and the degassed mold template. The entrapped air bubbles between the precursor and the micromold are removed through gas uptake by the degassed micromolds, which are then filled with the precursor. With the cover compressed onto the mold template, UV irradiation was applied to polymerize the precursor. (c) An image of fluorescent particles synthesized by the DML method. Source: Kim et al. [13].

0.1 atmospheric pressure on 70 °C heating. Until a solidified PDMS micromold was formed, the degassed mold template was taken out from the vacuum. The PDMS micromold was separated from the master mold and contacted with the hydrogel precursor. In this process, the micromold would strongly attract the precursor into the designed microwells with specific shapes on the micromold by absorbing the entrapped air bubbles between the micromold and the precursor. The microdroplets in the micromold were then cross-linked by UV irradiation. The cross-linked microgels were treated by the recovery solution composed of 70% ethanol, 30% water, and 0.05% Tween-20 and the microgels released from the PDMS micromold into the recovery solution. For demonstration, the authors prepared PEGDA hydrogel precursor and produced microgels with the proposed DML approach. The entering process of the prepolymer and the final static state were researched in detail. The filling duration would increase with the decreasing of the degassing procedure. The PEGDA microgels with different shapes were successfully printed.

8.4.2 Discrepant Wettability

Li et al. presented a convenient approach to assemble heterogeneous 3D cellular microenvironment in microgel arrays by establishing discrepant wettability [14]. To build the surface with different wettability micropatterns, a rapid surface modification way based on PDMS stamping method was proposed (Figure 8.22). The PDMS stamp with an ideal micropattern was made with conventional soft lithography technology. With oxygen plasma, the PDMS stamp and a washed glass slide were temporarily bonded together. After that, the PDMS stamp was detached from the glass slide, leaving a hydrophobic nanofilm of PDMS oligomers. Based on this, template with different wettability patterns was successfully built, in which the glass surface without PDMS oligomers was hydrophilic. According to the results of attenuated total reflectance Fourier-transform infrared spectroscopy (ATR-FTIR) absorption spectra, the PDMS nanofilm contained hydrophobic methyl groups while the glass surface owned hydrophilic hydroxyl groups. When the hydrogel prepolymer solution was poured onto the template surface, aqueous liquid tended to soak the hydrophilic areas and repelled by the hydrophobic areas, so that a microdroplet array with a specific arrangement could be formed. The prepolymer solution was treated with UV irradiation so that it would further form a microgel array. As the testing results, this facile strategy could generate microgel arrays with individual microgels sized from 20~2000 μm. The microgel's height was determined by the contact angle and microdroplet diameter. Taking microgel with the size of 100~800 μm as an example, which was with high feasibility and controllability, the height was tested as 5~45 μm. Moreover, the researchers also extend the applications of this method, including the assembly of microgels with spatially organized multiple components, heterogeneous microenvironment arrays with chemical/cell gradients, and complex cell microenvironment arrays with orthogonal gradients, displaying potentials to be widely applied in more biomedical applications in the future.

Besides, Le et al. proposed another facile approach to form a microgel array with surfaces patterned with different wettability [15]. First, gold-coated glass slides

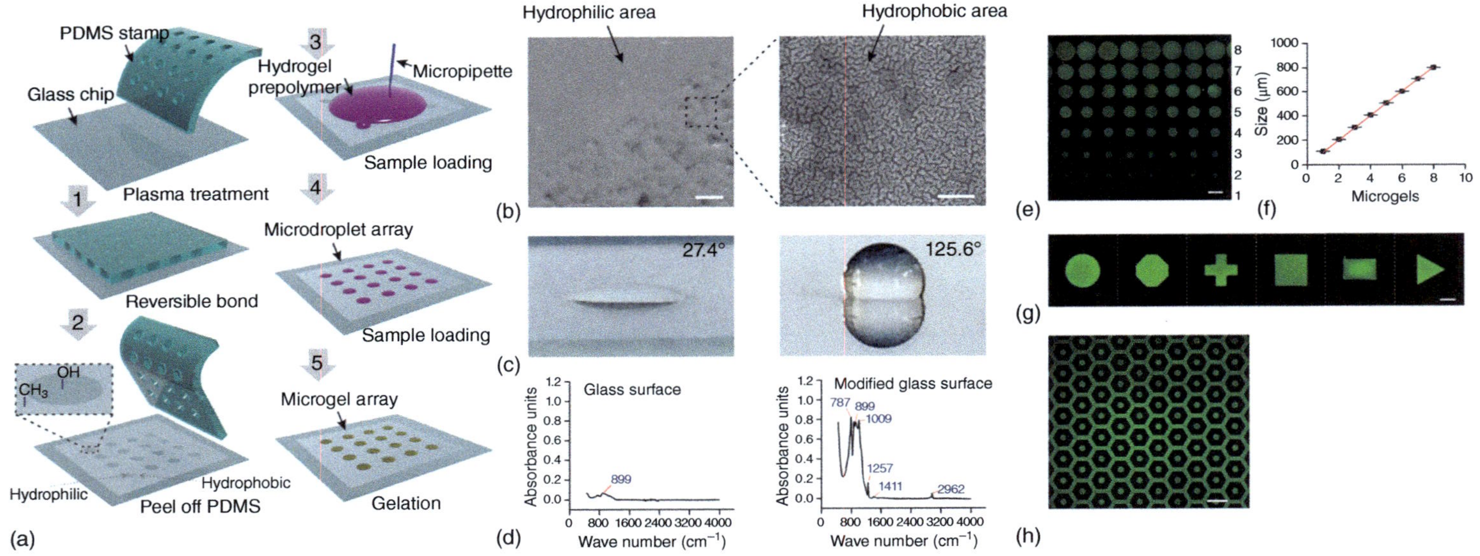

Figure 8.22 Surface-wettability-guided assembly of microgel arrays. (a) Schematics of wettability-guided assembly of a microgel array. (b) SEM of PDMS nanofilm. Scale bar, 4 μm in the left image, 500 nm in the right image. (c) Water-contact-angle measurements for plasma-treated glass and PDMS nanofilm; (d) ATR-FTIR analysis of unmodified glass and PDMS-nanofilm-stamped glass. (e) Assembly of size-varied PEG microgels. Scale bar, 600 μm. (f) Quantitative analysis of size variation of assembled microgels. The data represent mean ± standard deviation (SD) ($n = 10$). (g) Assembly of shapevaried PEG-microgel arrays. Scale bar, 300 μm. (h) Assembly of a complex PEG-hydrogel architecture. Scale bar, 600 μm. Source: Li et al. [14].

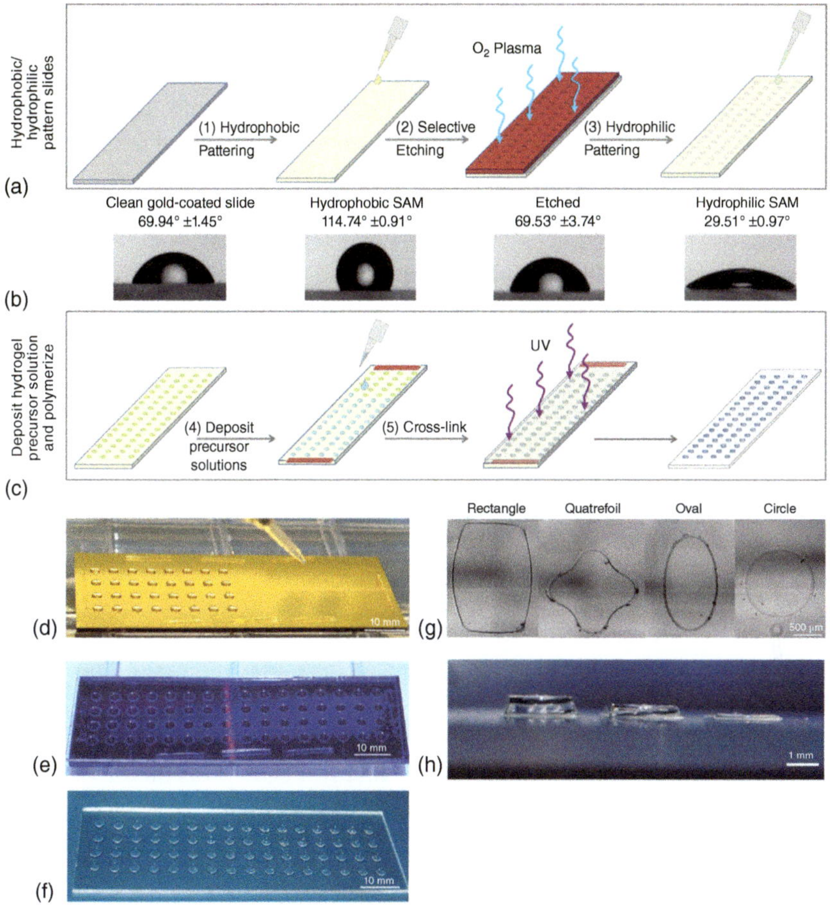

Figure 8.23 Hydrogel array formation procedure and outputs. (a and b) Hydrogel arrays were formed on gold-coated glass slides patterned with SAMs with differential wettability. (c–e) Hydrogel precursor solutions deposited onto the hydrophilic SAM regions of the patterned slide were cross-linked via UV-initiated photopolymerization to form hydrogel spots. (f) The resulting array is composed of hydrogel spots immobilized on a glass slide. (g and h) The array formation procedure allows for the formation of hydrogel spots of various size, shape, and height. Source: Le et al. [15].

were immersed in ethanol and washed by sonication for one minute (Figure 8.23). Then, the slides were taken out and after the evaporation of ethanol on the surface, hydrophobic self-assembled monolayer (SAM) generation on the gold-coated glass slides was carried out by submerging fluorinated alkanethiols in ethanol. PDMS elastic templates to determine the pattern for selective etching of hydrophilic SAM regions were built by soft lithography technology. The cleaned and dried gold-coated glass slides with the hydrophobic SAM were covered with the PDMS elastic templates. The areas that were not covered by the PDMS template were etched by oxygen plasma treatment to form the hydrophilic alkanethiol SAM layer in the etched areas so that a template with discrepant wettability was prepared.

Then, hydrogel precursor solutions containing PEG-NB, PEG-D, peptides, and irgacure 2959 photoinitiator were made up and diluted to ideal concentrations with PBS. For cell encapsulation studies, cells were suspended at the ideal density in the medium and added to the hydrogel precursor. All encapsulating research applied the MMP-degradable cross-linker where the nondegradable PEG-DT cross-linker. The hydrogel precursor solution was deposited onto the hydrophilic areas of the patterned gold-coated slides with a pipette. With the DTT-treated silanized glass coverslip, the hydrogel precursor solution was restricted between the coverslip and the gold-coated slide and cured by UV irradiation. After curing, the glass-immobilized microgels could be easily detached from the gold-coated slide by removing it. The experimental results confirmed the feasibility of this strategy to establish a microgel array. Moreover, the microgels generated by this way could be applied for stem cell culturing and screening for the influence of various chemical and mechanical substrate properties to guide the growth states of cells. The results showed that the microgel array system could be applied to test and optimize substrate features and culture conditions for stem cell culture, which is of great significance in stem cell engineering.

8.4.3 Photomask Film

In Gulfam et al.'s previous study, a bottom-up method was introduced as an important way for the assembly of cell-laden microgels [16]. Independent PEG microgels with square and hexagonal shapes were fabricated by using lithography technology based on the photomask film (Figure 8.24). Briefly, PEGDA was mixed with photoinitiator irgacure 2959 to form a curable prepolymer solution. The photomask thin film was designed in square shape (500 μm × 500 μm) and hexagonal shape (500 μm in length). Two spacer shims were placed on the bottom glass slide and the upper glass slide was placed on top of the poured hydrogel prepolymer solution in order to get the solution distributed uniformly in the cave. Then, a photomask thin film was placed on the upper glass slide, and UV light was switched on. The UV irradiation was reflected or absorbed by the opaque areas of the photomask and went through from the transparent areas so that the prepolymer was cross-linked according to the mask pattern arrangement and formed specific microgels and then they were further assembled in a bioreactor.

Alexander Revzin also used the method of lithography based on photomask film to generate microgels [17]. Apart from a glass basement, the microgels were also cross-linked on a silicon substrate (Figure 8.25). The surface of the silicon/glass was treated with 3-(trichlorosilyl)propyl methacrylate to form a SAM with pendant acrylate groups, which was verified with ellipsometry and time-of-flight secondary ion mass spectrometry. A solution consisted of acrylated or methacrylated poly(ethylene glycol) derivative and photoinitiator (2,2-dimethoxy-2-phenylacetophenone) was spin-coated onto the treated substrate and was cross-linked by the UV irradiation through a photomask film, and developed with either toluene, water, or supercritical carbon dioxide. The 3D PEG microgels, according to the designed photomask, were formed on the treated surface. Researchers demonstrated that with different

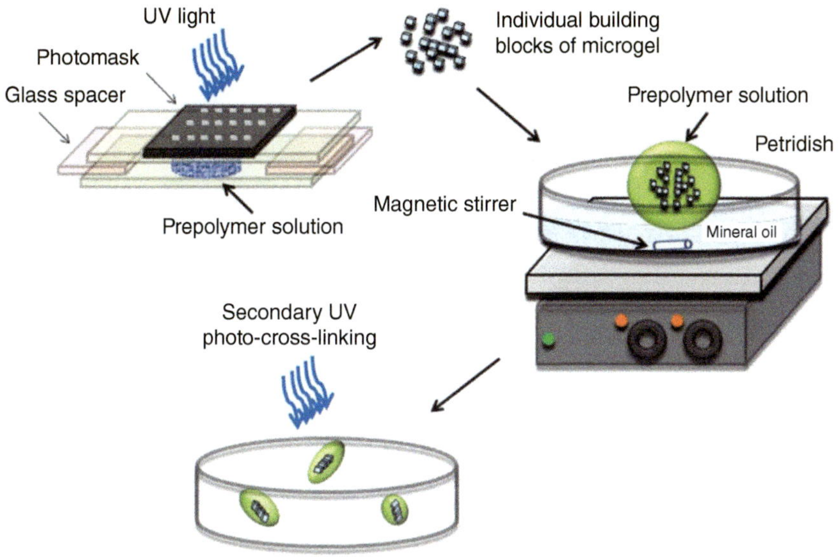

Figure 8.24 Schematic representation of the hydrogel assembly process. Individual building blocks of PEG microgels were fabricated by using a photolithography technique. Individual building blocks were subsequently assembled on the two-phase oil-aqueous interface in a bioreactor with a magnetic stirrer. To stabilize the hydrogel assembly, secondary photo-cross-linking was performed by using a UV light. Source: Gulfam et al. [16].

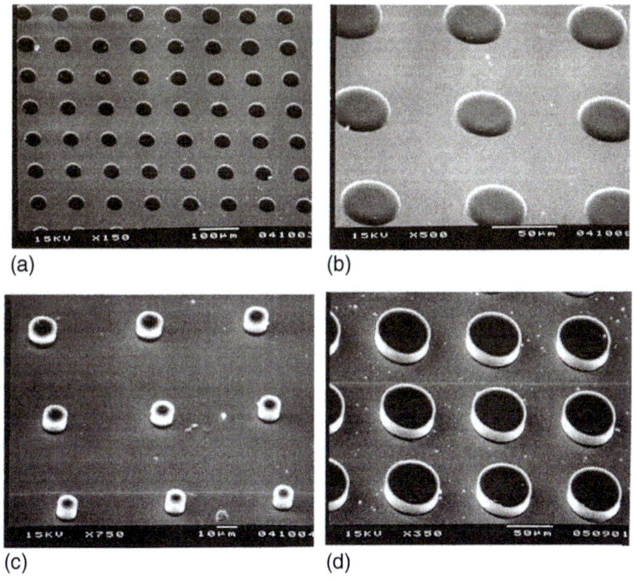

Figure 8.25 Scanning electron micrographs of PEG-DA (MW 575) gel microstructures on silicon: (a) 50 μm diameter microstructures; (b) higher magnification view of 50 μm diameter microstructures, (c) 7–10 μm diameter microstructures. The gels in images (a)–(c) were developed in toluene. (d) 50 μm diameter microstructures developed in supercritical carbon dioxide. Source: Alexander Revzin et al. [17].

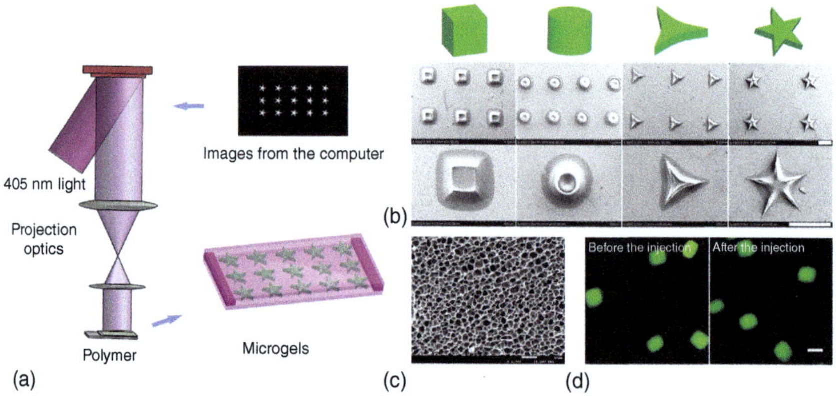

Figure 8.26 (a) Schematic representation of DLP-3D printer for microgels customization. After designing the digital 3D images in the computer, the microgels with the precise 3D structure are constructed. (b) The digital design and scanning electron microscopy (SEM) images of 3D printing microgels treated with gradient dehydration. Scale bars, 500 μm. (c) Representative SEM images of lyophilized microgel samples. Scale bar, 10 μm. (d) The morphology of fluorescein isothiocyanate (FITC)-labeled microgels (before and after the injection). Scale bar, 200 μm. Source: Liu et al. [18].

photomasks, the range of cylindrical microgels diameters was 7∼600 μm and the range of thickness was 3∼12 μm, and the microgels could keep high water content for three weeks on the substrate.

8.4.4 Digital Light Processing

Liu et al. described a novel approach based on advanced scalable digital light processing (DLP-3D) printing technology to fabricate microgels with different shapes and sizes [18]. GelMA prepolymer solution mixed with photoinitiator LAP was prepared. Two narrow silicon membranes in the printer were applied to form a narrow gap, where the cross-linked microgels were formed and the height of the microgels could be easily tuned (Figure 8.26). Design the specific digital patterns on the computer and uploaded them to the printers. The specific microgels could be printed after two seconds of curing exposure. Researchers observed the microgels undergoing both gradient dehydration and lyophilization by a SEM and quantitatively analyzed the pore size and porosity. Furthermore, the microgels that could realize the controlled drug releasing were also fabricated by adding nanoparticle-laden drugs into the monomer GelMA prepolymer solution. Moreover, cells could also be printed into microgels and the experimental results displayed that the cells kept high survival and proliferation capability, which owns great significance in customizing functional microgels encapsulating drugs or cells with potential therapeutic applications.

8.5 Bulk Crushing

All of the methods introduced in Sections 8.1–8.4 to print microgels are based on the hydrogel prepolymer with a liquid state. Apart from these approaches, bulk

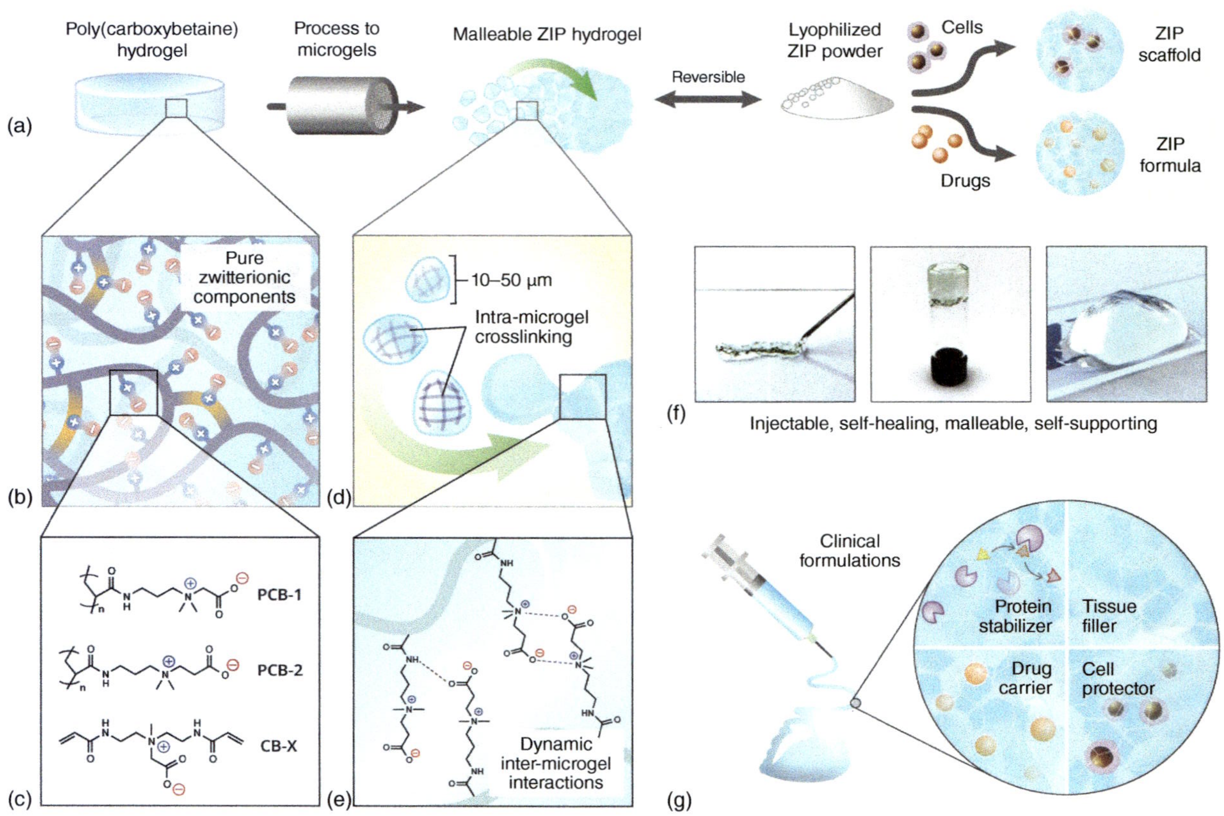

Figure 8.27 Production, properties, and applications of "ZIP" hydrogels. (a) Overview schematic showing the production of viscoelastic zwitterionic injectable pellet (ZIP) gels, which can be lyophilized for simple formulation and mixed with cells or therapeutics. (b) Hydrogel components are purely zwitterionic, consisting of (c) carboxybetaine acrylamide polymers (PCB-1 or PCB-2) with carboxybetaine diacrylamide cross-linker (CB-X). (d) Covalent cross-links inside each microgel enable bulk support and elasticity. (e) Dynamic zwitterionic fusion interactions enable the reconstruction of microgels into new viscoelastic ZIP material. (f) ZIP gels can be injected through needles, self-heal, and retain their shape. (g) Applications of ZIP gels include injectable soft tissue fillers, therapeutic carriers, and cell scaffolds for growth and protected injection. Source: Sinclair et al. [19].

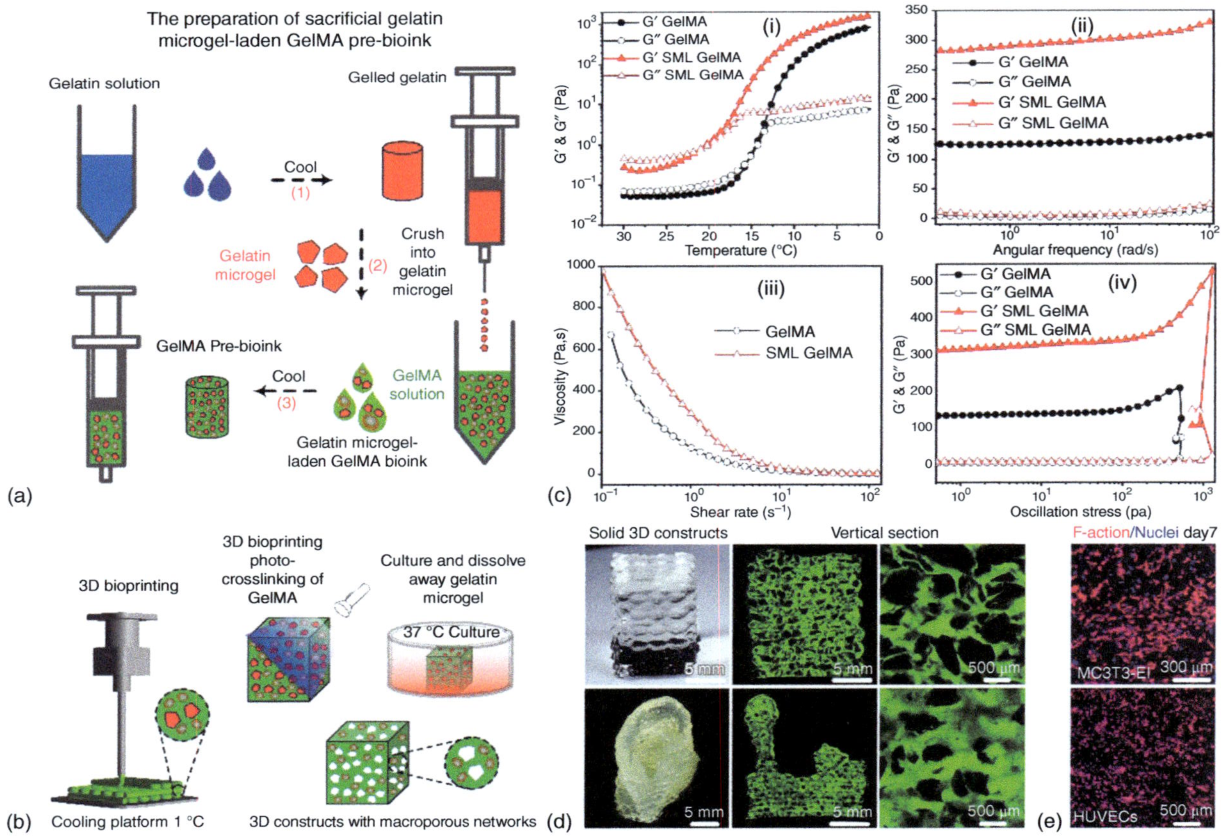

Figure 8.28 3D bioprinting large-scale tissue constructs with MPNs. (a) Schematic illustration of the preparation of sacrificial gelatin microgel-laden GelMA pre-bioink. (b) The process of 3D bioprinting constructs with MPNs. (c) Characterization of the GelMA pre-bioink and sacrificial microgel-laden GelMA (SML GelMA) pre-bioink; (i) G′ and G″ values as functions of temperature; (ii) mechanical spectra of the pre-bioinks at 22 °C; (iii) viscosity as a function of shear rate (22 °C); Source: Shao et al. [20].

crushing, which is based on the hydrogel prepolymer with solid state, is also a commonly used way to generate microgels.

Sinclair et al. reported a novel and scalable method to fabricate malleable and injectable zwitterionic polycarboxybetaine (PCB) microgels with bulk crushing principle, which displayed the properties of superhydrophilic, nonimmunogenic, and completely devoid of nonspecific interactions [19]. First, macroscopic PCB hydrogels were generated by the photo-cross-linking method (Figure 8.27). Then, researchers established an extruding outlet assembled with a micronic metal mesh. By repeatedly extruding the bulk hydrogel, the hydrogel integrity was crushed into tiny particles immediately by the effect of the mechanical force. To modify the general size of the microgels fabricated in this strategy, meshes with different pore sizes were applied (25~500 μm). The zwitterionic hydrogel formulations based on the generated microgels in this strategy displayed excellent shear-thinning and self-healing properties, demonstrating its potential to become an important bioink unit in more research.

To establish porous bulk tissue *in vitro*, Shao et al. also applied the bulk crushing method to generate gelatin microgel as the sacrifice material [20]. The gelatin solution was cooled down and was made to become fully gelled gelatin bulk at −20 °C temperature in the syringe by taking the advantage of the thermo-cross-linking property (Figure 8.28). Subsequently, the thermo-cross-linked gelatin bulk was directly squeezed out from the syringe needles and transformed into crushed gelatin microgels. The researchers mixed the gelatin microgels with GelMA prepolymer solution for further extrusion printing. The GelMA 3D structures encapsulating gelatin microgels were cross-linked together with UV irradiation. In the culturing atmosphere of 37 °C, the gelatin microgels would gradually melt into a liquid state, so that the micropores were formed inside the cured GelMA network.

References

1 Gao, Q., He, Y., Fu, J.-z. et al. (2015). *Journal of Sol-Gel Science and Technology* 77: 610.

2 Xu, C., Zhang, M., Huang, Y. et al. (2014). *Langmuir* 30: 9130.

3 López-Iglesias, C., Casielles, A.M., Altay, A. et al. (2019). *Chemical Engineering Journal* 357: 559.

4 Zhang, Z., Xiong, R., Corr, D.T., and Huang, Y. (2016). *Langmuir* 32: 3004.

5 Xie, M., Gao, Q., Zhao, H. et al. (2019). *Small* 15: 1804216.

6 Xie, M., Gao, Q., Qiu, J. et al. (2020). *Biomaterials Science* 8: 109.

7 Zhao, H., Chen, Y., Shao, L. et al. *Small* 0: 1802630.

8 Caldwell, A.S., Campbell, G.T., Shekiro, K.M.T., and Anseth, K.S. (2017). *Advanced Healthcare Materials*: 6.

9 Franco, C.L., Price, J., and West, J.L. (2011). *Acta Biomaterialia* 7: 3267.

10 Ahn, S. and Kim, G. (2015). *Journal of Materials Chemistry B* 3: 9132.

11 Choi, C.H., Wang, H., Lee, H. et al. (2016). *Lab on a Chip* 16: 1549.

12 Chu, L.-Y., Utada, A.S., Shah, R.K. et al. (2007). *Angewandte Chemie* 119: 9128.

13 Kim, H.U., Lim, Y.J., Lee, H.J. et al. (2020). *Lab on a Chip* 20: 74.
14 Li, Y., Chen, P., Wang, Y. et al. (2016). *Advanced Materials* 28: 3543.
15 Le, N.N.T., Zorn, S., Schmitt, S.K. et al. (2016). *Acta Biomaterialia* 34: 93.
16 Gulfam, M., Lee, J.M., and Chung, B.G. (2011). *Biotechnology Progress* 27: 466.
17 Alexander Revzin, R.J.R., Yadavalli, V.K., Koh, W.-G. et al. (2001). *Langmuir.*
18 Liu, X., Tao, J., Liu, J. et al. (2019). *ACS Applied Materials & Interfaces* 11: 12209.
19 Sinclair, A., O'Kelly, M.B., Bai, T. et al. (2018). *Advanced Materials* 30: e1803087.
20 Shao, L., Gao, Q., Xie, C. et al. (2020). *Bio-Design and Manufacturing* 3: 30.

9

Biomedical Applications of Microgels

9.1 Introduction

As the introduction in Chapter 8, lots of novel approaches have been developed to generate microgels, and a series of biomaterials have been applied as the basement. Compared to bulk hydrogels, microgels own their specific properties to match different requirements in the field of biomedical engineering. In terms of the practical application requirements, the advanced properties of microgels could be summarized as below.

9.1.1 Tiny Size

Microgels have extremely tiny size, which is controlled by the printing parameters. On the one hand, due to the tiny size, it will be easy to capture a small number of cells or even a single cell. When a single cell is encapsulated in the microgel, it could be easily separated from other cell clusters for further examinations. On the other hand, due to the tiny size of microgels, they have the capability to be injected or implanted in the narrow parts of a patients' body to recover the injury. Besides, compared to the bulk hydrogels with the same volume, microgels can expose a much bigger surface area, which extremely promotes the materials between the inner and outer environment of a hydrogel. This profile can also provide more growing space for the cells cultured based on a hydrogel system.

9.1.2 Hydrogel Network

Combining the tiny size feature and a hydrogel network, functional molecules, such as drugs and growth factors, could be loaded in the microgels. The network can act as a semiclosed carrier to achieve the controlled release of the molecules. Different types and concentrations of hydrogels own different pore density after cross-linking treatment. Thus, to meet different requirements of pharmacokinetics, the specific pore distribution could be easily built to release the promising molecular releasing curves. Furthermore, the hydrogel network has also a significant role to provide enough nutrients and 3D culturing microenvironment for the encapsulated cells.

Cell Assembly with 3D Bioprinting, First Edition. Yong He, Qing Gao, and Yifei Jin.

9.1.3 Complex Mechanical Properties

If microgels are placed together and regarded as a unit, the microgel system would show extremely special rheological properties. On the one hand, the system is composed of solid particles, which makes it display the profile of a stiff body or elastic body with specific Young's modulus. When a lower force is added to the microgel system, it can maintain the previous shape or show an elastic deformation. On the other hand, when higher force is added, the microgel system tends to show some fluid feature, such as shear-thinning phenomenon, which is known as Bingham plastics fluid. These special phenomenon makes it suitable for secondary bioprinting or supporting matrices of complicated 3D structures establishment.

In summary, these significant properties of microgels make them excellent self-adapted with self-healing behavior and suitable for many biomedical applications. Therefore, in this chapter, a series of representative applications using microgels would be introduced below, namely in vitro model, cell therapy, drug delivery, cell amplification, single-cell capture, supporting matrices, and secondary bioprinting.

9.2 In Vitro Model

In Pradhan et al.'s previous contribution, a facile manufacturing approach was proposed to generate high-throughput tumor models based on PEG-fibrinogen microgels. The tumor microspheres (TMs) were fabricated using the dual-photoinitiator, water-in-oil emulsion method proposed by Franco et al. with some modifications [1]. Researchers compared these TMs to the tumor spheroids (TS) built by suspended droplets, verifying the advantages of TM in this work (Figure 9.1). Because of the existence of the hydrogel network, the encapsulated cells owned more space to tune their pose in the 3D microenvironment, which could be difficult for the ones in TS for the intercellular restriction. The capability of encapsulating and keeping different kinds of cancer cell types in a promising growth state, namely MDA-MB-231, SK-BR-3, HT29, PC-3, and PC-3-Met cells, in these microgels for a long culturing duration was also confirmed. The high-throughput TM building strategy owned great potential to be applied for enhancing predictive and relevant data collection of the antitumor drug effect in preclinical pharmacokinetics assessment.

Moreover, combining 3D bioprinter and microdroplets generating system of electrohydrodynamic (EHD) principle, Xie et al. established a novel in vitro tumor models, namely "layer cake" 3D tumor array chip, as an antitumor drug screening system that could match the conventional drug-screening examinations based on 2D models [2]. Human breast cancer cells (MDA-MB-231) were encapsulated in a series of GelMA microgel arrays (Figure 9.2). Building tumor array on demand precisely and providing suitable microenvironment were verified to be feasible. The encapsulated tumor cells kept high viability after the printing process and blue light irradiation for cross-linking. The F-actin staining images displayed that

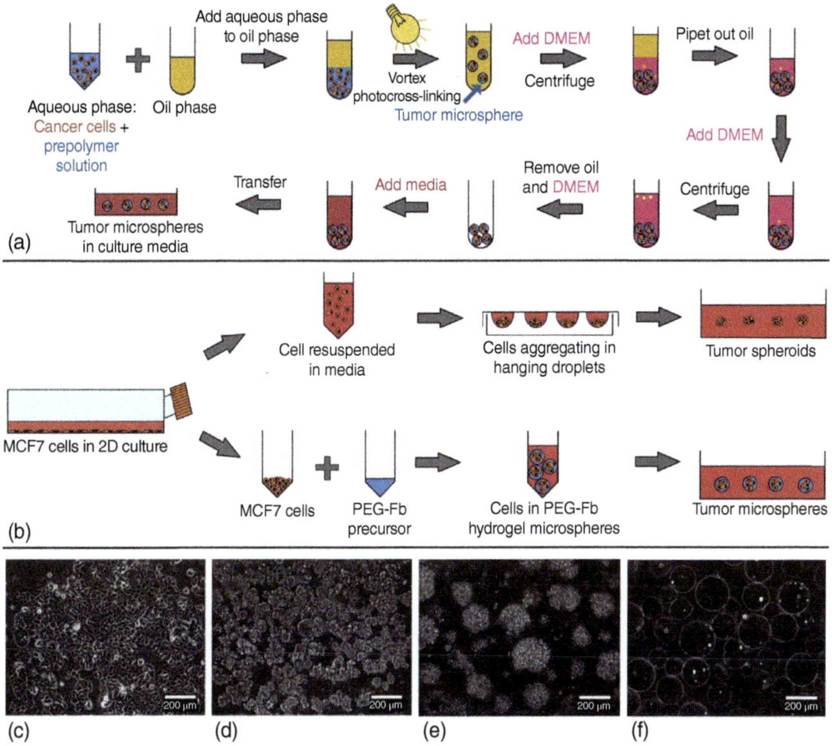

Figure 9.1 Schematic of fabrication processes for PEG-Fb tumor microsphere (TM) and self-aggregated tumor spheroid (TS) models. (a) Dual-phase, water-in-oil emulsion technique for the 3D encapsulation of cancer cells in PEG-fibrinogen hydrogel microspheres. (b) Comparative fabrication technique for TS and TM. MCF7 cells (c) in 2D culture, (d) cultured as TS, or (e) in TM. (f) PEG-fibrinogen microspheres formed without cells. Source: Pradhan et al. [1]. Reproduced with permission of Elsevier Ltd.

the encapsulated cells could spread in the 3D microenvironment. Importantly, because of the micropore network inside a GelMA system, MDA-MB-231s showed a large-scale migration at high speed. All of these research works indicated the rationality of the in vitro breast tumor models.

As a demonstration of drug screening, two common chemotherapy drugs, i.e. epirubicin and paclitaxel, were applied to treat the 3D tumor models. A series of biological characterization items were carried out to reveal the drug effect with different concentrations, such as laser scanning confocal fluorescence microscopy observation (live/dead testing, tumor F-actin morphology), enzyme-linked immunosorbent assay testing (NAD^+ examination for proliferation analysis, half-maximal inhibitory concentration [IC50], vascular endothelial growth factor [VEGF] accumulative expression), and cellular apoptosis, examined by flow cytometric analysis. This chapter demonstrated the application value of microgels in in vitro tumor model building and antitumor drug research.

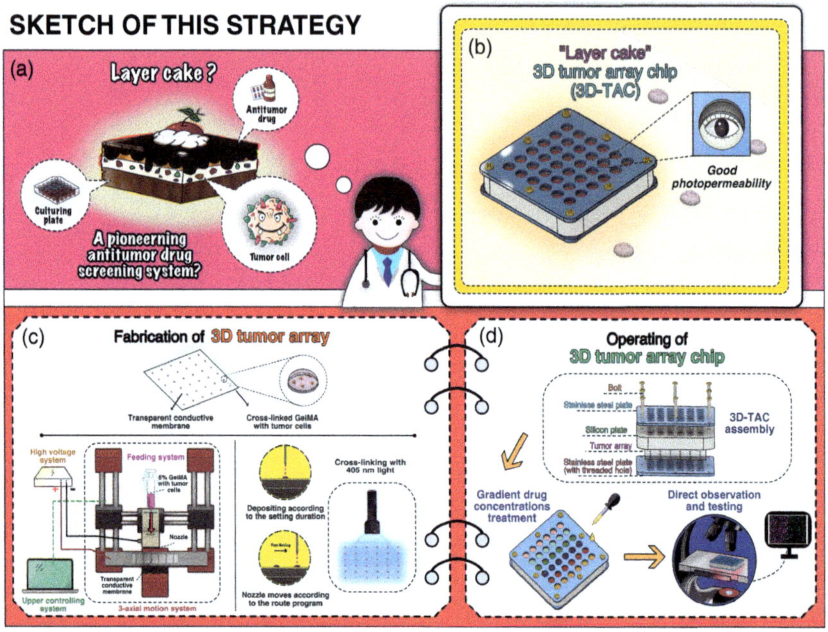

Figure 9.2 Sketch of the pioneering strategy. (a) Inspiration of this strategy. A layer cake structure is considered to be used to build a new kind of 3D drug screening system. (b) Sketch of the "layer cake" 3D tumor array chip (3D-TAC). The transparent conductive membrane of 3D-TAC owns good photopermeability for further examinations. (c) Fabrication protocol of 3D-TAC. (d) Assembly and using the method of 3D-TAC. Source: Xie et al. [2].

9.3 Cell Therapy

Zhao et al. published a paper in which the GelMA microgels encapsulating bone marrow stromal cells (BMSCs) and growth factors were injected into rabbit bone defects to enhance osteogenesis repairment [3]. Microdroplets were generated using the method of a microfluidic strategy (Figure 9.3). Based on the *analysis* in vitro, BMSCs were confirmed that they could keep high viability, obvious proliferation, and significant osteogenesis differentiation in the microgels. Then, New Zealand rabbits, which were six months old, were treated with anesthesia and were drilled to form a cavity (6-mm diameter and 10-mm depth) on the joint surface. After that, 250-μl GelMA microgels encapsulating BMSCs or BMP-2 were injected into the cavity, expecting to realize the bone regeneration in vivo. The rabbits were sacrificed after four weeks by overdose of thiopental sodium and harvested the defect samples. The histological examination with H&E and VG dying confirmed the realization of fresh bone production with the BMSCs or BMP-2 microgels compared to the control groups. Importantly, the microgels encapsulating BMSCs and BMP-2 simultaneously displayed the largest volume of the new bone (Figure 9.4). In summary, this strategy from the fabrication of cells/growth factors-laden microgels by microfluidic method to the osteogenesis differentiation in vitro/vivo showed the excellent effect of microgels in the field of tissue regeneration.

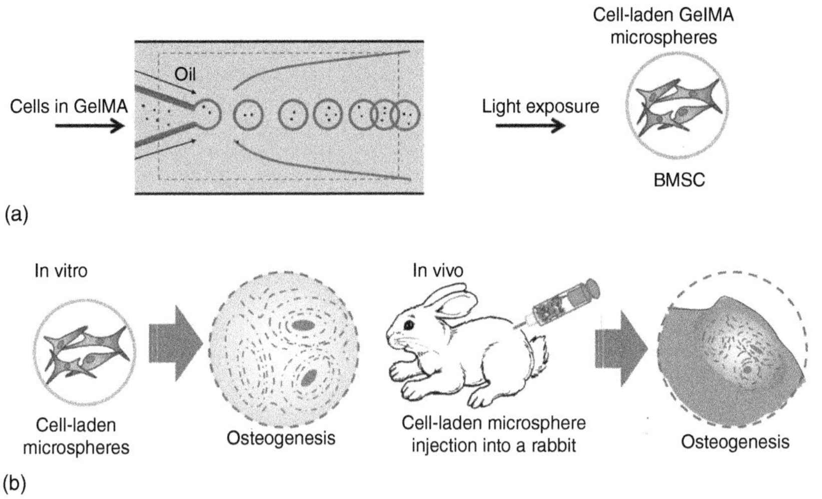

Figure 9.3 Schematic diagram of fabrication of BMSC-laden GelMA microspheres and its application for osteogenesis and regeneration of injured bones. (a) Photocross-linking-microfluidic fabrication of GelMA microspheres. Aqueous droplets containing GelMA gel precursors are produced from a microfluidic flow-focusing device and photopolymerized to form GelMA microspheres. (b) Osteogenesis of BMSC embedded in GelMA microspheres. BMSCs encapsulated in GelMA microspheres differentiate and regenerate bone in vitro and in vivo. Source: Zhao et al. [3].

Limb ischemia has become a global health problem because of the increasing elderly population and the prevalence of obesity, diabetes mellitus, and atherosclerotic disease. To realize the regeneration of vascular tissues, Shen et al. also proposed a perfect strategy that took advantage of microgels [4]. Researchers cultured a batch of HEK293T cells with stably integrated VEGF-A (VEGF type A) plasmids and encapsulated them in the alginate microgels, which could persistently release VEGF-A. The alginate microgels with HEK293T cells were prepared using the method of the EHD system (Figure 9.5). The releasing profiles were tested at 4, 12, and 24 hours in vitro with the medium that cultured the cell-laden microgels and with the whole blood in vivo. The concentration of VEGF-A increased from 115.3 ± 16.52 pg/ml at 4 hours, 229.9 ± 35.39 pg/ml at 12 hours, 680.0 ± 7.03 pg/ml at 24 hours, and arrived at a peak concentration of 746.3 ± 90.73 pg/ml at 72 hours (107 cells in each sample). The circulating VEGF-A concentration in the mouse serum was increasing during the first 72 hours (peaked at 177.8 ± 7.106 pg/ml) after implantation, verifying the continuous VEGF-A releasing capability of the microgels. Furthermore, the microgels were cocultured with HUVECs seeded on the Matrigel. The HUVECs morphology was observed with the optical microsphere. As expected, the medium containing enough VEGF-A was confirmed to promote the proliferation, migration, and sprouting of HUVECs. Guided by these previous research works, to assess the therapeutic effect of these microgels in promoting angiogenesis and decreasing muscle injury, the HEK293T-laden alginate microgels were implanted into the experimental mice. 10-week-old male C57BL/6 mice were kept at 23 ± 1 °C, 55 ± 5% humidity, and under a 12-hour dark/light cycle, and

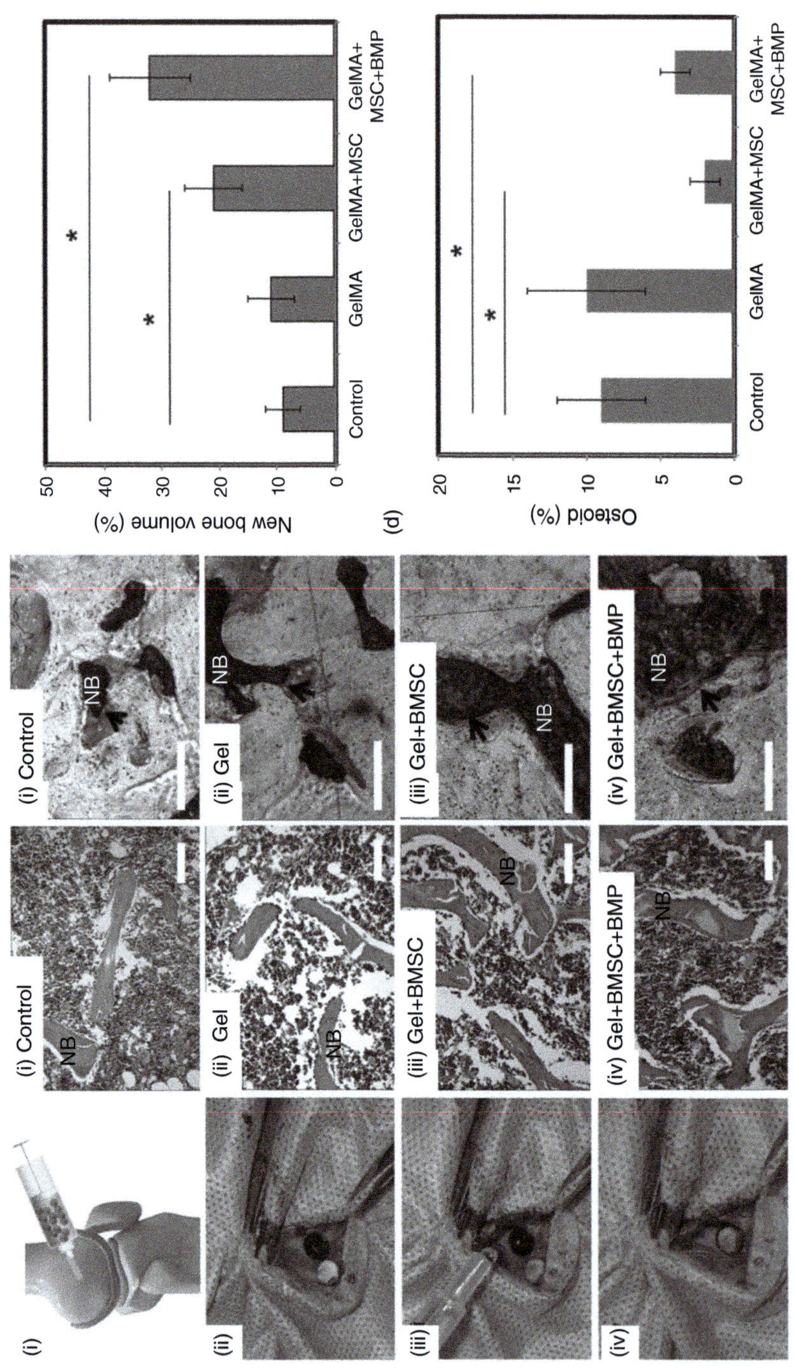
(i)
(ii)
(iii)
(iv)
(a)
(i) Control
NB
(ii) Gel
(iii) Gel+BMSC
(iv) Gel+BMSC+BMP
(b)
(c)
New bone volume (%)
Control
GelMA
GelMA+MSC
GelMA+ MSC+BMP
(d)
Osteoid (%)
(e)

Figure 9.4 Bone defect repair in vivo. (a) Operation processes. (i) Schematic illustration of rabbit femoral defect model. (ii) A circular defect is created on the lateral aspect of the distal part of a rabbit femur. (iii) Injection of normal saline (control), empty GelMA microspheres (blank), or GelMA microspheres containing BMSC or BMSC + BMP-2 into the wound bed. (iv) Closure of the femoral defect using the circular piece of bone originally removed to create the defect. (b) Histological sections of nonimplanted (i), implanted GelMA microspheres (ii), BMSC-laden microspheres (iii), and BMSC-laden microspheres containing BMP-2 (iv) in rabbit's femur after implantation for four weeks stained using hematoxylin and eosin (H&E) and (c) Van Gieson's Picro–Fuchsine staining. The BMSC-laden GelMA microspheres show more new bone formation than the control after four weeks postsurgery. "NB" indicates new bone. Scale bar = 200 μm. (d) Histomorphometrical analysis (%) of new bone formation and (e) osteoid (arrows) formation and total area in the defect zone (*$p < 0.05$). Source: Zhao et al. [3]. Reproduced with permission of WILEY-VCH Verlag GmbH & Co. KGaA, Weinheim.

allowed unlimited regular chow diet and water. By ligation of the femoral artery, the mice were treated as a hind-limb ischemia model. These mice in the treatment group received alginate microgels encapsulating VEGF-over-expressing HEK293T cells via subcutaneous grafting. Researchers noticed that the hind-limb blood flow in the experimental mice was obviously improved around the hind-limb ischemia injury area and the gastrocnemius muscle cells viability increased, which indicated the potentiality of the proposed alginate microgels as an effective approach of carrying out therapeutic angiogenesis for treating limb ischemia.

9.4 Drug Delivery

Cai et al. developed a 3D-printed concentration-controlled microfluidic chip to generate alginate microgels with a diffusion mixing pattern, which could deal with the problems of the rapid ionic gelation rate between alginate and free cations [5]. The diffusion mixing pattern in the microfluidic chip could undermine the ionic gelation process in the central stream (Figure 9.6). With this device, the calcium alginate microgels encapsulating Doxorubicin (Dox), which is a commonly used antitumor drug, were generated in large yield. The microgels containing Dox were placed in the dialysis bag (MW 14 000) and the gradual releasing property of Dox was tested in PBS of pH 7.4 and pH 6.5, respectively, by UV measurement (absorbance at 480 nm). The accumulative releasing curve confirmed that Dox-loaded alginate microgels had accelerated drug release in the PBS at lower pH 3. Moreover, drug screening studies were implemented by adding Dox-loaded microgels (different concentrations of Dox) into the well-plates seeded MCF-7 cells. Forty-eight hours later, the seeded cells displayed extremely low viability of 10.5% in Dox-loaded microgels groups, while the cells with free Dox of the same concentration showed a cell viability of 60.2%. This difference could verify the promising drug delivery efficiency of the calcium alginate microgels. This proof-of-concept about 3D-printed concentration-controlled microfluidic chip with a diffusion mixing pattern would be improved as a significant synthesis technology to fabricate other kinds of functional microgels for drug delivery.

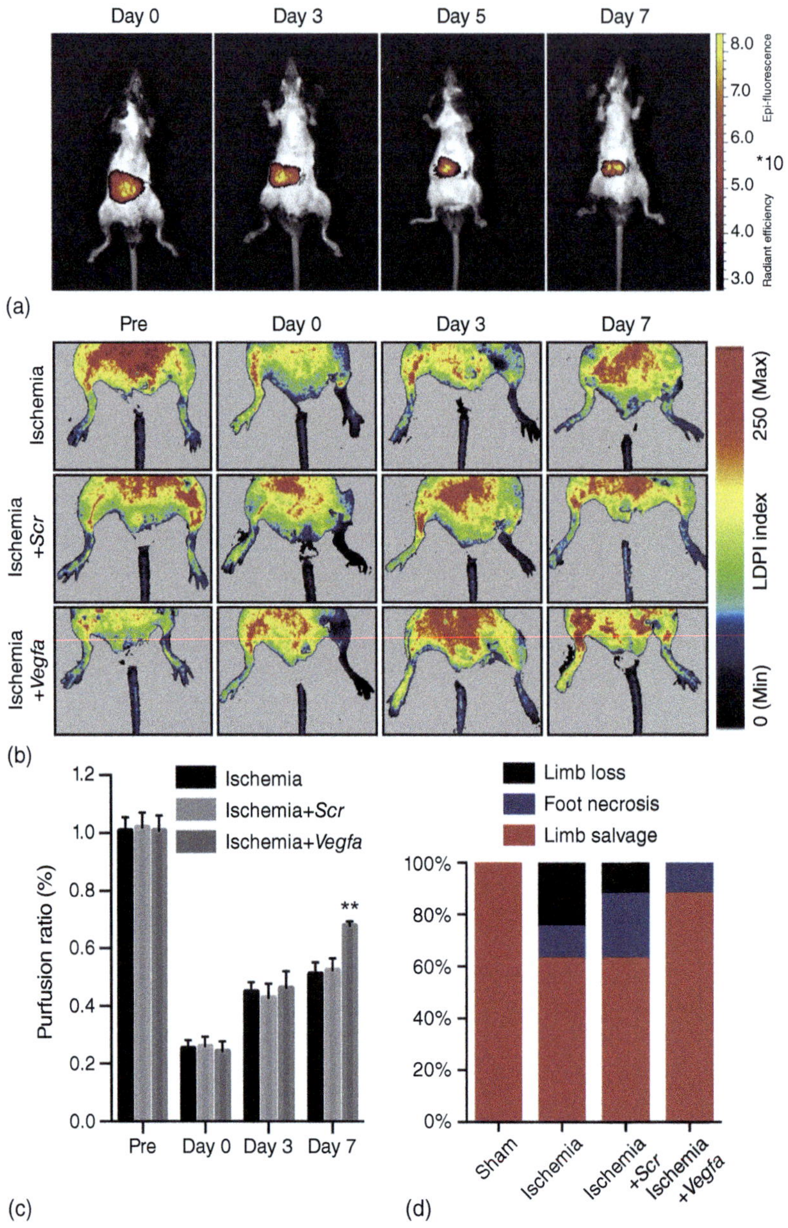

Figure 9.5 Generated microspheres containing VEGF-overexpressing HEK293T cells increased blood reperfusion and limb salvage. (a) IVIS spectrum imaging system detected the cell survival and location after the implantation of VEGF microspheres.
(b) Representative images of blood flow of the lower limbs from different treated mice at different time points. (c) Perfusion ratio among mice with different treatments. The perfusion ratio was calculated as the ratio of ischemic (left) side to nonischemic (right) side before and after ischemia ($n = 8$ in each group). $^{**}P < 0.01$ vs. scramble group.
(d) Pathological status of the ischemic hind limbs seven days after transplantation ($n = 24$ for Sham, 8 for Ischemia, 8 for Ischemia+Scr, and 8 for Ischemia+Vegfa). Source: Shen et al. [4]. Reproduced with permission of Elsevier B.V.

Additionally, a novel core–shell alginate microgel was developed with microfluidic technology for protein encapsulating and controlled releasing in Yu et al.'s previous paper [6]. To observe the releasing behavior in this microgel system, FITC-dextran (40 kDa) was loaded in the core–shell structure first rather than protein (Figure 9.7). From the fluorescent images, the signal intensity was extremely low in that the loaded FITC-dextran molecules could release from the core–shell structures immediately when they touched the calcium chloride bath. To solve this problem, researchers utilized polymers with positive charge, namely poly(ethyleneimine) (PEI), to quickly adsorb to the surface of alginate microgels according to alginate's negative charge profile to avoid premature releasing. After that, 42.7 kDa ovalbumin from chicken egg white (OVA) was applied as a model protein or antigen for delivery. As a result, the OVA loading efficiency of the PEI-modified alginate microgels arrived at 88%, which was much higher than the highest efficiency (58.6%) in others' reported contributions. Furthermore, an alternative material, chitosan, was applied to coat the microgels to prevent the premature releasing thanks to its mucoadhesive profile. Chitosan is insoluble at neutral pH and only soluble in acid conditions, making it a kind of pH-sensitive material. As we know, microgel-loaded drugs always enter human body from mouth so that the microgels would experience different pH environment and have to go through the gastrointestinal area to the small intestine. Therefore, to examine the protein-releasing kinetics in different pH environments, the chitosan-coated alginate microgels were immersed at pH 3 than at pH 7. In the earlier stage, less OVA was released from the microgels in that the barrier function of chitosan was much stronger in an acid solution. However, when the microgels were immersed in the environment at pH 7, the releasing rate of OVA was much higher due to the weakened chitosan strength. This pH-responsive release of proteins would potentially be used in more applications.

Figure 9.6 3D-printed concentration-controlled microfluidic chip for Dox delivered CaA microgels (CaA@Dox). (a) Particle sizes distribution of CaA microgels and CaA@Dox microgels 3 by DLS. (b) Left figure shows particle sizes of CaA@Dox microgels 1–5. Right shows the drug delivery rate of CaA microgels 1–5. (c) Evaluation of the release rate of Dox in CaA microgels 3 under different pH conditions. (d) Toxicity assay of CaA microgels 3 under different concentrations by a CCK-8 kit. The drug administration time was 48 h. (e) Cytotoxicity of CaA@Dox microgels 1–5 and free Dox with different Dox concentrations against MCF-7 cells for 48 h. Data are shown as mean ± SEM, $n = 3$, $^{*}p < 0.05$, $^{**}p < 0.01$, $^{***}p < 0.001$. (f) Fluorescence paragraph for live and dead cell assay of CaA@Dox microgels 3, Dox, CaA microgels, and PBS under Dox concentration of 20 μg/l for 24 h. The live cells were dyed by FDA (Ex/Em = 494/520 nm), whereas the dead cells were dyed by PI (488/630 nm). Scale bar was 50 μm. (g) Live cell rate calculated by fluorescence by fluorescence paragraph. Data are shown as mean ± SEM, $n = 3$, $^{*}p < 0.05$, $^{**}p < 0.01$, $^{***}p < 0.001$. (h) Cell survival test of CaA@Dox microgels 3 for 24, 48, and 72 h on MCF-7. Data are shown as mean ± SEM, $n = 3$, $^{*}p < 0.05$, 3, $^{**}p < 0.01$, $^{***}p < 0.001$. Source: Cai et al. [5]. https://www.ncbi.nlm.nih.gov/pmc/articles/PMC6835883/ CC-BY 4.0.

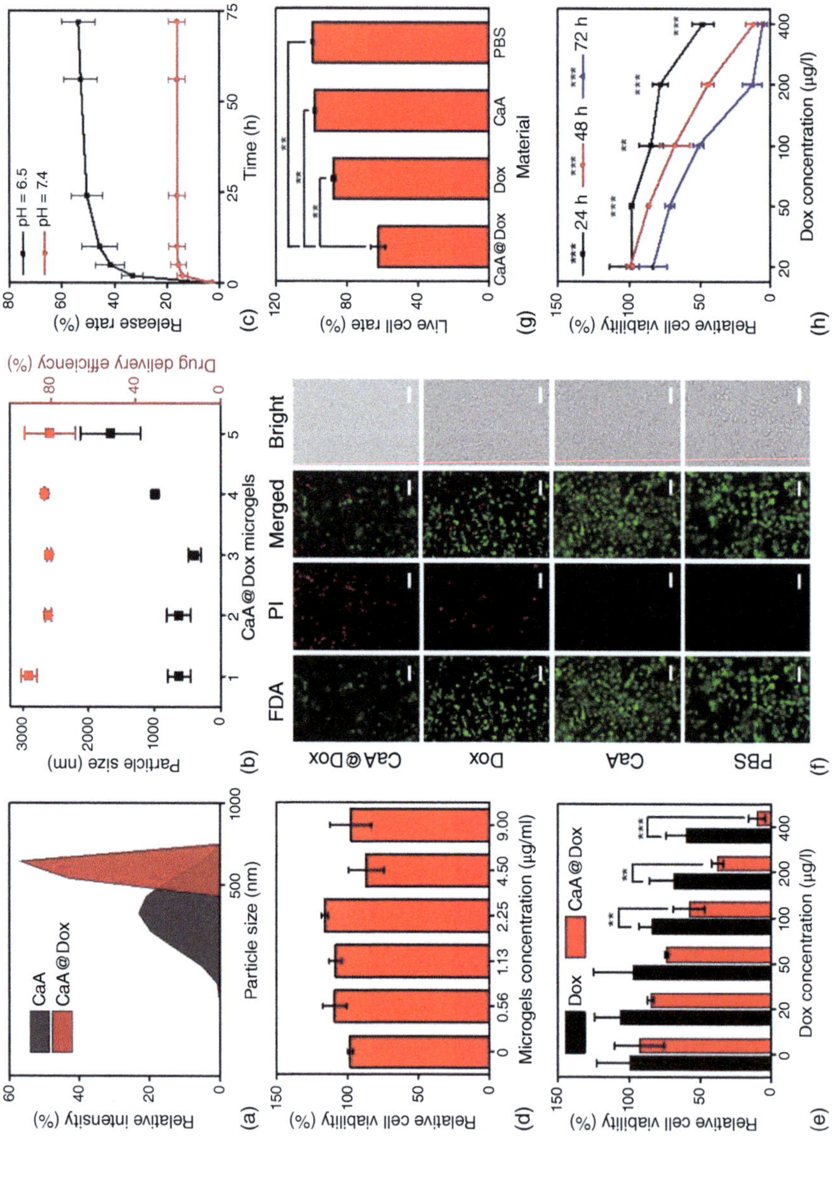
(a)
Relative intensity (%)
CaA
CaA@Dox
Particle size (nm)
(b)
Particle size (nm)
CaA@Dox microgels
Drug delivery efficiency (%)
(c)
Release rate (%)
pH = 6.5
pH = 7.4
Time (h)
(d)
Relative cell viability (%)
Microgels concentration (μg/ml)
(e)
Relative cell viability (%)
Dox
CaA@Dox
Dox concentration (μg/l)
(f)
FDA
PI
Merged
Bright
CaA@Dox
Dox
CaA
PBS
(g)
Live cell rate (%)
Material
(h)
Relative cell viability (%)
24 h
48 h
72 h
Dox concentration (μg/l)

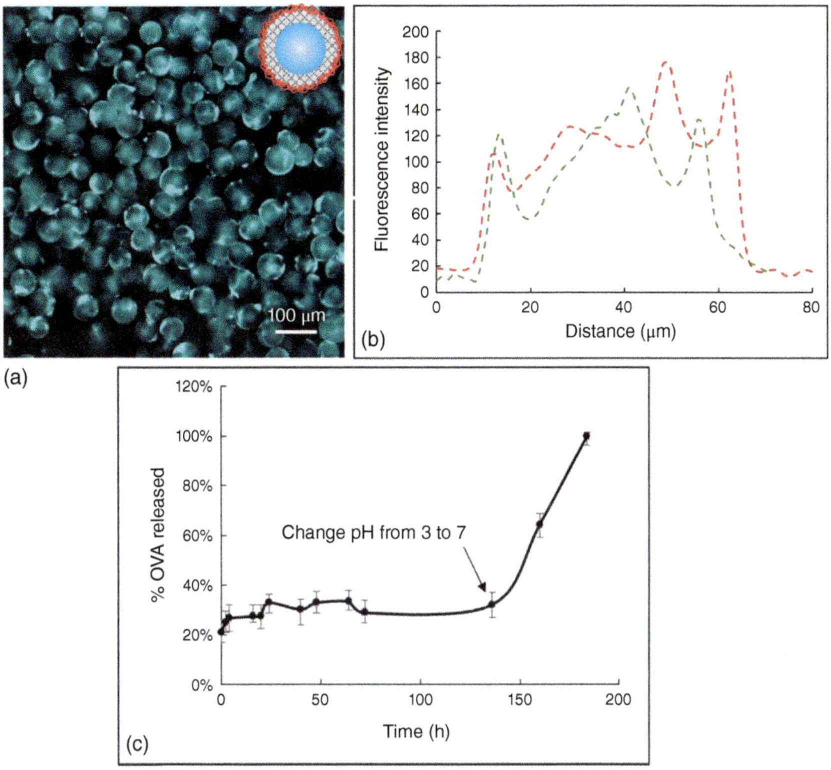

Figure 9.7 OVA encapsulated core–shell alginate microparticles coated with chitosan. (a) Confocal image of fluorescent OVA encapsulated alginate microparticles. (b) Fluorescent intensity distribution along the diameters of two different alginate microparticles (shown in green and red dot lines). (c) pH triggered release of OVA from the microparticles. (For interpretation of the references to color in this figure legend, the reader is referred to the web version of this article.) Source: (a) Yu et al. [6]. Reproduced with permission of Elsevier Inc, (b, c) Yu et al. [6].

9.5 Cell Amplification

He et al. established an "all-in-one" microgel system for human adipose-derived mesenchymal stem cells (hADSCs) based on DLP technology to achieve 3D cell culture with promoting cell amplification efficiency, facilitated workflow, and building of macrotissues with even cell distribution and specific function [7]. Due to the short exposure duration of each layer, a large number of each printing, and convenient subsequent process, thousands of microgels could be printed in one step (Figure 9.8). The property of the printed GelMA microgels for hADSC amplification was analyzed comparing to Cytodex-1microsphere, which is a commonly used commercial structure for cell expansion. From the viability testing results, the viability of cells on the printed GelMA microgels was much higher than the one on Cytodex-1 microspheres in one-week culturing. In terms of the attaching efficiency and proliferation speed, almost all the cells get attached to the printed GelMA microgels (106.20 ± 10.62%),

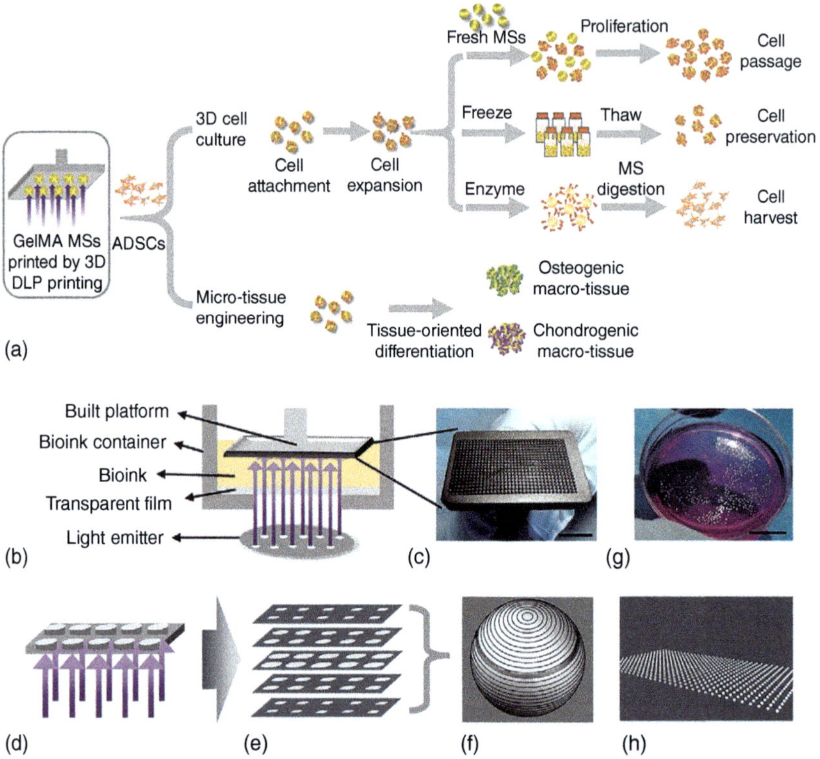

Figure 9.8 Summary of this study and schematic diagram of 3D DLP technology. (a) GelMA MSs are fabricated by 3D DLP technology, and used for hADSCs culture and macrotissues construction. (b) Schematic diagram of the self-made 3D DLP printer. (c) Images of the printed MSs placed on a built platform. (d)–(f) Schematic diagram of layer-by-layer printing. (g) Images of the printed MSs in a petri dish. (h) The CAD file of MSs for printing. Source: (a, b, d–f) He et al. [7], (c, g) He et al. [7]. Reproduced with permission of WILEY-VCH Verlag GmbH & Co. KGaA, Weinheim.

whereas only $83.85 \pm 13.94\%$ cells get attached to Cytodex-1 microspheres after seeding for 24 hours. Additionally, the metastasis capability, spreading morphology, and functionalization degree on the printed GelMA microgels were all superior to the ones on the Cytodex-1 microspheres, which fully confirmed the superiority of the former structures.

In the cell passage process, unlike the conventional digestion method in 2D culturing, which is time and financially consuming, cells growing on the 3D surface of GelMA microgels did not need to be digested (Figure 9.9). To realize the further amplification of cells, a researcher could directly add more microgels without cells to provide more area for cell proliferation because the cells could transfer among microgels. To harvest the cells, GelMA microgels with a diameter of 400 μm were tested to be digested by collagenase within five minutes utilizing good biodegradability of GelMA. Moreover, in terms of the preservation and transportation of amplified cells, the printed GelMA microgels with cells could be frozen thawed for

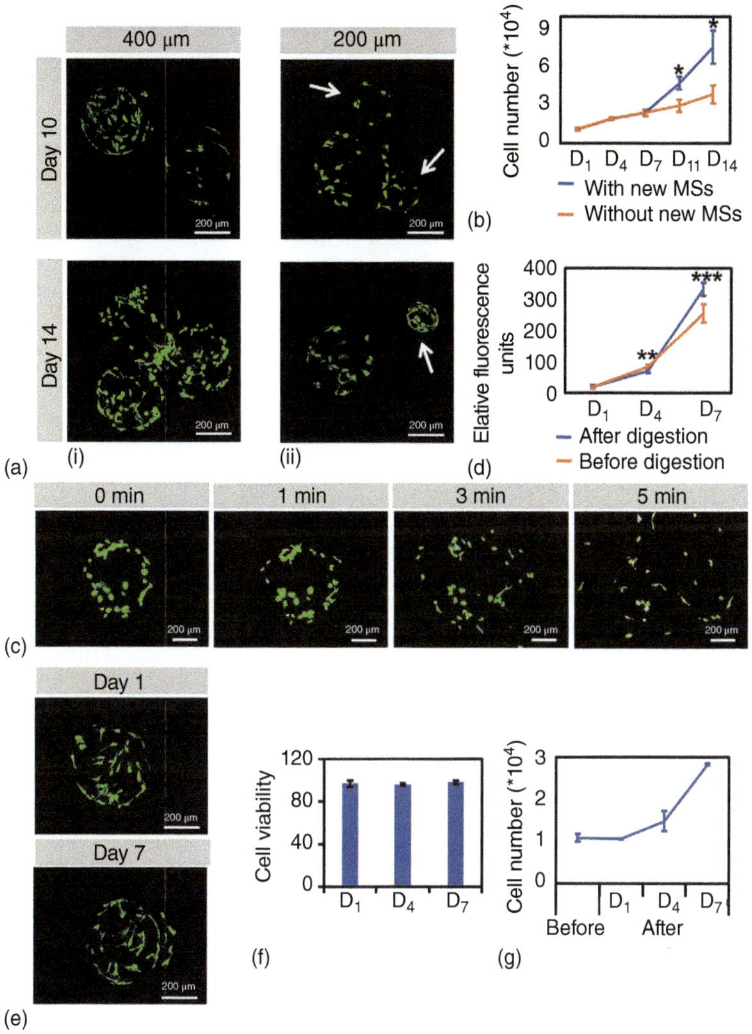

Figure 9.9 The printed GelMA MSs for cell passage, cell harvest, and cell cryopreservation. (a) and (b) Confocal fluorescence images of live/dead staining and cell number of cells after adding fresh MSs. The newly added MSs were marked with white arrow. (c) Fluorescence images of live/dead staining of cells during MS digestion. (d) Prestoblue testing of cells before and after digestion. (e)–(g) Confocal fluorescence images of live/dead staining, cell viability, and the cell number of cells after freezing and thawing process. $^{*}p < 0.5$, $^{**}p < 0.01$, $^{***}p < 0.001$, $N = 3$, $n = 6$. Source: He et al. [7]. Reproduced with permission of WILEY-VCH Verlag GmbH & Co. KGaA, Weinheim.

further cell culture. The cells undergoing the freezing and thawing process still kept high viability ($97.26 \pm 0.74\%$), and the proliferation speed still remained high.

Besides, in Zhang et al.'s previous research work, the mass production of fluorescent chitosan (CS)/graphene oxide (GO) hybrid microgels for the amplification of human umbilical cord mesenchymal stem cells (HUMSCs) was also introduced [8].

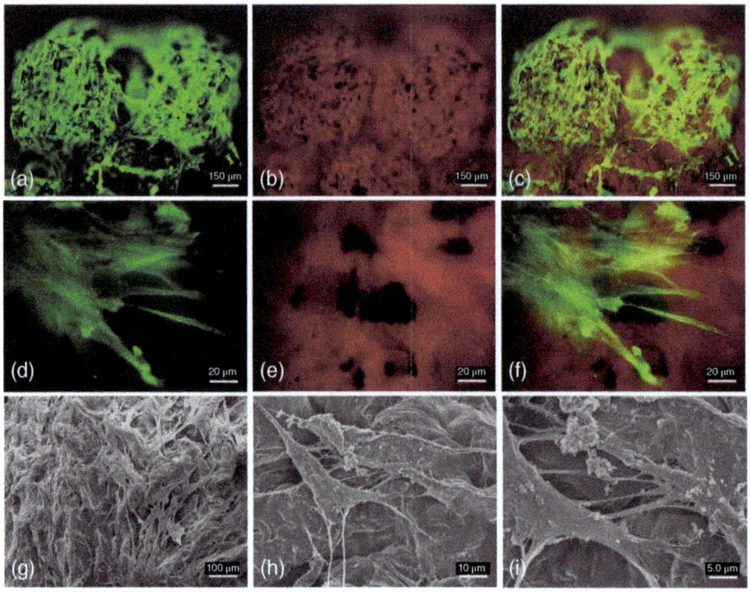

Figure 9.10 (a)–(c) The fluorescent images of the overall HUMSC-seeded GCS/GO microspheres after seven days' culture. (d)–(f) The high magnification images of HUMSC seeded on GCS/GO microspheres. (g)–(i) SEM images of the HUMSC-seeded GCS/GO microspheres after seven days' culture. Source: Zhang et al. [8]. Reproduced with permission of Elsevier B.V.

The CS/GO microgels were prepared by directly dripping the hybrid solution with CS and GO with a syringe needle with 360-μm inner diameter into NaOH aqueous solution. After resting overnight, the microgels were washed to pH-neutral immersed in genipin aqueous solution for 12 hours. After being washed, they were dried in an oven at 60 °C for 48 hours. The prepared microgels were placed in a bacteriological well plate, where cells are difficult to attach, and presoaked with α-modified minimum essential media (α-MEM). HUMSCs suspension was dropped onto the microgel surface and then incubated for 20 minutes for cell adhesion and then more medium was added (Figure 9.10). As shown by cell attachment and proliferation results, the seeded cells displayed excellent morphology and independent locations of the cells were observed, which verifies the promising biocompatibility of the hybrid microgels. Furthermore, marker proteins, namely CD29 and CD44 normally expressed, confirm the functionalization of HUMSCs on the microgel system. In summary, the proposed strategy to produce hybrid microgels with a high yield could achieve the amplification of stem cells based on a 3D growing environment in vitro, which was potentially to be introduced into industrial production and utilized in more biomedical applications (Figure 9.11).

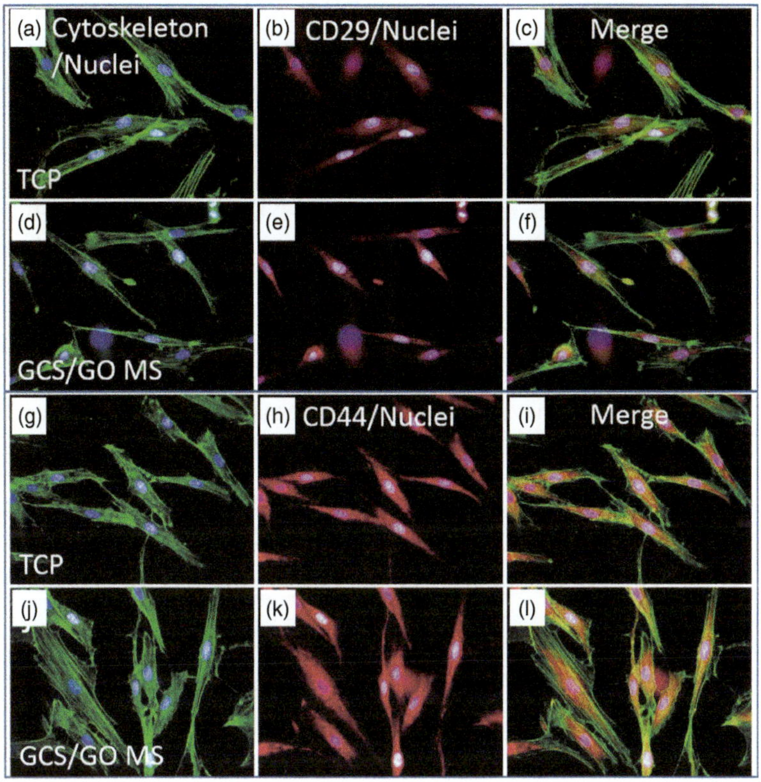

Figure 9.11 HUMSCs cytoskeleton and stromal markers immunofluorescence staining after cultured for seven days. (a)–(c) The fluorescent images of cytoskeleton staining and CD29 staining of HUMSC digested from GCS/GO microspheres. (d)–(f) The fluorescent images of cytoskeleton staining and CD29 staining of HUMSC cultured on TCP. (g)–(i) The fluorescent images of cytoskeleton staining and CD44 staining of HUMSC digested from GCS/GO microspheres. (j)–(l) The fluorescent images of cytoskeleton staining and CD44 staining of HUMSC cultured on TCP. Source: Zhang et al. [8]. Reproduced with permission of Elsevier B.V.

9.6 Single-Cell Capture

Circulating tumor cells (CTCs) have been confirmed to be important to build metastasis, which can provide predictive and prognostic information in cancer diagnosis and precision medicine. However, the extraordinarily low density of CTCs in patients' blood makes it hard to examine. The previous research work discovered that CD44 was always overexpressed in lots of cancer cells, which is the main site of HA binding on cell surface for tumor migration. Therefore, CD44 has become a significant biomarker for early cancer examination and drug therapy. Based on these, Li et al. published a contribution to propose a novel strategy to capture CTCs that overexpressed CD44 by a dopamine-functionalized hyaluronic acid (HA-DA) microgel with the aid of microfluidic technology [9]. The diameter of HA-DA microgels was around 45 μm, effectively enlarging the cell size and differentiating the CTCs from white blood cells (15–20 μm, for monocytes, 10–15 μm, for granulocytes and 6–15 μm, for lymphocytes). Furthermore, a microfluidic filter with double layers (DLMF) containing many slit openings (15 μm in height) was manufactured for dynamic cell isolation, with which the HA-DA microgels encapsulating cancer cells were intercepted and separated from blood cells (Figure 9.12). As a result, the capture efficiency of HA-$DA_{2.0}$ microgels for CD44+HeLa and HepG2 cells were 85.94 ± 1.98% and 88.73 ± 3.66%, respectively. However, the percentage for CD44-NIH-3T3 cells was just 5.85 ± 0.96%. Most importantly, the method displayed promising capture efficiency when the cell density was extremely low (10 cells/ml), demonstrating that this approach could be potentially used in cancer patients for prognosis monitoring and treatment in the future.

Nan et al. also developed a microfluidic system for one-chip harvesting of single-cell-laden microgels in an aqueous medium [10]. The microfluidic channel combined with an optical setup to classify the microgels encapsulating single cells

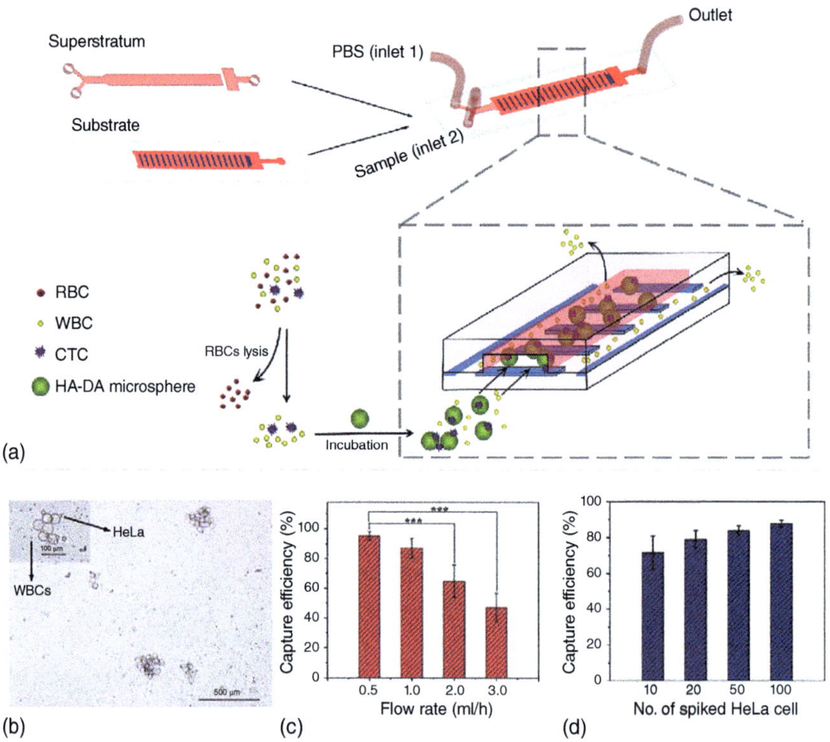

Figure 9.12 (a) Schematic diagram of the DLMF for the separation of CTCs. The pink area on the pattern represented the outflow path of the sample. (b) Bright-field image of HeLa cell attached to HA-DA 2.0 microspheres in RBC-lysed blood. The capture efficiency of HeLa cells in the DLMF device at (c) different flow rates and (d) different numbers of spiked CTCs samples. Error bars represent the standard deviations ($n \geq 3$). ***$p < 0.001$. Source: (a, c, d) Li et al. [9], (b) Li et al. [9]. Reproduced with permission of Elsevier B.V.

into aqueous flow. After the real-time UV irradiation, cross-linked GelMA microgels were guided into a sorting chamber to collect the ones encapsulating cells into the medium. When microgel flowed to the examination area, the fluorescent intensity would be immediately detected by an optical setup. If a single cell was captured inside the microgel, the fluorescent signal was triggered, which produced an output pulse to the sorting electrode and guide the microgels with single cells into the medium.

With this system, researchers realized the capturing and collection of two types of cells. All the working steps were carried out without complicated operation, which were possible to cause the cell damage. This system could be explored as a meaningful tool for single-cell analysis, including immune responses, monitoring of cell differentiation, and revealing the role of cell-to-cell variations, and had the potential to be applied as building blocks for bottom-up tissue establishment and antibody production.

9.7 Supporting Matrices

Hinton et al. reported the development of a 3D bioprinting method named freeform reversible embedding of suspended hydrogels (FRESH) [11]. In this strategy, thermoreversible microgels were applied as supporting matrices to support the deposition of hydrogels on complex structures. In this work, the support bath was constituted with gelatin microgels, which were like a Bingham plastic in the print process. The bioink was extruded from the bioprinting nozzle inserting into the supporting bath and moved controlled by 3D bioprinter and upper computer. After the designed 3D structure was successfully printed, the supporting microgels could be easily removed by heating the printing system in 37 °C. The temperature is raised to a cell-friendly 37 °C, causing the gelatin support bath to melt in a nondestructive manner. A series of complex 3D structures were printed by the FRESH method. To create complex external surface structures, researchers utilized an MRI image of the human brain and embryonic heart and formed 3D models in detail. Both the embryonic heart and brain 3D structures were perfectly printed, indicating the excellent performance of the FRESH method to rebuild complex internal and external structures.

Similarly, in Jeon's research work, the strategy of individual cell-only bioink and photocross-linkable supporting bath for 3D printing structures with complex geometries was proposed [12]. Human bone marrow-derived mesenchymal stem cells were directly printed inside the supporting bath composed of alginate microgels as Bingham plastic fluids, which were generated by ionic cross-linking of dual-cross-linkable, oxidized and methacrylated alginate (OMA) (Figure 9.13). With the decrease in the microgel size, it could be realized to print tissue constructs with complicated structures because of the shear-thinning profile of supporting bath with the movement of the nozzle and the self-healing profile when the external strain was withdrawal. The cells undergoing the printing process displayed high viability after the treatment of UV irradiation. Different from the other bioprinting methods based on external solid materials for keeping structures or auxiliary procedures to prepare cell aggregates, the photocross-linked OMA microgels supporting

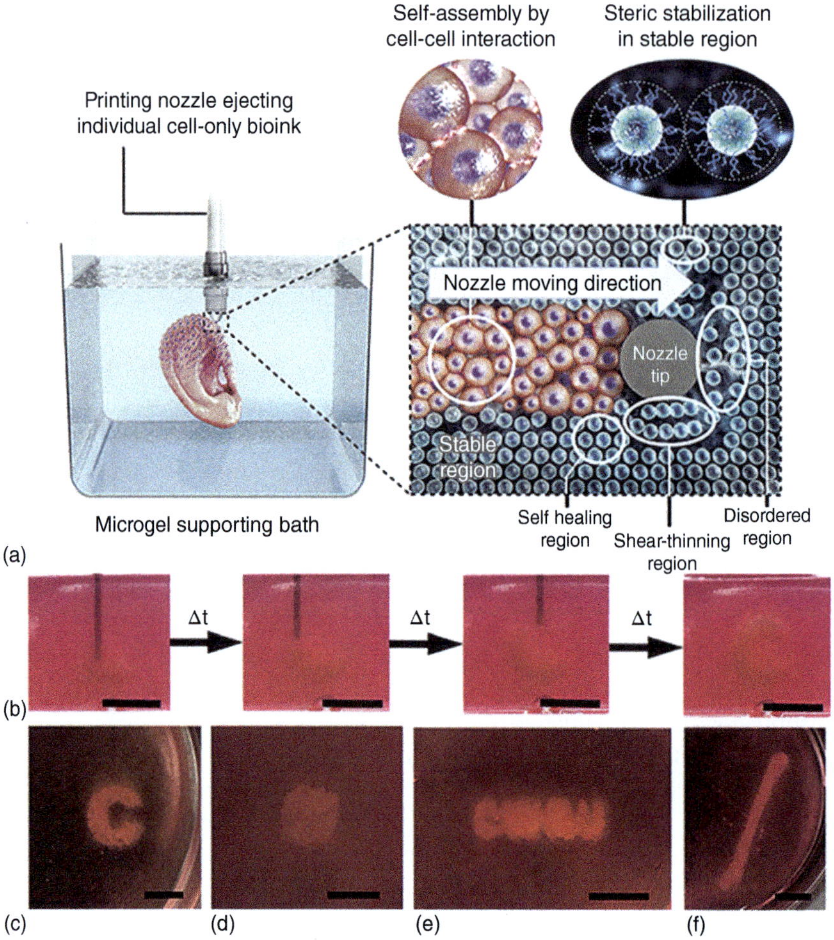

Figure 9.13 Shear-thinning and self-healing alginate microgel supporting medium for 3D bioprinting of living individual stem cells. (a) A schematic of 3D printing of cells within the alginate microgel supporting medium. OMA microgels in the supporting medium fluidize via their shear-thinning properties when stress is applied by the motion of the printing needle and cell-only bioink (shear-thinning region) and rapidly fill in after the needle passes by self-healing properties (self-healing region) without creating crevasses. Microgel supporting medium without shear stress presents solid-like properties, which provide mechanical stability for the printed cell construct (stable region). (b) Captured images at different times during bioprinting of the letter "C" using living stem cell-only bioink into the alginate microgel supporting medium. As the printing progresses, cells are arranged into the letter "C" shape in 3D without disturbing previously printed regions, which is achieved as a result of the shear-thinning and self-healing properties of the alginate microgel supporting medium. Images of the 3D bioprinted structures of (c) a letter "C," (d) a cube, (e) letters comprising the acronym "CWRU," and (f) a femur in alginate microgel supporting medium. Scale bars indicate 5 mm. Source: Jeon et al. [12]. Reproduced with permission of The Royal Society of Chemistry.

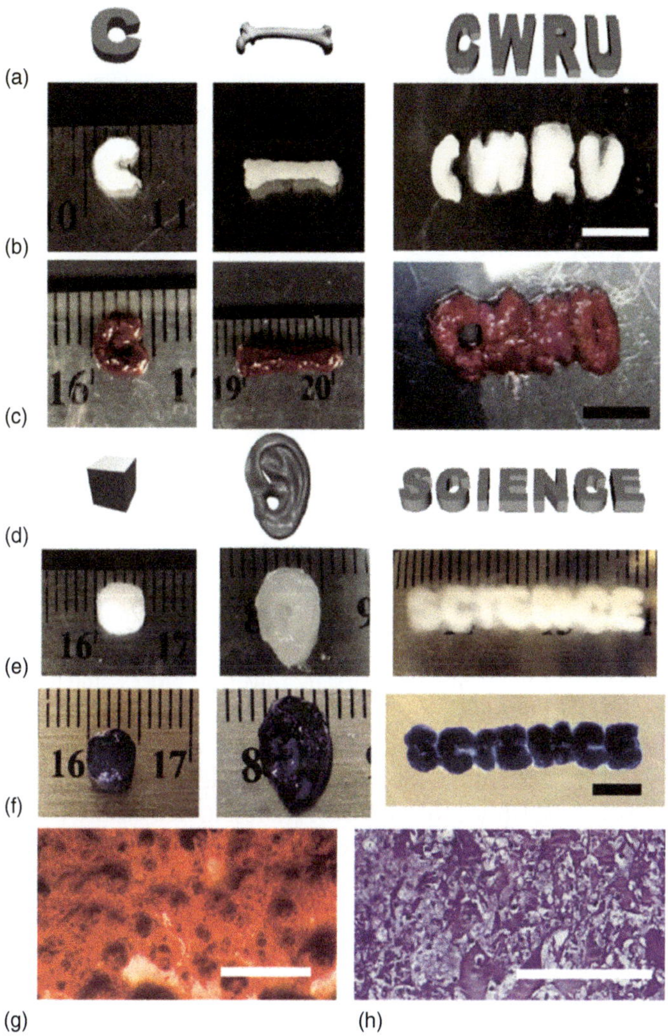

Figure 9.14 Differentiation of 3D bioprinted hMSC constructs. (a) Digital images and photographs of osteogenically differentiated 3D printed individual hMSC-only bioink construct morphology (b) before and (c) after Alizarin red S staining. Scale bars indicate 5 mm. (d) Digital images and photographs of chondrogenically differentiated 3D printed hMSC construct morphology (e) before and (f) after Toluidine blue O staining. The constructs presented well-preserved structures after long-term four weeks culture without evidence of construct deformation due to cellular contraction or proliferation, and generation of specific tissue types (i.e. bone and cartilage) with desired geometries. Scale bar indicates 5 mm. Photomicrographs of (g) Alizarin Red S and (h) Toluidine blue O stained construct sections. The images demonstrate hMSC differentiation and deposition of lineage-specific ECM in the cell-only bioink printed constructs. Scale bars indicate 200 mm. Source: Jeon et al. [12]. Reproduced with permission of The Royal Society of Chemistry.

bath could realize ECM applying and long-term culture. After the printed tissue got mature, the OMA supporting bath could be removed by simple agitation or degradation so that the printed structures could be directly harvested (Figure 9.14). This novel method has the potential to be widely used in the field of regenerative medicine.

9.8 Secondary Bioprinting

The combination of multiscale micro- and macroenvironments has been described as extremely significant in rebuilding biomimetic scaffolds in vitro for peripheral nerve tissue regeneration although it remained to be a great challenge so far. To achieve this goal, Chen et al. successfully applied GelMA/chitosan microgels as bioprinting units to print multiscale scaffolds, which contained multifunctional GelMA/chitosan microgels and GelMA shell to mimic the 3D extracellular matrix of the native nerve tissue [13]. GelMA/chitosan microgels were generated with the method of microfluidic technology (Figure 9.15). Neurotrophins, especially nerve growth factor (NGF), is a significant component to promote neurite outgrowth and extension during peripheral nerve regeneration, which is also commonly applied to induce the growth and neuron-like differentiation of PC12 cells in vitro. In this work, researchers loaded NGF in the microgels first by a swelling process and tested the controlled release behavior and the corresponding influence on neuron-like differentiation of PC12 cells. The qualitative and quantitative examinations of the controlled release of NGF from the microgels were carried out by analyzing the fluorescent intensity of the labeled NGF release. As a result, NGF loaded on the microgels sustained release from the inside to the outside of the hydrogel network obviously. Furthermore, PC12 cells were seeded on the microgels to analyze the effects of microgels and microgel-loaded NGF on the morphology and physiological function of nerve cells. The experimental results showed that the cells on both types of microgels kept an excellent growing state. The neurite growth was detected by SEM after three-day culturing. PC12 cells on microgels without NGF displayed the polygon outline, whereas the ones on microgel-loaded NGF were observed with clear neurite growth. The axon length of PC12 cells on microgels with NGF was 14.51 ± 6.86 mm, and the percent of PC12 cells with axon was performed $40 \pm 14.3\%$. These results indicated that microgel-loaded NGF could promote the axon outgrowth and elongation of PC12 cells.

After the basic cell experiments, the multifunctional modular bioink was made with the microgels and GelMA prepolymer solution for the following bottom-to-up 3D bioprinting (Figure 9.16). Based on extrusion bioprinting method, the bioink was extruded via printing nozzle and establish a 3D hybrid scaffold. A lower microgel density was applied in order to provide enough space for the RSC96 cells in the GelMA shell of the scaffolds. In this work, a scaffold with four layers was printed. Because of the ability of RSC96 cells to continuously release various neurotrophins, including NGF, research results showed that the coculture of RSC96 cells and PC12 cells owned an obvious effect on enhancing neurite growth and elongation of PC12

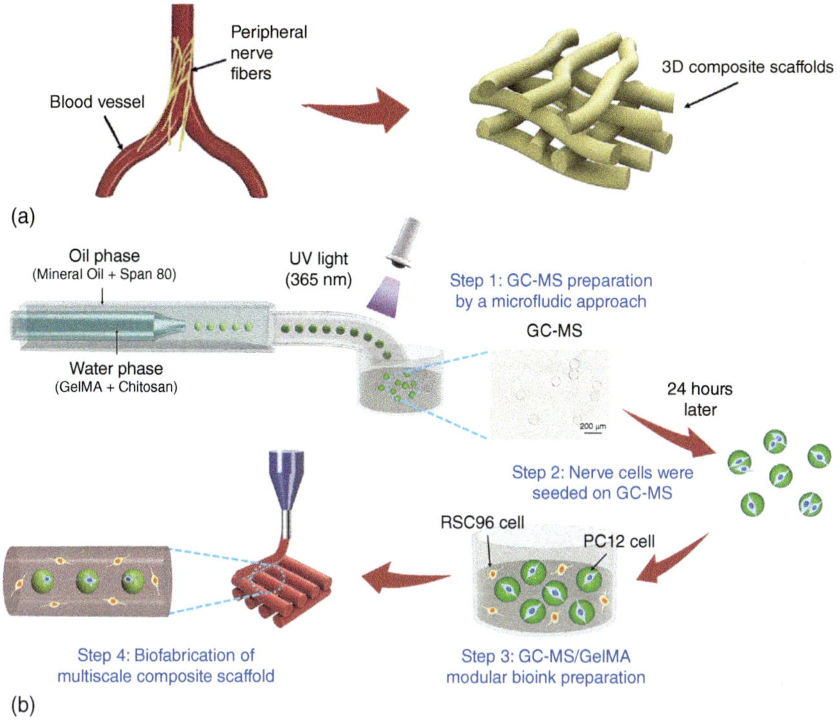

Figure 9.15 The design and preparation of a multiscale composite scaffold mimic the native structure of the peripheral nerve tissue. (a) The native peripheral nerve tissue network shows the 3D hierarchical structure, and the 3D composite scaffolds would be beneficial for nerve tissue regeneration. (b) The scheme shows the multiscale composite scaffold preparation and neuronal outgrowth within 3D composite scaffolds: GC–MS preparation by a microfluidic approach (Step 1), nerve cells seeded on GC–MS (Step 2), GC–MS/GelMA modular bioink preparation (Step 3), and biofabrication of 3D composite scaffold performed by extruding bioink with the 3D printer (Step 4). Source: Chen et al. [13]. Reproduced with permission of Elsevier Inc.

cells, which indicated that the proposed 3D bioprinting method using microgels would have a great potential application in the field of peripheral nerve tissue regeneration.

In 2018, Highley et al. developed the strategy where the ink was composed of jammed microgels, which were generated by microfluidic technology [14]. Unlike the low-density microgels in Chen et al.'s work above, the main body of the bioink for extrusion printing was microgels rather than a hydrogel prepolymer solution. This designation could simultaneously meet the requirements of rheological profiles for successfully printing and appropriate mechanical properties for cell growth. Microgels in the bioink system were densely packed and restricted by each other. When the applied force was lower, the bioink system would display a solid feature. The volume of microgels could elastically deform below the yield stress without motion. With the increase in the additive force, the microgels tended to produce relative motion to each other. However, if the added force was repealed,

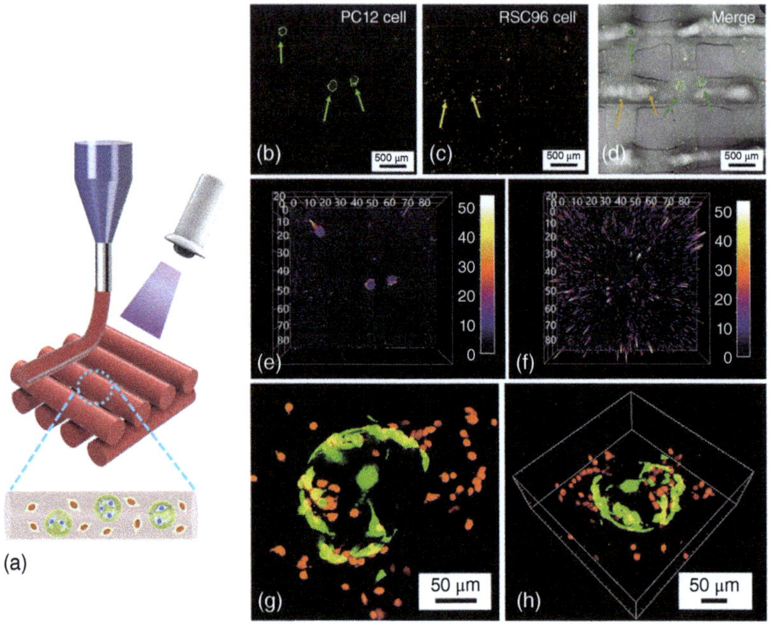

Figure 9.16 The bioprinting of the multiscale GC–MS+NGF/GelMA composite scaffold within PC12 cells and RSC96 cells. (a) The schematic representation showed the assembly of the encapsulated GC–MSC+NGF inside the bioprinted 3D GelMA structure. (b)–(d) The 2D views of confocal fluorescent microscopy images showed the cross-sectional view of a two-layer scaffold at one day, indicating the formation of a GC–MSC+NGF/GelMA scaffold, and PC12 cells and RSC96 cells were stained with CellTracker green (b) and CellTracker orange (c), respectively. Scale: 500 μm. The gray-level analyses of surface plot 3D images transformed from Figure 9.8b,c shows the location of PC12 cells (e) and RSC96 cells (f) (g, h). The top view and 3D view of confocal fluorescent microscopy images of a single GC–MS+NGF seeded with PC12 cells in the GC–MS+NGF/GelMA scaffold encapsulated with RSC96 cells, where PC12 cells were stained with CellTracker green and RSC96 cells were stained with CellTracker orange, respectively. Scale: 50 μm. (For interpretation of the references to color in this figure legend, the reader is referred to the web version of this article.) Source: Chen et al. [13]. Reproduced with permission of Elsevier Inc.

the bioink system would recover to the initial state as a solid. Based on this promising property, the jammed microgels performed outstanding printability during the extrusion bioprinting process. To test the jammed microgels' behavior in practical printing, researchers generated the microgels bioink composed of norbornene-modified hyaluronic acid (NorHA), poly(ethylene glycol)diacrylate, and agarose. The aqueous medium among microgels was removed via centrifugation over a filter or via vacuum filtration to form a jammed bioink with extremely high density of microgels. The rheological profiles of jammed NorHA were tested. The bioink showed obvious shear-thinning behavior. Additionally, when the jammed microgels were at a lower strain, the bioink tended to display the elastic feature, while rapidly transited to a liquid-like viscous state at a higher strain due to disrupted contacts among microgels, which was a suitable state for extruding printing.

References

1 Pradhan, S., Clary, J.M., Seliktar, D., and Lipke, E.A. (2017). *Biomaterials* 115: 141.

2 Xie, M., Gao, Q., Fu, J. et al. (2020). *Bio-Design and Manufacturing* 3: 175.

3 Zhao, X., Liu, S., Yildirimer, L. et al. (2016). *Advanced Functional Materials* 26: 2809.

4 Shen, J., Ji, Y., Xie, M. et al. (2020). *Materials Science & Engineering. C, Materials for Biological Applications* 112: 110896.

5 Cai, S., Shi, H., Li, G. et al. (2019). *Nanomaterials (Basel)*: 9.

6 Yu, L., Sun, Q., Hui, Y. et al. (2019). *Journal of Colloid and Interface Science* 539: 497.

7 He, Q., Liao, Y., Zhang, J. et al. (2020). *Small* 16: e1906539.

8 Zhang, S., Ma, B., Wang, S. et al. (2018). *Chemical Engineering Journal* 331: 675.

9 Li, X., Cui, T., Zhang, W. et al. (2020). *Colloids and Surfaces. B, Biointerfaces* 196: 111281.

10 Nan, L., Yang, Z., Lyu, H. et al. (2019). *Advanced Biosystems* 3: e1900076.

11 Hinton, T.J., Jallerat, Q., Palchesko, R.N. et al. (2015). *Science Advances* 1: e1500758.

12 Jeon, O., Lee, Y.B., Jeong, H. et al. (2019). *Materials Horizons* 6: 1625.

13 Chen, J., Huang, D., Wang, L. et al. (2020). *Journal of Colloid and Interface Science* 574: 162.

14 Highley, C.B., Song, K.H., Daly, A.C., and Burdick, J.A. (2019). *Advanced Science (Weinheim)* 6: 1801076.

10

Microfiber-Based Organoids Bioprinting for In Vitro Model

10.1 Introduction

10.1.1 Significance and Challenge

Microfiber-based organoids (MFOs) attract increasing attention because the basic structures of the organisms contain various types of fiber-shaped three-dimensional (3D) cellular constructs such as muscle fibers, blood vessels, and neural pathways [1, 2]. Hereby, as building blocks, MFOs can directly construct fibrous tissue, such as muscle fibers, blood and lymph vessels, nerve networks, ligaments, and tendons, which are essential for functional tissue/organ system. Specifically, constructing a blood vessel network in 3D tissues in vitro is a tough challenge for achieving the long-term culture of engineered macroscopic tissues [3–6]. Vascular organoids as building blocks could greatly assist in overcoming the bottleneck in forming 3D tissues with vascular networks. Coincidentally, as a building block, the characteristic advantages of MFOs are easy to prepare, handle, and assemble to meet engineering and application scenarios. These characteristics enable rapid assembly of MFOs to form higher-order macroscopic tissue construction through 3D assembly technologies, such as 3D bioweaving and 3D bioprinting. Additionally, another fascinating function of MFOs is transplantation therapy. As an advantage of its thinness, the MFO for implantation can remain in the implanted site; as an advantage of its continuity, it can easily be identified during the whole treatment process and removed at any time.

Due to its biomimetic fibrous geometry and ease for nutrients/oxygen diffusion, cell-laden hydrogel microfiber (CLHM, approximately 10–1000 μm in diameter) is considered as the most attractive medium to build MFOs in vitro [7–10]. Yet, to fabricate MFOs that are close to real tissues, CLHMs need to mimic the biological, morphological, and mechanical properties of the natural tissues. To meet these requirements, CLHMs must be constructed of hydrogel that (i) is easily processed into microfibers; (ii) offers a cell-favorable environment for cell growth; (iii) offers a tunable mechanical strength for functionalization of different cells. Thus, to fabricate MFOs, it is necessary to find or develop a suitable hydrogel that balances processability and biological functionality for engineering bioactive CLHMs.

Cell Assembly with 3D Bioprinting, First Edition. Yong He, Qing Gao, and Yifei Jin.

10.1.2 Hydrogel Materials

As the basic and the most important composition of CLHMs, hydrogels used to play a decisive role in building functional microenvironments for different cells. Hydrogels can be composed of natural biomaterials or synthetic biomaterials alone, or a combination of both to form mixed hydrogels. For natural biomaterials, such as collagen, although they have superior biological properties, it is difficult to directly manufacture microfiber structures due to their difficulty in gelation and poor mechanical properties of gelation [11–45]. To meet the requirements of fiber-forming, synthetic biomaterials were mixed with natural biomaterials to obtain the prescriptions of mixed hydrogels with compromised forming and biological properties. For mixed materials, alginate is the most commonly used basic composition due to its fast ionic cross-linking and strong mechanical properties [46–53]. However, because alginate components could not provide cell adhesion sites, the biological properties of the whole mixed system were not conducive to MFOs formation. So far, due to the challenge of balancing the processability and biological functionality, hydrogels for the construction of bioactive CLHMs are limited, despite great developments in hydrogel design. With the continuous development of biomaterials, artificial modification of protein with superior processing and biological properties may be the most promising biomaterials for manufacturing bioactive CLHMs. Therein, chemical functionalization of proteins, such as remarkable photo-cross-linking modification, is proposed to improve the usability of natural biomaterials. Gelatin methacrylate (GelMA) hydrogel, as a photocurable hydrogel representative, is synthesized by adding methacrylate groups to the amine-containing side-groups to enable to have the common advantages of both natural and synthetic biomaterials [54–65]. The biological and mechanical properties of GelMA can be precisely and purposefully adjusted by adjusting the degree of substitution of amino. Additionally, GelMA hydrogel will not evoke noticeable host inflammatory responses due to its low immunogenicity. Accordingly, GelMA hydrogels have great potential as the material candidate for manufacturing MFOs. However, due to its generally low initial viscosity and easy to block nozzle after curing, it is difficult to directly manufacture GelMA microfibers.

10.2 Coaxial Bioprinting of Bioactive Cell-laden Microfiber

To fabricate MFOs, various CLHM methods have been developed, such as wetspinning [9], microfluidic spinning, [2, 66–77] and coaxial bioprinting [78, 79]. Each of these methods is suitable for different hydrogel prescriptions. For wetspinning and microfluidic spinning, by virtue of its fast ionic cross-linking properties, an alginate-based hydrogel is the most common prescription. Unfortunately, these alginate-based cell-laden microfibers are incapable of providing a microenvironment conducive to cell growth similar to that in the body, and thus unable to construct MFOs similar in morphologies and functions to that in vivo. As a result, other hydrogels, such as collagen, fibrin, and GelMA, due to their better biological performance, gradually became material candidates for manufacturing CLHMs. As described earlier, for collagen and fibrin, these materials require a relatively long gelation time and remain mechanically fragile after gelation; for GelMA, limited by initial low viscosity and easy to block nozzle after curing, it is very difficult to directly fabricate GelMA microfibers. To overcome this limitation for engineering MFOs, a coaxial bioprinting strategy was developed to fabricate bioactive CLHMs, such as microfluidic coaxial bioprinting, coaxial nozzle-assisted bioprinting.

The basic principle of coaxial bioprinting is to use shell flow assisted core flow gelation and smooth continuous extrusion without clogging the nozzle. Coaxial bioprinting mainly depends on a coaxial device, which can construct coaxial laminar flow (core/shell). To be clear, the coaxial laminar core/shell flows in coaxial devices are affected by many parameters, including the composition of different fluid phases and their viscosities. Accordingly, to construct coaxial laminar flow, the shell flow generally chooses the material with adjustable viscosity and fast gelation, such as alginate. The core flow is the expected CLHM material. The outer protective sleeve of rapid gelation can maintain the structure of the inner hydrogel fiber and assist its further gelation. Alternatively, the non-cross-linked outer shell fluid assisted the in situ gelled inner fiber to be continuously extruded without blocking the nozzle.

10.2.1 Microfluidic Coaxial Bioprinting

Originally, for ECM proteins, such as collagen and fibrin, microfluidic coaxial bioprinting was developed to manufacture core–shell CLHMs and then form MFOs [1]. To apply microfluidic coaxial bioprinting, a coaxial laminar flow microfluidic device was well-designed to print core–shell microfiber, which was composed of inner core fiber of ECM proteins covered by a tubular outer layer of mechanically stable and rapidly gelating hydrogel (Ca-alginate). Specifically and as shown in Figure 10.1a, to fabricate core–shell hydrogel microfiber, three solutions were prepared: (i) pre-gel cell-laden ECM protein solution for the core flow, (ii) Na-alginate solution for the shell flow, and (iii) $CaCl_2$ solution for the sheath flow to rapidly cross-link Na-alginate solution. *Caution:* Before loading solutions into the coaxial microfluidic device, the device and all tubing were filled with 70% ethanol for one hour for sterilization. Additionally, it is important to note that careful handling is required or it may lead to faults in the manufacturing process, such as clogging. After sterilization, CLHMs are manufactured as follows: (i) Pump core and shell solutions into the corresponding device channels, and introduce saline to the sheath channel instead of the $CaCl_2$ solution to avoid blocking at the exit where $CaCl_2$ and alginate solution meet. (ii) After the stable formation of core–shell laminar flow, switch the saline to $CaCl_2$ solution. Subsequently, the core–shell CLHM could be continuously generated and collected in a saline bath. (iii) Finally, after forming the desired length of the CLHMs, switch the $CaCl_2$ solution to the saline again; after several seconds, stop all pumps and clean the microfluidic equipment for the next use. The diameters of the core/shell can be adjusted by changing the flow rates of the core/shell flows.

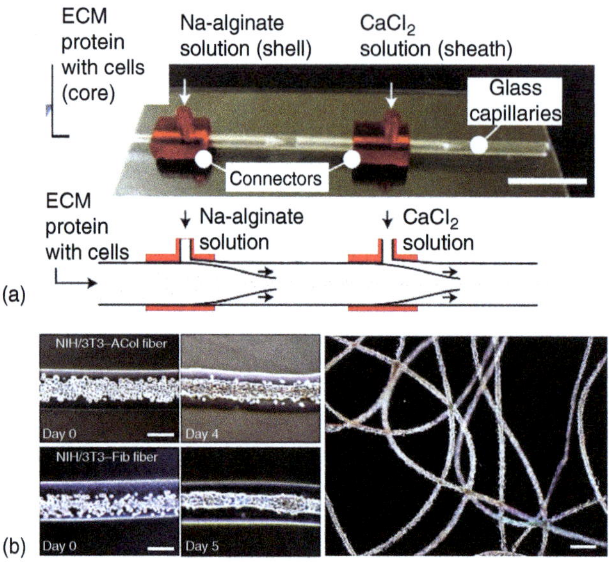

Figure 10.1 Coaxial laminar flow microfluidic device (a); core–shell cell-laden microfibers (b, left); microfiber-based organoids (MFOs, b, right). Source: Onoe et al. [1]. Reproduced with permission of Nature Publishing Group.

As shown in Figure 10.1b, left, the core of the CLHMs is cell-laden ECM proteins in the pre-gel state and is encapsulated by the shell of the mechanically stable Ca-alginate hydrogel. The shell could prevent the ECM proteins from diffusing away from the core until they completely gelate to form a cell-laden ECM proteins microfiber. After a period of culturing, the cells in this core can form a fiber-shaped cellular construct. The gelated ECM proteins in the core can regulate cell migration and proliferation inside the tubular Ca-alginate shell, then direct their morphological organizations and physiological functions (Figure 10.1b, left). Finally, the Ca-alginate shell can be selectively digested by alginate lyase and removed to complete the creation of the MFOs (Figure 10.1b, right).

10.2.2 Coaxial Nozzle-Assisted Bioprinting

Despite advances in microfluidic coaxial bioprinting technology, the sophisticated microfluidic devices inevitably bring inconvenience to operation and instability to the manufacturing process. And that, for nonviscous hydrogel materials, such as GelMA, due to the tubular shell layer cannot cover the inner flow resulting in escape, the microfluidic coaxial bioprinting technique does not work very well. Thus, to fabricate GelMA microfibers, our research group developed a coaxial nozzle-assisted bioprinting strategy (Figure 10.2a), and it is a simple fabrication method based on easy to use apparatus [78, 79].

The key of the coaxial nozzle-assisted bioprinting strategy is to rely on an easily made coaxial needle extrusion device (Figure 10.2b), in which a coaxial nozzle was connected to a transparent capillary for cross-linking GelMA and immersed

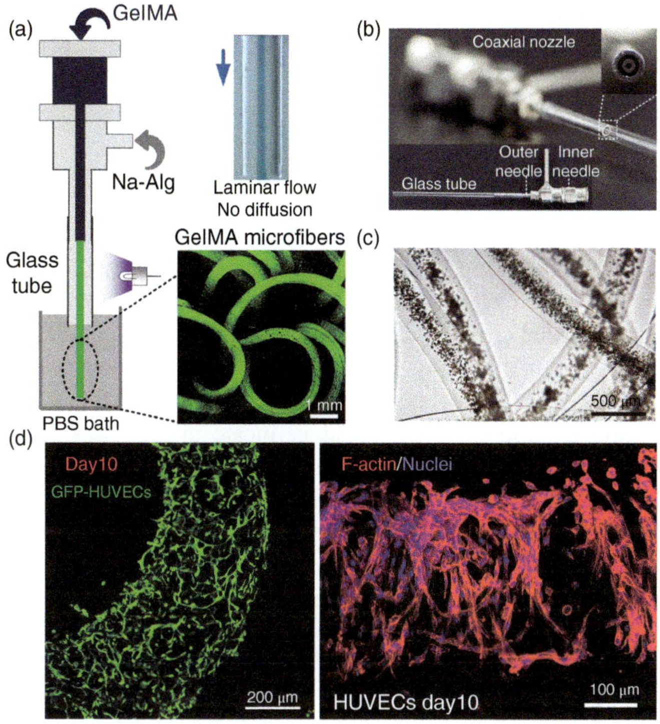

Figure 10.2 (a) Mechanism of coaxial nozzle-assisted bioprinting. Source: Shao et al. [79]. Reproduced with permission of WILEY-VCH.(b) Coaxial needle extrusion device. Source: Shao et al. [79]. Reproduced with permission of WILEY-VCH. (c) Cell-laden GelMA microfiber. Source: Shao et al. [79]. Reproduced with permission of WILEY-VCH. (d) Vascular microfiber-based organoids. Source: Shao et al. [79]. Reproduced with permission of WILEY-VCH.

in collection bath for collecting GelMA microfibers. Specifically, the nonviscous GelMA solution containing photoinitiator was pumped to the internal needle, while a viscous solution, such as sodium alginate (Na-Alg), was pumped to the external needle. As these flows had a low Reynolds number, they merge in the transparent capillary channel and form a core–shell laminar flow. Meanwhile, with the blue light penetrating through the transparent capillary to cross-link GelMA, the GelMA microfibers could be generated in situ (Figure 10.2c). To show the potential of making MFOs from these fibers, cell-laden GelMA microfibers were printed, and these encapsulated cells gradually stretched, migrated, and connected to form MFOs within a culture period (Figure 10.2d).

10.3 Heteromorphic/Heterogeneous Microfiber Bioprinting

Heteromorphic microfibers are desirable for tissue engineering applications because they allow mimicking the natural structures of organisms. Meanwhile, heterogeneous microfibers with tunable compositions are highly needed in tissue engineering applications, because they allow mimicking the multicomponent properties of natural organisms, such as multiple cell types and the extracellular matrix (ECM).

10.3.1 Heteromorphic Microfiber

Based on the coaxial nozzle-assisted bioprinting method, inspired by the "liquid rope-coil effect," a coaxial bioprinting method was developed for scalable fabrication of morphology-controllable GelMA microfibers. Based on the coflow rope-coil (CRC) effect and the sequential cross-linking strategy, straight, wavy, and helical GelMA microfibers could be achieved by adjusting the flow rates, as shown in the schematic Figure 10.3 [78]. In this method, a coaxial nozzle was connected with a transparent capillary and UV light could permeate to cross-link inner photo-cross-linkable hydrogel (GelMA, Q_{in}). Meanwhile, the coaxial nozzle was immersed in a calcium chloride ($CaCl_2$) bath to rapidly solidify outer sodium alginate (Na-Alg, Q_{out}) for directing inner GelMA microfibers morphology change. As Na-Alg and GelMA solutions had low Reynolds numbers, when the two solutions merged in the transparent capillary, a hierarchical coaxial laminar flow formed. When the coaxial laminar flow flowed through the transparent capillary, the GelMA

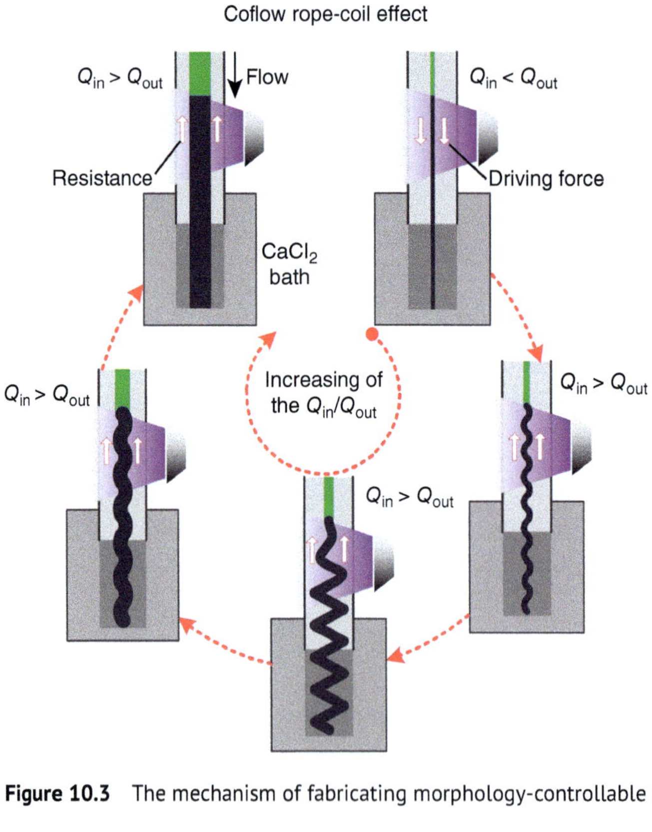

Figure 10.3 The mechanism of fabricating morphology-controllable GelMA microfibers.

could be UV-cross-linked in situ forming elastic GelMA microfibers and flowed with Na-Alg. Subsequently, at the end of the nozzle, when Na-Alg enters the calcium chloride bath, progressive ion cross-linking immediately occurs from the outside to the inside. When the velocity of GelMA was greater than that of Na-Alg, due to the resistance of the inner semi-cross-linked Na-Alg, the GelMA microfibers began to coil inside. Then, with more extensive cross-linking of Na-Alg, helical GelMA microfibers were fixed in the gelled Na-Alg (Ca-Alg).

Taking advantage of this effect, with an increase of the flow rate ratio (Q_{in}/Q_{out}), the morphology of the GelMA microfibers can stably change from straight to wavy to helical to thick wavy to thick straight (Figure 10.4). Specifically, the transformation is attributed to the force of the outer fluid (Na-Alg) acting on the inner GelMA microfibers. When the velocity of inner GelMA microfibers is smaller than that of outer fluid (Na-Alg), the driving force from the outer fluid would push GelMA microfiber out with straight morphology. When the velocity of inner GelMA microfibers is larger than that of outer fluid (Na-Alg), the force of the outer fluid would resist the inner GelMA microfibers, resulting in wavy morphology. As the Q_{in}/Q_{out} increases to a certain value, the inner GelMA microfiber starts to coil into helical morphology. With increasing Q_{in}/Q_{out}, the diameter and elastic modulus of the inner GelMA microfibers increases, and the GelMA microfiber fails to coil, resulting in thick wavy morphology. With further increase of Q_{in}/Q_{out}, the GelMA microfiber is unable to deform, resulting in thick straight morphology. *Caution:* When Q_{in}/Q_{out} is too large, GelMA would stack rapidly and solidify, forming a block glued to the inner wall of the outlet and blocking the nozzle.

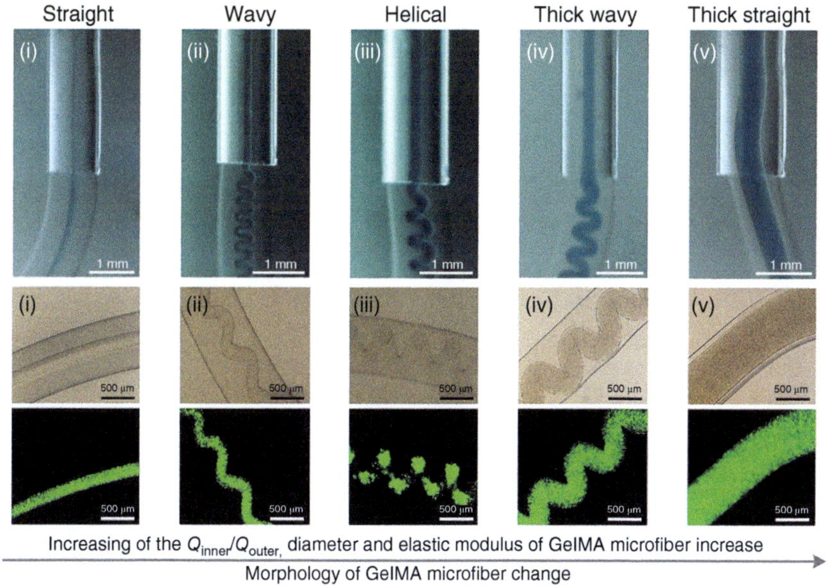

Figure 10.4 With increase of the flow rate ratio (Q_{in}/Q_{out}), the morphology of the GelMA microfibers can stably change from straight to wavy to helical to thick wavy to thick straight. Source: Shao et al. [78]. Reproduced with permission of WILEY-VCH.

10.3.2 Heterogeneous Microfiber

To fabricate desired heterogeneous GelMA microfibers, two-in-one coaxial nozzle, three-layer coaxial nozzle, and one pack two coaxial nozzles were designed, as shown in Figure 10.5 [78, 79]. Instead of a single inner GelMA injection, bicomponent laminar GelMA flow was used for the inner injection, as shown in Figure 10.6a [78, 79]. Two GelMA solutions were simultaneously pumped into the inner injection channel. As these solutions had low Reynolds numbers, the in situ cross-linked GelMA microfibers are consistent with the laminar fluid form. Similar to the formation of single-component GelMA microfibers, multicomponent GelMA microfibers with straight or helical morphology can be achieved using these specially designed nozzles. Meanwhile, using a two-in-one coaxial nozzle, the

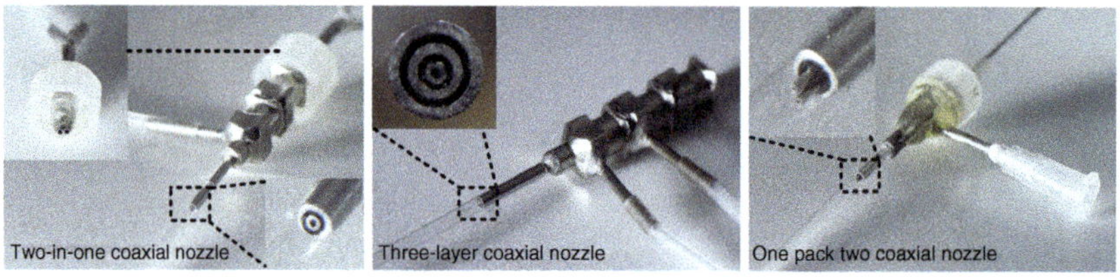

Figure 10.5 Two-in-one coaxial nozzle, three-layer coaxial nozzle and one pack two coaxial nozzle. Source: Shao et al. [78]. Reproduced with permission of WILEY-VCH.

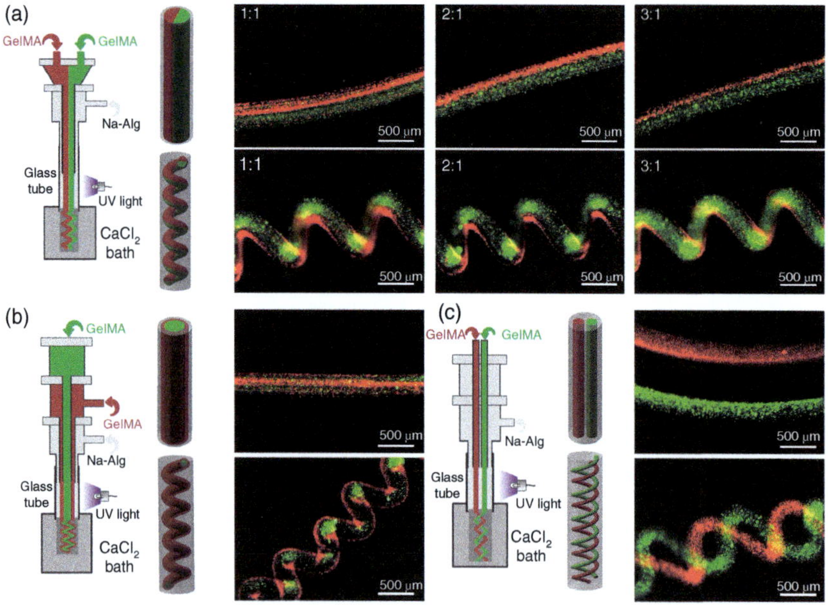

Figure 10.6 (a) Different proportions of Janus GelMA microfibers. Source: Shao et al. [78]. Reproduced with permission of WILEY-VCH. (b) Multilayered GelMA microfibers. Source: [78]. Reproduced with permission of WILEY-VCH. (c) Double paratactic straight or double helical GelMA microfibers. Source: Shao et al. [78]. Reproduced with permission of WILEY-VCH.

volume ratio of each component of the GelMA microfibers could be modulated by controlling the different flow rates of GelMA injections, as shown in Figure 10.6a. Additionally, using a three-layer coaxial nozzle, multilayered GelMA microfibers can be simply fabricated, as shown in Figure 10.6b. As illustrated in Figure 10.6b, three fluids including a core fluid of GelMA solution, a middle fluid of GelMA solution, and an outer sheath fluid of Na-Alg solution were simultaneously pumped into the corresponding channels, and double-layered GelMA microfibers with concentric straight or helical morphology can be generated. Inspired by the DNA double helix structure and using one pack two coaxial nozzles, two GelMA solutions were separately pumped into the two parallel inner injection channels and Na-Alg solution was pumped into the outer channel, generating double paratactic straight or double helical GelMA microfibers, as shown in Figure 10.6c. The successful generation of these multicompartmental GelMA microfibers with controllable morphology offers potential applications in creating multicellular encapsulations.

10.4 3D Assembly of Microfibers

Apparently, these versatile microfibers can be used as templates for constructing MFOs. Meaningfully, to more accurately mimic thick tissues or organs, complex and high-cell-density 3D tissue constructs are desired in tissue engineering. To apply these MFOs or CLHMs as building blocks for higher-order assembly, strategies were developed to handle these bioactive microfibers, such as 3D bioweaving, 3D bioprinting, etc.

10.4.1 3D Bioweaving

To apply these functional cell fibers as building blocks for higher-order cellular assembly, 3D bioweaving was developed to handle the CLHMs or MFOs [1]. As shown in Figure 10.7, a microfluidic weaving machine can be constructed by thin capillaries and fluid flow. And the CLHMs or MFOs can be withdrawn, ejected, bridged, and clamped through the microfluidic weaving machine. Using the 3D bioweaving technique, high-order assembly of various fiber-based 3D macroscopic cellular constructs can be fabricated, as shown in Figure 10.7. Furthermore, a macroscopic fabric-like cellular structure containing different CLHMs or MFOs can be woven through the 3D bioweaving technique. Accordingly, this weaving-based approach could be effective in creating pre-designed patterned cellular constructs, which would be useful in mimicking fabric-like network tissues in vivo such as neuronal pathways in the brain.

10.4.2 3D Bioprinting

For more effectively manipulating the CLHMs to construct higher-order assembly, a 3D bioprinting strategy was developed, as displayed schematically in Figure 10.8 [79]. To bioprint GelMA microfiber based macroscopic constructs through coaxial 3D bioprinting strategy, the core flow (GelMA solution containing $CaCl_2$), and the

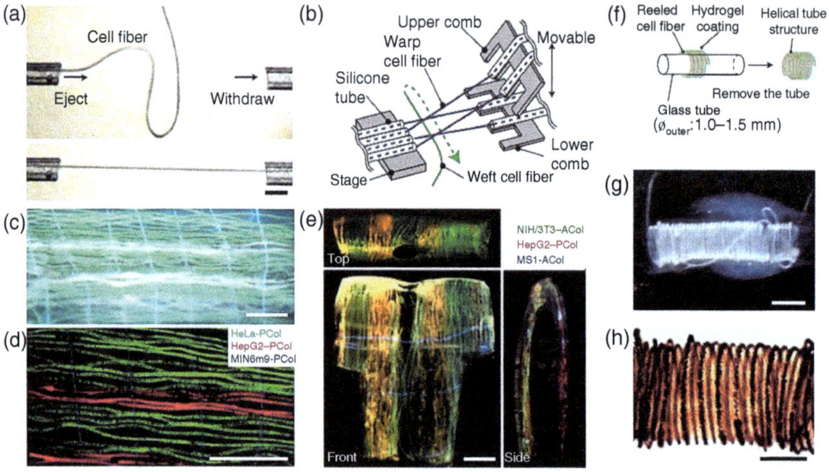

Figure 10.7 3D bioweaving of CLHMs or MFOs. Source: Onoe et al. [1]. Reproduced with permission of Nature Publishing Group.

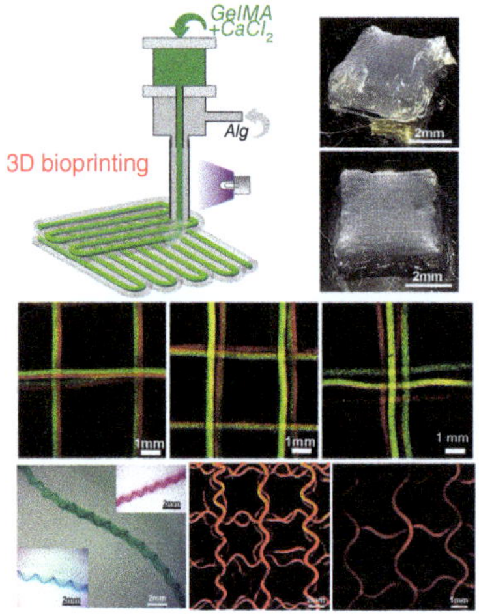

Figure 10.8 3D bioprinting of CLHMs. 3D heterogeneous constructs and 3D helical structure. Source: Shao et al. [79]. Reproduced with permission of WILEY-VCH.

sheath flow (Na-Alg solution) were injected into the inner and outer channels of the coaxial nozzle respectively. Specifically, when the core and sheath flows meet, Ca^{2+} diffused into the Na-Alg solution and progressive cross-linking started from inside to outside. Subsequently, as they flow through the lighting region and GelMA solution begins photo-cross-linking, the GelMA/alginate core–shell microfibers were generated. Then, through a 3D printer, the adjacent filament can be deposited in the precise location and fused into 3D constructs. When the entire structure was printed, the 3D structure was immersed in the $CaCl_2$ solution for complete cross-linking and a macroscopic 3D construct was fabricated, as shown in Figure 10.8.

Additionally, similar to the formation process of the multicomponent GelMA microfibers, by selecting the type of GelMA bioink sent to the inner channel of a coaxial nozzle, 3D heterogeneous constructs containing designed compositions in different layers can be constructed, as shown in Figure 10.8. Additionally, when the velocity of core flow (GelMA, $CaCl_2$) was greater than that of sheath flow (Na-Alg) during the bioprinting process, due to the resistance of the cross-linked Na-Alg, the GelMA microfibers could coil into a helical structure, as shown in Figure 10.8. Obviously, these helical GelMA microfibers can also be printed into a macroscopic 3D structure using the aforementioned 3D bioprinting strategy, as shown in Figure 10.8. It is observed that this novel 3D bioprinting strategy minimizes the bioprinting requirements for the core hydrogels, and facilitates the fabrication of cell-laden GelMA constructs at low concentrations. In view of the above advantages, spatially organized and heterogeneous 3D tissue constructs could be generated using the above 3D bioprinting strategy.

10.5 Microfiber-Based Organoids Bioprinting for In Vitro Mini Tissue Models

Apparently, these CLHMs can be directly used as templates for the reconstruction of MFOs, such as fiber-shaped mini tissues that mimic muscle fibers, blood vessels, or nerve networks in vitro.

10.5.1 Vascular Organoid

During the development of many functional tissue/organ systems, vascularization presents one of the most critical steps, because blood vessels enable the transport of nutrients, oxygen, and wastes to/from the tissues. To demonstrate the feasibility of using CLHMs in the application of vascularization, blood vessels were built by culturing HUVEC-laden hydrogel microfibers, as shown in Figure 10.9 (microfluidic coaxial bioprinting [1]) and Figure 10.10 (coaxial nozzle-assisted bioprinting [78, 79]). First, HUVECs were dispersed into corresponding hydrogels to fabricate cell-laden microfibers. With further culture, the HUVECs gradually migrated, stretched, and connected to each other, ultimately, potentially driven by the nature of HUVECs, the cells migrated toward the peripheries of the microfibers and formed endothelial tubular structures, as shown in Figures 10.9 and 10.10. As almost all the cells are encapsulated in the center of the fiber and there is little hydrogel in the fiber center, this endothelial tubular structure is hollow, which can be perfused after cell migration.

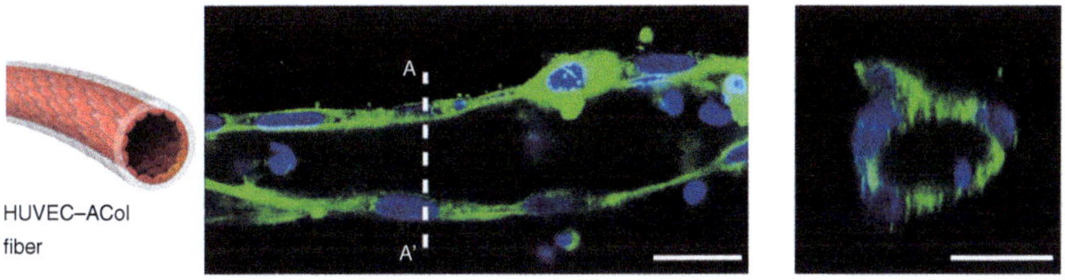

Figure 10.9 Microfluidic coaxial bioprinting of vascular organoids. Source: Onoe et al. [1]. Reproduced with permission of Nature Publishing Group.

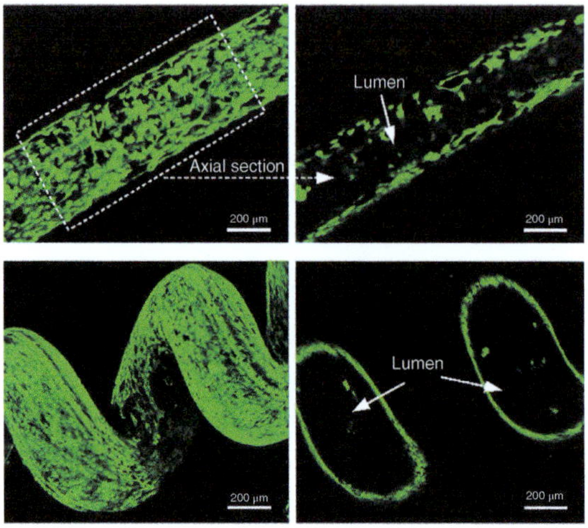

Figure 10.10 Coaxial nozzle-assisted bioprinting of vascular organoids. Source: Shao et al. [78]. Reproduced with permission of WILEY-VCH.

10.5.2 Myocyte Fiber

To mimic the "native" muscle structure, using microfluidic coaxial bioprinting strategy and collagen, Shinako Bansai et al. fabricated C_2C_{12} cell-laden core–shell hydrogel microfiber, which allows the cells to grow, migrate, promote cell–cell interaction, and form muscle MFOs, as shown in Figure 10.11 [80]. To evaluate the effect of the three-dimensional culture condition on the myogenesis of C_2C_{12} cells, the cytoskeletons of the MFOs were evaluated. The cytoskeleton in the MFOs aligned to the cylindrical axis of the fiber. It was revealed that the core–shell hydrogel fiber structure could simulate "native" muscle fibrous structure to maintain the cell and muscle fiber alignment.

10.5.3 Nerve Fiber

Using microfluidic coaxial bioprinting strategy and neural stem cells (NSCs)-laden pepsin-solubilized type-I collagen (PCol) hydrogel, Onoe et al. fabricated NSCs-laden MFOs [1]. Confocal laser scanning microscopy (CLSM) observation of the NSC–PCol fibers showed that the differentiated NSC–PCol fibers expressed a

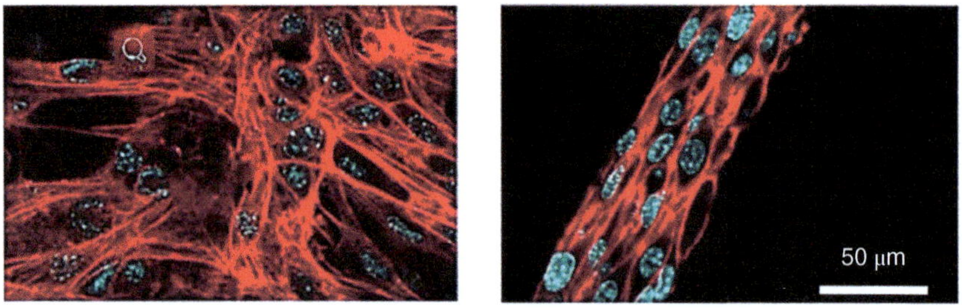

Figure 10.11 Microfluidic coaxial bioprinting of myocyte fiber. Source: Bansai et al. [80]. https://www.ncbi.nlm.nih.gov/pmc/articles/PMC6630375 CC-BY 4.0.

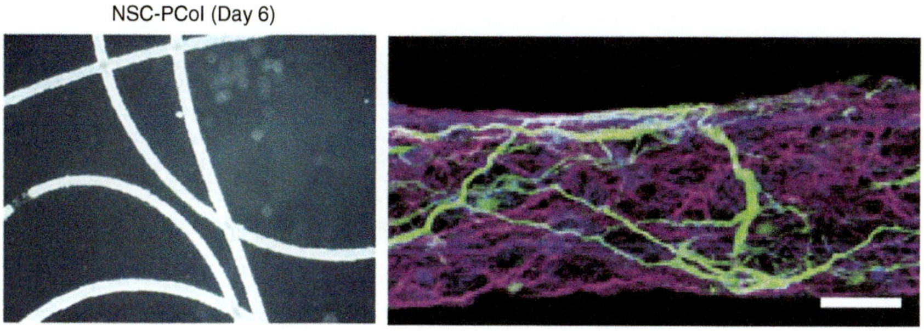

Figure 10.12 Microfluidic coaxial bioprinting of nerve fiber. GFAP, magenta; Tuj1, green. Source: Onoe et al. [1]. Reproduced with permission of Nature Publishing Group.

neuronal cell marker (Tuj1, green) and a glial cell marker (GFAP, magenta). The MFOs maintained shape even after the Ca-alginate shell was completely removed. Immunofluorescence staining of the MFOs showed that during the formation of the MFOs, cell-to-cell contacts occurred. Also, this study revealed that the initial ECM proteins gradually disappeared and that cell-derived ECM was newly produced (Figure 10.12).

10.5.4 Cardiomyocyte Fiber

Furthermore, using microfluidic coaxial bioprinting strategy and cardiomyocyte-laden fibrin hydrogel, Onoe et al. fabricated primary rat cardiomyocyte-fibrin fibers, which can contract spontaneously at 0.5–1 Hz after three days of culture, and the contracting motion actuated the whole fiber structure, as shown in Figure 10.13 [1]. These results suggest that the microenvironment provided by the ECM proteins could regulate cell migration and cell–cell interactions so that the MFOs exhibit intrinsic morphologies and functions.

10.5.5 Co-cultured Multi-organoids Interactions

Except for the individual MFOs, multicellular or multiple tissues/organoids interactions in an engineered environment are highly desired in vitro. So, to further demonstrate the potential of these microfibers, CLHM co-culture models were also researched [79]. As displayed in Figure 10.14, HUVEC-laden

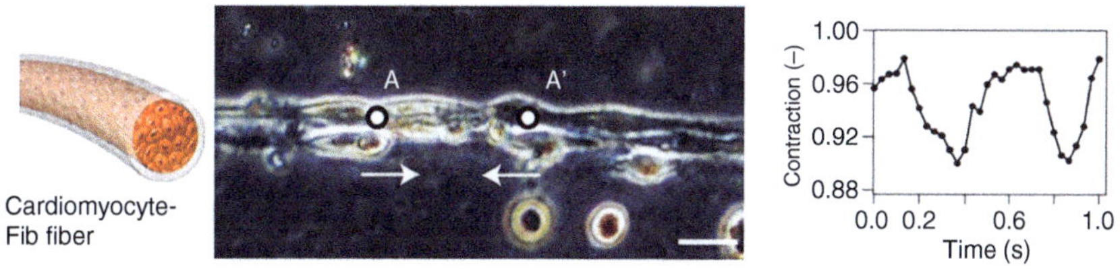

Figure 10.13 Microfluidic coaxial bioprinting of cardiomyocyte fiber. Source: Onoe et al. [1]. Reproduced with permission of Nature Publishing Group.

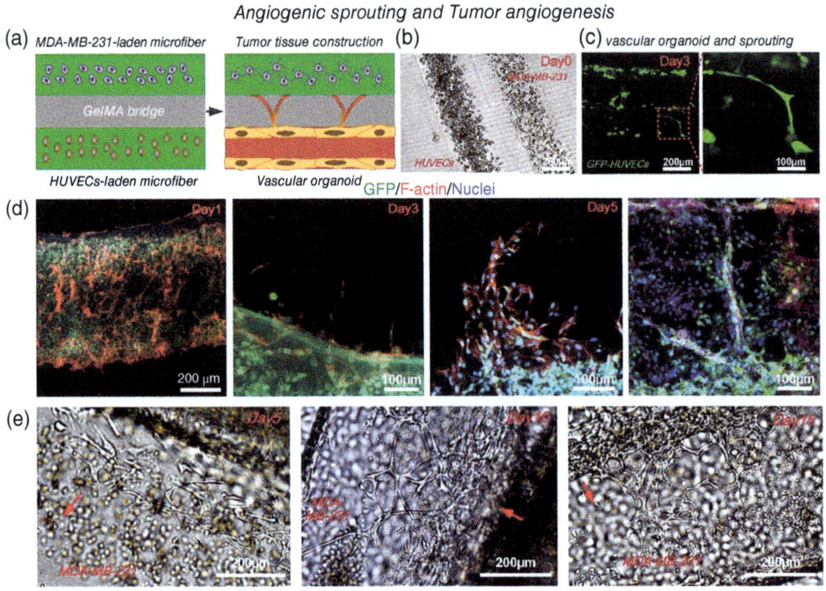

Figure 10.14 Co-cultured multi-organoids interactions by CLHMs. Angiogenic sprouting and tumor angiogenesis. Source: Shao et al. [79]. Reproduced with permission of WILEY-VCH.

microfiber and MDA-MB-231-laden microfiber are closely arranged in parallel and encapsulated in GelMA. After a period of culture, macroscopic endothelial tubular structures formed, and then angiogenic sprouts were generated and primarily elongated toward the increasing concentrations of angiogenic factors secreted by MDA-MB-231. Meanwhile, the sprouts grew longer with increasing co-culture time. These results indicate that GelMA microfibers provided a permissive 3D environment for cell functionalization, and complex intrinsic interactions of living tissues can also be reconstituted by co-culturing different CLHMs.

10.6 Discussion and Outlook

CLHMs can be used to directly build MFOs and complex co-culture tissue models in vitro. These in vitro functional biological models have many significant values, such as cell/tissue/organ development, pathologic/pharmacological study, drug screening, and tissue/organ repair in vivo. Moreover, organ-on-a-chip is another important research direction of MFOs.

Compared to the conventional 2D culture, CLHMs can provide a 3D culture environment, which is almost similar to the real physiological environment of an organism; thus, MFOs are closer to the morphology and function of tissues in vivo, and their related application results are also considered to be relatively more accurate. Specially, in the fields of stem cell biology and regenerative medicine, CLHMs are of great significance, such as stem cell amplification for stem cell therapy, controlling the differentiation of cells with the designed microenvironment, and mimicking stem cell niches. Additionally, by utilizing the shape of the fiber, various types of external mechanical stimuli, such as stretching, bending, and twisting, could be applied to the CLHMs to research the relationship between mechanical cues and tissue formation, maturation, and development.

Apparently, CLHMs and their 3D assembled structures could provide a meaningful 3D cell culture platform to construct healthy or diseased tissue models for a variety of related applications. Particularly, for drug screening, MFOs can be manufactured in large quantities and used for high-throughput screening and can replace some of the animal pharmacokinetic tests, which can eliminate ethical problems in animal experiments and reduce the cost of new drugs development. For tissue/organ repair in vivo, CLHMs can be easily used for implantation therapy. Because CLHM is thin and long, it can be implanted through a microcatheter, leading to minimally invasive implantation of artificially constructed tissues, and the implanted fiber can remain in the implanted site and can easily be recognized and even removed. However, to facilitate the clinical use of CLHMs, CLHM technology should simplify the fabrication method and the device so that it can be translated into medical practice.

Simple and easy-to-use devices could enable operators who are not specialized in CLHM technology to encapsulate micro-tissues from the patient in microfiber for producing medical implantation at the clinical site.

Although CLHMs have much important significance, there are still some difficult issues to solve. Specifically, for transplantation therapy of CLHMs, except for regulatory requirements and ethical issues, hydrogel materials should avoid or reduce immune responses. Meanwhile, for the use in drug discovery or medical implantation fields, due to the requirement of large batch and small difference between batches, the long-term preservation, transportation, and the maintenance of sterile environments are problems that need to be solved. In the near future, these problems could be solved if the CLHM-based tissues have the ability to cure patients who cannot be cured by existing medical technologies or to mimic the functions of real tissues in vivo for drug testing in vitro.

References

1 Onoe, H., Okitsu, T., Itou, A. et al. (2013). Metre-long cell-laden microfibres exhibit tissue morphologies and functions. *Nature Materials* 12 (6): 584–590.

2 Onoe, H. and Takeuchi, S. (2015). Cell-laden microfibers for bottom-up tissue engineering. *Drug Discovery Today* 20 (2): 236–246.

3 Carmeliet, P. and Jain, R.K. (2000). Angiogenesis in cancer and other diseases. *Nature* 407 (6801): 249–257.

4 Jain, R.K., Au, P., Tam, J. et al. (2005). Engineering vascularized tissue. *Nature Biotechnology* 23 (7): 821–823.

5 Rouwkema, J. and Khademhosseini, A. (2016). Vascularization and angiogenesis in tissue engineering: beyond creating static networks. *Trends in Biotechnology* 34 (9): 733–745.

6 Rouwkema, J., Rivron, N.C., and van Blitterswijk, C.A. (2008). Vascularization in tissue engineering. *Trends in Biotechnology* 26 (8): 434–441.
7 Jun, Y., Kang, E., Chae, S. et al. (2014). Microfluidic spinning of micro-and nano-scale fibers for tissue engineering. *Lab on a Chip* 14 (13): 2145–2160.
8 Yamada, M., Sugaya, S., Naganuma, Y. et al. (2012). Microfluidic synthesis of chemically and physically anisotropic hydrogel microfibers for guided cell growth and networking. *Soft Matter* 8 (11): 3122–3130.
9 Lee, B.R., Lee, K.H., Kang, E. et al. (2011). Microfluidic wet spinning of chitosan-alginate microfibers and encapsulation of HepG2 cells in fibers. *Biomicrofluidics* 5 (2): 022208.
10 Chung, B.G., Lee, K.H., Khademhosseini, A. et al. (2012). Microfluidic fabrication of microengineered hydrogels and their application in tissue engineering. *Lab on a Chip* 12 (1): 45–59.
11 Pati, F., Jang, J., Ha, D.H. et al. (2014). Printing three-dimensional tissue analogues with decellularized extracellular matrix bioink. *Nature Communications* 5 (1): 1–11.
12 Pati, F. and Cho, D.W. (2017). *Bioprinting of 3D Tissue Models Using Decellularized Extracellular Matrix Bioink[M]//3D Cell Culture*, 381–390. New York, NY: Humana Press.
13 Jang, J., Kim, T.G., Kim, B.S. et al. (2016). Tailoring mechanical properties of decellularized extracellular matrix bioink by vitamin B2-induced photo-crosslinking. *Acta Biomaterialia* 33: 88–95.
14 Lee, H., Han, W., Kim, H. et al. (2017). Development of liver decellularized extracellular matrix bioink for three-dimensional cell printing-based liver tissue engineering. *Biomacromolecules* 18 (4): 1229–1237.
15 Kim, B.S., Kim, H., Gao, G. et al. (2017). Decellularized extracellular matrix: a step towards the next generation source for bioink manufacturing. *Biofabrication* 9 (3): 034104.
16 Choudhury, D., Tun, H.W., Wang, T. et al. (2018). Organ-derived decellularized extracellular matrix: a game changer for bioink manufacturing? *Trends in Biotechnology* 36 (8): 787–805.
17 Kim, B.S., Kwon, Y.W., Kong, J.S. et al. (2018). 3D cell printing of in vitro stabilized skin model and in vivo pre-vascularized skin patch using tissue-specific extracellular matrix bioink: a step towards advanced skin tissue engineering. *Biomaterials* 168: 38–53.
18 Han, W., Singh, N.K., Kim, J.J. et al. (2019). Directed differential behaviors of multipotent adult stem cells from decellularized tissue/organ extracellular matrix bioinks. *Biomaterials* 224: 119496.
19 Jang, J., Park, H.J., Kim, S.W. et al. (2017). 3D printed complex tissue construct using stem cell-laden decellularized extracellular matrix bioinks for cardiac repair. *Biomaterials* 112: 264–274.
20 Kim, J., Shim, I.K., Hwang, D.G. et al. (2019). 3D cell printing of islet-laden pancreatic tissue-derived extracellular matrix bioink constructs for enhancing pancreatic functions. *Journal of Materials Chemistry B* 7 (10): 1773–1781.

21 Sasikumar, S., Chameettachal, S., Cromer, B. et al. (2019). Decellularized extracellular matrix hydrogels-cell behavior as a function of matrix stiffness. *Current Opinion in Biomedical Engineering* 10: 123–133.

22 Das, S., Kim, S.W., Choi, Y.J. et al. (2019). Decellularized extracellular matrix bioinks and the external stimuli to enhance cardiac tissue development in vitro. *Acta Biomaterialia* 95: 188–200.

23 Kabirian, F. and Mozafari, M. (2020). Decellularized ECM-derived bioinks: prospects for the future. *Methods* 171: 108–118.

24 Ma, X., Yu, C., Wang, P. et al. (2018). Rapid 3D bioprinting of decellularized extracellular matrix with regionally varied mechanical properties and biomimetic microarchitecture. *Biomaterials* 185: 310–321.

25 Heath, D.E. (2019). A review of decellularized extracellular matrix biomaterials for regenerative engineering applications. *Regenerative Engineering and Translational Medicine* 5 (2): 155–166.

26 Dzobo, K., Motaung, S., and Adesida, A. (2019). Trends in decellularized extracellular matrix bioinks for 3D printing: inspiring a new kind of medicine. *International Journal of Molecular Sciences* 20: 4268.

27 Saldin, L.T., Cramer, M.C., Velankar, S.S. et al. (2017). Extracellular matrix hydrogels from decellularized tissues: structure and function. *Acta Biomaterialia* 49: 1–15.

28 Taylor, D.A., Sampaio, L.C., Ferdous, Z. et al. (2018). Decellularized matrices in regenerative medicine. *Acta Biomaterialia* 74: 74–89.

29 Bejleri, D. and Davis, M.E. (2019). Decellularized extracellular matrix materials for cardiac repair and regeneration. *Advanced Healthcare Materials* 8 (5): 1801217.

30 Rhee, S., Puetzer, J.L., Mason, B.N. et al. (2016). 3D bioprinting of spatially heterogeneous collagen constructs for cartilage tissue engineering. *ACS Biomaterials Science & Engineering* 2 (10): 1800–1805.

31 Lee, A., Hudson, A.R., Shiwarski, D.J. et al. (2019). 3D bioprinting of collagen to rebuild components of the human heart. *Science* 365 (6452): 482–487.

32 Mazzocchi, A., Devarasetty, M., Huntwork, R. et al. (2018). Optimization of collagen type I-hyaluronan hybrid bioink for 3D bioprinted liver microenvironments. *Biofabrication* 11 (1): 015003.

33 Osidak, E.O., Karalkin, P.A., Osidak, M.S. et al. (2019). Viscoll collagen solution as a novel bioink for direct 3D bioprinting. *Journal of Materials Science: Materials in Medicine* 30 (3): 1–12.

34 Yang, X., Lu, Z., Wu, H. et al. (2018). Collagen-alginate as bioink for three-dimensional (3D) cell printing based cartilage tissue engineering. *Materials Science and Engineering: C* 83: 195–201.

35 Lee, C., Abelseth, E., de la Vega, L. et al. (2019). Bioprinting a novel glioblastoma tumor model using a fibrin-based bioink for drug screening. *Materials Today Chemistry* 12: 78–84.

36 Zhang, K., Fu, Q., Yoo, J. et al. (2017). 3D bioprinting of urethra with PCL/PLCL blend and dual autologous cells in fibrin hydrogel: an in vitro evaluation of biomimetic mechanical property and cell growth environment. *Acta Biomaterialia* 50: 154–164.

37 Xu, Y. and Wang, X. (2015). Fluid and cell behaviors along a 3D printed alginate/gelatin/fibrin channel. *Biotechnology and Bioengineering* 112 (8): 1683–1695.

38 Chameettachal, S., Midha, S., and Ghosh, S. (2016). Regulation of chondrogenesis and hypertrophy in silk fibroin-gelatin-based 3D bioprinted constructs. *ACS Biomaterials Science & Engineering* 2 (9): 1450–1463.

39 Zheng, Z., Wu, J., Liu, M. et al. (2018). 3D bioprinting of self-standing silk-based bioink. *Advanced Healthcare Materials* 7 (6): 1701026.

40 Fan, R., Piou, M., Darling, E. et al. (2016). Bio-printing cell-laden Matrigel–agarose constructs. *Journal of Biomaterials Applications* 31 (5): 684–692.

41 Zhang, Y.S., Duchamp, M., Oklu, R. et al. (2016). Bioprinting the cancer microenvironment. *ACS Biomaterials Science & Engineering* 2 (10): 1710–1721.

42 Swaminathan, S., Hamid, Q., Sun, W. et al. (2019). Bioprinting of 3D breast epithelial spheroids for human cancer models. *Biofabrication* 11 (2): 025003.

43 Peng, W., Unutmaz, D., and Ozbolat, I.T. (2016). Bioprinting towards physiologically relevant tissue models for pharmaceutics. *Trends in Biotechnology* 34 (9): 722–732.

44 Knowlton, S., Onal, S., Yu, C.H. et al. (2015). Bioprinting for cancer research. *Trends in Biotechnology* 33 (9): 504–513.

45 Schmidt, S.K., Schmid, R., Arkudas, A. et al. (2019). Tumor cells develop defined cellular phenotypes after 3D-bioprinting in different bioinks. *Cells* 8 (10): 1295.

46 Axpe, E. and Oyen, M.L. (2016). Applications of alginate-based bioinks in 3D bioprinting. *International Journal of Molecular Sciences* 17 (12): 1976.

47 Tabriz, A.G., Hermida, M.A., Leslie, N.R. et al. (2015). Three-dimensional bioprinting of complex cell laden alginate hydrogel structures. *Biofabrication* 7 (4): 045012.

48 Tan, E., Y.S. and Yeong, W.Y. (2015). Concentric bioprinting of alginate-based tubular constructs using multi-nozzle extrusion-based technique. *International Journal of Bioprinting* 1 (1).

49 Jeon, O., Lee, Y.B., Hinton, T.J. et al. (2019). Cryopreserved cell-laden alginate microgel bioink for 3D bioprinting of living tissues. *Materials Today Chemistry* 12: 61–70.

50 Zhu, K., Chen, N., Liu, X. et al. (2018). A general strategy for extrusion bioprinting of bio-macromolecular bioinks through alginate-templated dual-stage crosslinking. *Macromolecular Bioscience* 18 (9): 1800127.

51 Kim, M.H., Lee, Y.W., Jung, W.K. et al. (2019). Enhanced rheological behaviors of alginate hydrogels with carrageenan for extrusion-based bioprinting. *Journal of the Mechanical Behavior of Biomedical Materials* 98: 187–194.

52 Duin, S., Schütz, K., Ahlfeld, T. et al. (2019). 3D bioprinting of functional islets of Langerhans in an alginate/methylcellulose hydrogel blend. *Advanced Healthcare Materials* 8 (7): 1801631.

53 Rastogi, P. and Kandasubramanian, B. (2019). Review of alginate-based hydrogel bioprinting for application in tissue engineering. *Biofabrication* 11 (4): 042001.

54 Yue, K., Trujillo-de Santiago, G., Alvarez, M.M. et al. (2015). Synthesis, properties, and biomedical applications of gelatin methacryloyl (GelMA) hydrogels. *Biomaterials* 73: 254–271.

55 Shirahama, H., Lee, B.H., Tan, L.P. et al. (2016). Precise tuning of facile one-pot gelatin methacryloyl (GelMA) synthesis. *Scientific Reports* 6: 31036.

56 Noshadi, I., Hong, S., Sullivan, K.E. et al. (2017). In vitro and in vivo analysis of visible light crosslinkable gelatin methacryloyl (GelMA) hydrogels. *Biomaterials Science* 5 (10): 2093–2105.

57 Yin, J., Yan, M., Wang, Y. et al. (2018). 3D bioprinting of low-concentration cell-laden gelatin methacrylate (GelMA) bioinks with a two-step cross-linking strategy. *ACS Applied Materials & Interfaces* 10 (8): 6849–6857.

58 Stratesteffen, H., Köpf, M., Kreimendahl, F. et al. (2017). GelMA-collagen blends enable drop-on-demand 3D printability and promote angiogenesis. *Biofabrication* 9 (4): 045002.

59 Ying, G., Jiang, N., Yu, C. et al. (2018). Three-dimensional bioprinting of gelatin methacryloyl (GelMA). *Bio-Design and Manufacturing* 1 (4): 215–224.

60 Pepelanova, I., Kruppa, K., Scheper, T. et al. (2018). Gelatin-methacryloyl (GelMA) hydrogels with defined degree of functionalization as a versatile toolkit for 3D cell culture and extrusion bioprinting. *Bioengineering* 5 (3): 55.

61 Sun, M., Sun, X., Wang, Z. et al. (2018). Synthesis and properties of gelatin Methacryloyl (GelMA) hydrogels and their recent applications in load-bearing tissue. *Polymers* 10 (11): 1290.

62 Hribar, K.C., Soman, P., Warner, J. et al. (2014). Light-assisted direct-write of 3D functional biomaterials. *Lab on a Chip* 14 (2): 268–275.

63 Chen, Y.C., Lin, R.Z., Qi, H. et al. (2012). Functional human vascular network generated in photocrosslinkable gelatin methacrylate hydrogels. *Advanced Functional Materials* 22 (10): 2027–2039.

64 Bertassoni, L.E., Cardoso, J.C., Manoharan, V. et al. (2014). Direct-write bioprinting of cell-laden methacrylated gelatin hydrogels. *Biofabrication* 6 (2): 024105.

65 Liu, W., Heinrich, M.A., Zhou, Y. et al. (2017). Extrusion bioprinting of shear-thinning gelatin methacryloyl bioinks. *Advanced Healthcare Materials* 6 (12): 1601451.

66 Kang, E., Jeong, G.S., Choi, Y.Y. et al. (2011). Digitally tunable physicochemical coding of material composition and topography in continuous microfibres. *Nature Materials* 10 (11): 877–883.

67 Cheng, Y., Zheng, F., Lu, J. et al. (2014). Bioinspired multicompartmental microfibers from microfluidics. *Advanced Materials* 26 (30): 5184–5190.

68 Yu, Y., Fu, F., Shang, L. et al. (2017). Bioinspired helical microfibers from microfluidics. *Advanced Materials* 29 (18): 1605765.

69 Shang, L., Fu, F., Cheng, Y. et al. (2017). Bioinspired multifunctional spindle-knotted microfibers from microfluidics. *Small* 13 (4): 1600286.

70 Yu, Y., Wen, H., Ma, J. et al. (2014). Flexible fabrication of biomimetic bamboo-like hybrid microfibers. *Advanced Materials* 26 (16): 2494–2499.

71 Shin, S.J., Park, J.Y., Lee, J.Y. et al. (2007). "On the fly" continuous generation of alginate fibers using a microfluidic device. *Langmuir* 23 (17): 9104–9108.

72 Lee, K.H., Shin, S.J., Park, Y. et al. (2009). Synthesis of cell-laden alginate hollow fibers using microfluidic chips and microvascularized tissue-engineering applications. *Small* 5 (11): 1264–1268.

73 Xu, P., Xie, R., Liu, Y. et al. (2017). Bioinspired microfibers with embedded perfusable helical channels. *Advanced Materials* 29 (34): 1701664.

74 Xie, R., Xu, P., Liu, Y. et al. (2018). Necklace-like microfibers with variable knots and perfusable channels fabricated by an oil-free microfluidic spinning process. *Advanced Materials* 30 (14): 1705082.

75 Wang, J., Zou, M., Sun, L. et al. (2017). Microfluidic generation of Buddha beads-like microcarriers for cell culture. *Science China Materials* 60 (9): 857–865.

76 Um, E., Nunes, J.K., Pico, T. et al. (2014). Multicompartment microfibers: fabrication and selective dissolution of composite droplet-in-fiber structures. *Journal of Materials Chemistry B* 2 (45): 7866–7871.

77 Headen, D.M., Aubry, G., Lu, H. et al. (2014). Microfluidic-based generation of size-controlled, biofunctionalized synthetic polymer microgels for cell encapsulation. *Advanced Materials* 26 (19): 3003–3008.

78 Shao, L., Gao, Q., Zhao, H. et al. (2018). Fiber-based mini tissue with morphology-controllable GelMA microfibers. *Small* 14 (44): 1802187.

79 Shao, L., Gao, Q., Xie, C. et al. (2019). Bioprinting of cell-laden microfiber: can it become a standard product? *Advanced Healthcare Materials* 8 (9): 1900014.

80 Bansai, S., Morikura, T., Onoe, H. et al. (2019). Effect of cyclic stretch on tissue maturation in myoblast-laden hydrogel fibers. *Micromachines* 10 (6): 399.

11

Large Scale Tissues Bioprinting

11.1 Introduction

Due to the complexity of the composition and structure of tissues or organs in vivo, the ability to construct artificial volumetric tissues for research or therapeutic use remains a daunting challenge [1–6]. In vivo, solid living tissues/organs rely on the circulation of the blood to supply the cells with nutrients and oxygen, and remove metabolic byproducts, thus supporting cell survival and development [7–10]. Consequently, to engineer volumetric tissues, first and foremost, pervasive nutrient networks should be built to continuously supply nutrients to billions of cells for survival and subsequent functionalization. Without a prefabricated nutrient network, engineered tissues are often limited to several hundred micrometers in thickness. As a matter of fact, for a tissue to grow beyond the diffusion limit of nutrients or oxygen (100–200 μm), new blood vessel formation is required, and this is also the same case for artificial tissue constructs. Specifically, due to nutrient deficiencies or hypoxia, insufficient nutrient/oxygen can lead to improper cell integration or cell death in large-scale (≥1 mm) man-made tissue constructs. To promote the bio-function of large-scale or organ-level constructs, building effective nutrient networks within the engineered tissue constructs is crucial. However, despite significant advancements in tissue engineering, the creation of large-scale complex tissue constructs with nutrient networks on demand has remained a great challenge.

11.1.1 Challenges in Bioprinting Large Scale Tissues

Due to its advantages over other biofabrication methods, three-dimensional (3D) bioprinting has attracted increasing attention as an emerging technology for the fabrication of 3D tissues or organs [11–20]. Certainly, for bioprinted functional tissue constructs, in addition to requiring bio-inks to balance printability and biological performance [21–38], long-term perfusion culture is also required for promoting the growth and maturity of the tissue through exchanging nutrients/oxygen and removing metabolic byproducts. Accordingly, to bioprinting large-scale functionalized tissue constructs, there are two crucial challenges, (i) printing effective nutrient networks; (ii) long-term perfusion culture for functionalization and maturation of the bioprinted tissue through direct perfusion or perfusion-on-chip.

Cell Assembly with 3D Bioprinting, First Edition. Yong He, Qing Gao, and Yifei Jin.

11.1.2 Strategies in Bioprinting Large Scale Tissue with Nutrient Networks

With the development of bioprinting, a number of distinctive ideas and strategies for constructing nutrient networks have been put forward. Generally speaking, the structure forms of the nutrient network include porous network structure, hollow channel network, and biomimetic vascular network. In this case, the corresponding bioprinting strategies have been proposed and optimized successively.

11.1.2.1 Porous Network Printing

Generally, 3D bioprinting is usually performed by depositing bioinks layer-by by-layer to generate well-designed 3D tissue constructs, such as extrusion-based bioprinting. Obviously, bioinks, most of which are cell-laden biomaterials, are vital parts of 3D bioprinting. To ensure the structural fidelity and bio-functionality of the bioprinted tissues, an ideal bioink should possess proper mechanical property and superior biological performance. Due to their intrinsic micro-porosity and capacity for high nutrient loading, hydrogels are the most common candidates for bioink design. Whereas, to maintain adequate mechanical strength and structural fidelity during 3D bioprinting, the commonly used bioinks would inevitably form dense hydrogel networks with low porosity. And in turn, these dense gels have an inadequate supply of nutrients/oxygen, limiting the development of encapsulated cells and the functionalization of bioprinted tissue constructs. Thus, it is highly desired to design hydrogel prescriptions with the ability of bioprinting tissue constructs containing multiscale porous networks that ensure effective nutrient/oxygen diffusion and structural fidelity, facilitating the generation of functional tissues or organs.

In this context, emulsion templating has emerged as a strategy for the bioprinting of porous hydrogel structures [39, 40]. However, based on this strategy, the pore size can only reach several tens of micrometers, and they are not large enough to meet the nutrient transport requirements for large-scale tissue constructs. Given this issue, solid particles as sacrificial templates were used to create mesoscale porous hydrogel constructs, and the pore size can be up to hundreds of micrometers [41]. However, as large solid particles tend to clog printer nozzles, 3D bioprinting of hydrogel encapsulating solid particles does not work in practice. Accordingly, we developed a practical method for the production of a tunable and self-adapting microgel as a sacrificial template for 3D bioprinting structures with macro-pore networks without blocking nozzle (see below for details) [42].

Except for the original bioink design, the structural design is also a fascinating strategy for building nutrition networks, such as directly print porous cell-laden hydrogel scaffolds. However, to ensure the normal growth and development of encapsulated cells, the printing of porous scaffolds is often unable to use excessively strong hydrogels. As a result, due to the low strength of hydrogel fibers, the printed porous structures are often not ideal. To print an effective porous structure, we proposed a synchronous 3D bioprinting strategy, in which the extruded filament to be deposited layer-by-layer is composed of a half cell-laden ink and a half sacrificial ink, and effective nutrient pores could be constructed after the removal of the sacrificial inks (see below for details) [43].

11.1.2.2 Hollow Channel Network Printing

In addition to porous structures, the manufacture of biomimetic vascular structures has always been fascinating. To create biomimetic vascular structure, the corresponding bioprinting strategies of a hollow channel network are proposed. Most of these strategies are based on: (i) sacrificial bioprinting [1, 44, 45], where a sacrificial ink is first deposited and molded in a hydrogel matrix, and then removed to create channel networks; (ii) coaxial bioprinting [46–55], where the cross-linker/hydrogel is pumped into the inner/outer needles to generate hydrogel tube-based structures. However, due to the required molding process during sacrificial bioprinting or the continuity of the hydrogel tubes in coaxial bioprinting, one-step creation of large complex tissue constructs with hollow channel networks is still not easy. In this sense, we proposed one-step coaxial/sacrificial printing of large-scale vascularized tissue constructs (see below for details).

11.1.2.3 Advanced Bioprinting Techniques-Enabled Printing Highly Biomimetic Vascular Network

Of course, the ideal nutrient network is the highly biomimetic vascular network system, which can simulate the whole blood circulation effect. To create a highly biomimetic vascular network system, advanced bioprinting techniques with high precision are indispensable. With the development of bioprinting technology, some 3D bioprinting technologies created a great sensation, such as support bath-assisted bioprinting and light-based bioprinting, which can create highly biomimetic tissue/organ constructs with the directly perfusable vascular system. Due to their strong manufacturing capacity, these advanced technologies have great potential for creating large scale functional tissues or organs, yet, in addition to the vascular system, due to the complexity of the whole biological system of organisms, there is still a long way to go in the field of tissues or organs engineering.

11.2 Large Scale Cell-laden Porous Structures Printing

For porous structures, according to whether the pores are interconnected or not, they can be divided into independent porous structures and interconnected porous structures. In the independent pore structure, the pores are independent and disconnected from each other. In the interconnected porous structure, the pores are connected with each other, which can transport nutrients more effectively.

11.2.1 Independent Porous Structure Printing

Independent pore structures are generally fabricated using sacrificial microtemplates, such as gelatin microparticles. However, due to clogging printhead, a solid microparticle sacrificial template is not suitable for extrusion-based 3D bioprinting of independent pore structures. To avoid clogging the nozzle, we proposed the elastic sacrificial microgel template, which has great potential in producing an independent pore structure [42]. Specifically, the preparation process of the tunable

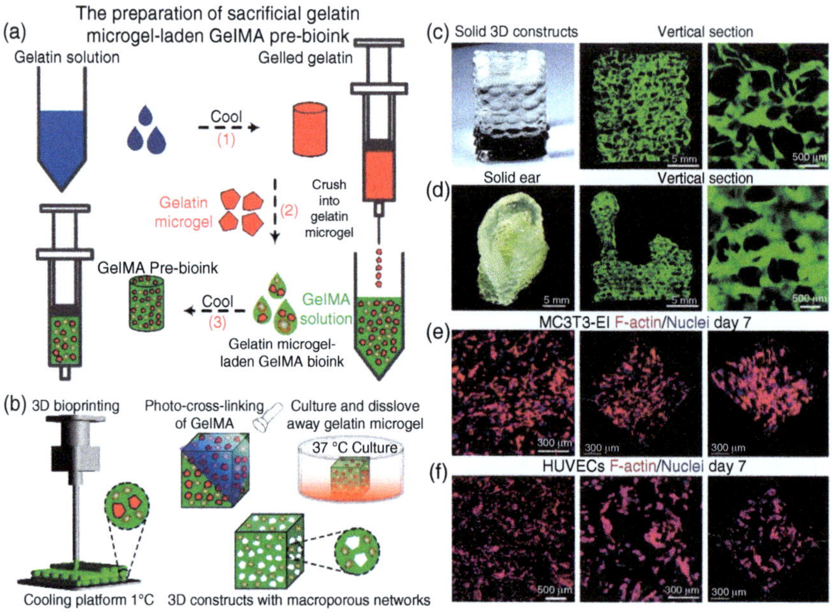

Figure 11.1 (a) The preparation of sacrificial gelatin microgel-laden GelMA pre-bioink. Source: Shao et al. [42]. Reproduced with permission of Nature Publishing Group. (b) The process of bioprinting of independent porous structures. Source: Shao et al. [42]. Reproduced with permission of Nature Publishing Group. (c, d) The printed complex soft constructs with independent pores. Source: Shao et al. [42]. Reproduced with permission of Nature Publishing Group. (e, f) Osteoblasts and HUVECs encapsulated in the bioprinted porous constructs could stretch, migrate, and connect during a period of culture. Source: Shao et al. [42]. Reproduced with permission of Nature Publishing Group.

sacrificial microgel-laden bioink (gelatin microgel-laden bioink) is shown in Figure 11.1a. First, the gelatin solution was completely cooled and gelatinized by the condensation mechanism of the gelatin solution. Then, the cooled gelatin gel was crushed into gelatin microgel through the printhead and mixed with GelMA solution directly. Finally, through pre-condensation again, printable sacrificial gelatin microgel-laden bioink was prepared.

Systematically, as shown in Figure 11.1b, the specific steps of bioprinting can be divided into three steps: (i) the preparation of the sacrificial gelatin microgel-laden GelMA pre-bioink according to the above preparation process; (ii) 3D bioprinting temporary structure via a cooling process; and (iii) permanent photo-cross-linking of GelMA and dissolving away gelatin microgel to create the final independent porous structure. Summarily, this strategy used a reversible thermo-cross-linking mechanism (gelatin and GelMA) and irreversible photo-cross-linking mechanism (GelMA) to create a porous structure, which allowed for effective oxygen/nutrient, facilitating the generation of functional tissue constructs.

Subtly, the relatively uniform and controllable sacrificial microgel was prepared through the common print nozzle, which was especially suitable for extrusion-based 3D printing without blocking the print nozzle. More importantly, as the cell-laden bioink and sacrificial microgel work together to promote their printability, complex soft constructs with independent pores can easily be printed, as shown in Figure 11.1c,d. Moreover, the osteoblasts and human umbilical cord vein endothelial cells (HUVECs) encapsulated in the bioprinted constructs could stretch, migrate, and connect during a period of culture, as shown in Figure 11.1e,f. In view of this, our sacrificial gelatin microgel-laden bioink strategy could partly balance physical printability and biological functionality.

Additionally, the sacrificial microgel-laden bioink can be expanded to be applicable to other biomaterials, such as collagen, matrigel, etc. It is a simple and practical bioink strategy for 3D bioprinting of porous structures, which can be more convenient for the development of related fields. This simple and practical sacrificial microgel-laden bioink has significant potentials in building functionalized tissue toward applications in tissue engineering, perhaps even in organ repair.

11.2.2 Interconnected Porous Structure Printing

Because the independent pores are not connected with each other, the nutrient/oxygen exchange between the pores is very limited. Obviously, the interconnected pore structure can improve the efficiency of nutrient transport. The common interconnected pore structure is a cell-laden scaffold. The strength of the commonly used cell-laden hydrogel is relatively weak, and the porous efficiency of traditional direct printing cell-laden scaffold is limited. For this reason, we developed a strategy to synchronously print a half sacrificing ink and a half cell-laden bioink, and after the sacrificial ink is removed, the interconnected porous structure is constructed [43].

11.2.2.1 Directly Cell-laden Scaffold Printing

The traditional porous structure printing is generally considered as the direct printing of cell-laden scaffold. It is relatively simple to directly print the porous scaffolds. But there is an inevitable problem, to provide a suitable soft tissue environment for the encapsulated cells, soft gelled bioinks with weak mechanical properties are usually necessary to enable biological function, paradoxically, to maintain high fidelity of printing structure, it is necessary for bioink to have high strength. In this dilemma, compromised bioink designs for both biological and mechanical properties are common candidates. Even so, due to the insufficient mechanical strength to support the weight of the bioink fiber itself, it inevitably leads to the collapse of the structure and pore failure, as shown in Figure 11.2. Because of this, it is impossible to print complex porous structures. Because of this, the printability of bioink is usually verified by simple, porous scaffold printing.

11.2.2.2 Synchronous Bioprinting (Bioink and Sacrificial Ink Half and Half)

To bioprint interconnected pore structure for more effective oxygen/nutrient transport, synchronous bioprinting strategy was proposed, which could simultaneously print a half cell-laden bioink and a half sacrificial ink to build a complex, self-supporting solid structure, and once the sacrificial ink was removed, interconnected pore structure could be achieved [43]. Specifically, synchronous printing

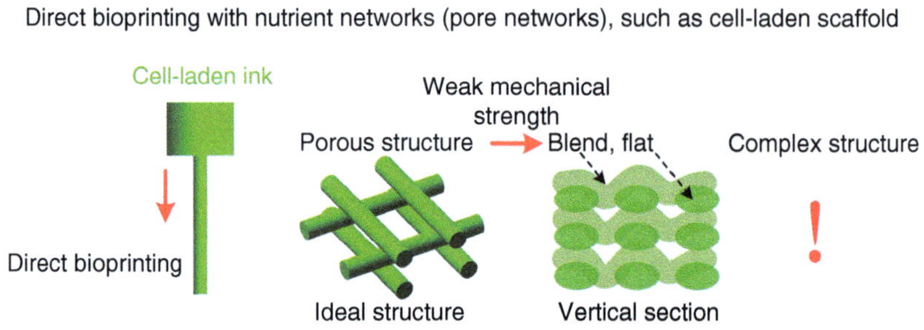

Figure 11.2 Direct bioprinting of cell-laden scaffold. Due to the insufficient mechanical strength to support the weight of the bioink fiber itself, it inevitably leads to the collapse of the structure and pore failure, let alone complex porous structures. Source: Shao et al. [43]. Reproduced with permission of WILEY-VCH.

is mainly divided into three steps (Figure 11.3a): (i) preparing printable gelatin and GelMA pre-bioink by pre-condensation and pumping side-by-side into the nozzle for 3D bioprinting temporary self-supporting solid structure; (ii) permanent photo-cross-linking of GelMA to obtain stable structure; (iii) culturing at 37 °C, and dissolving away gelatin to make porous nutrient network. Based on this strategy, due to the synergistic interaction between the cell-laden bioink and the sacrificial ink and the self-supporting effect of the solid structure, simple and complex engineering large-scale constructs with interconnected pore networks can easily be bioprinted, as shown in Figure 11.3b. Furthermore, the cells (osteoblast and HUVECs) encapsulated in the bioprinted large-scale constructs with interconnected pore networks could stretch, migrate, and connect within a period of culture, as shown in Figure 11.3c. The interconnected pore structure not only improves the efficiency of nutrient transport but also facilitates the dynamic perfusion culture, which has great potential and value in the field of tissue engineering and regenerative medicine.

11.2.3 Heterogeneous Independent/Interconnected Porous Structure Printing

As native tissues are composed of multicellular or multicellular extracellular matrix, to construct the heterogeneity, heterogeneous constructs with tunable compositions are highly desirable. Multi-material or multi-cell printing has always been the focus and difficulty in the field of bioprinting. To generate multicomponent and multicellular constructs, we designed a practical all-in-one nozzle that allowed the rapid bioprinting of constructs using different materials or cells, as displayed schematically in Figure 11.4 [42]. The all-in-one nozzle includes multiple material inlet channels, so operators can selectively pass through the pre-designed material components and print the expected multi-component structure through a single outlet channel. In fact, the all-in-one nozzle can be expanded to print as many as seven materials at the same time. As an example, using two-in-one nozzles, different types of gelatin microgel-laden GelMA pre-bioinks were selected to pass into the

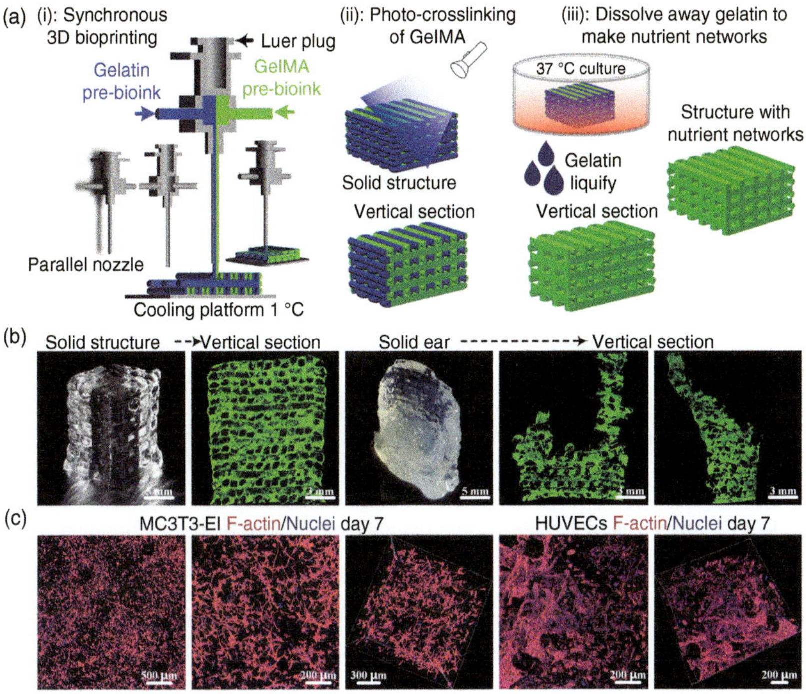

Figure 11.3 (a) Synchronous bioprinting strategy and specific steps. Source: Shao et al. [43]. Reproduced with permission of WILEY-VCH. (b) Bioprinted simple and complex engineering large-scale constructs with interconnected pore networks. Source: Shao et al. [43]. Reproduced with permission of WILEY-VCH. (c) Osteoblast and HUVECs encapsulated in the bioprinted large-scale constructs with interconnected pore networks could stretch, migrate, and connect within a period of culture. Source: Shao et al. [43]. Reproduced with permission of WILEY-VCH.

nozzle respectively, and finally, heterogeneous structures with independent pore could be generated. As shown in Figure 11.4, 2D patterns and 3D heterogeneous structures suggest that these all-in-one nozzles can rapidly switch between different bioinks in a fully pre-designed or programmable manner. Meanwhile, the vertical section views confirmed the presence of independent pores within 3D constructs (Figure 11.4). This strategy may open possibilities in the creation of freeform and heterogeneous constructs on demand. The adaptability of this bioprinting system provides a new level of convenience in fabricating complex constructs with independent pores, which allows the selection from multiple materials as necessary without the need for replacing the nozzles. Together, this enables the production of constructs that closely mimic the composition of real tissues and organs.

Similarly, for heterogeneous constructs with interconnected pores, the all-in-one nozzle is also suitable [43]. As an example, as shown in Figure 11.5, using three-in-one nozzles, one of the inlet channels is continuously pumped with pre-cooled gelatin ink, while the other two inlet channels are selectively pumped with pre-cooled GelMA bioinks according to the pre-designed scheme. According

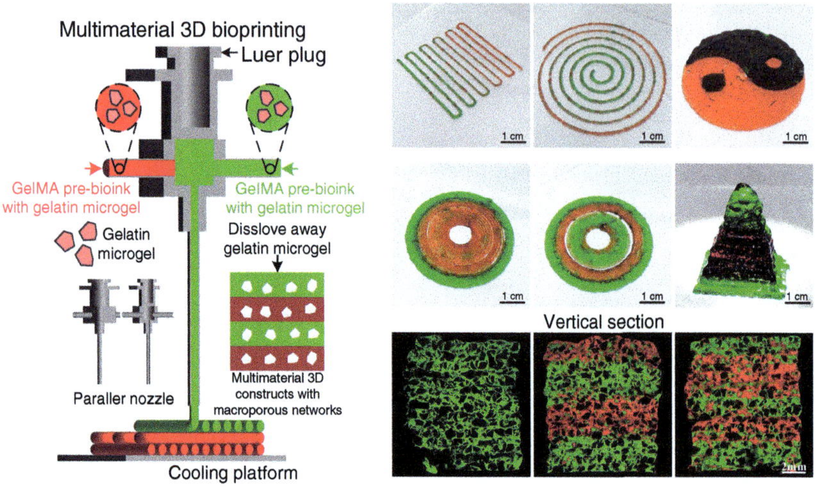

Figure 11.4 Heterogeneous 3D bioprinting of independent pore structure through an all-in-one nozzle. Source: Shao et al. [42]. Reproduced with permission of Nature Publishing Group.

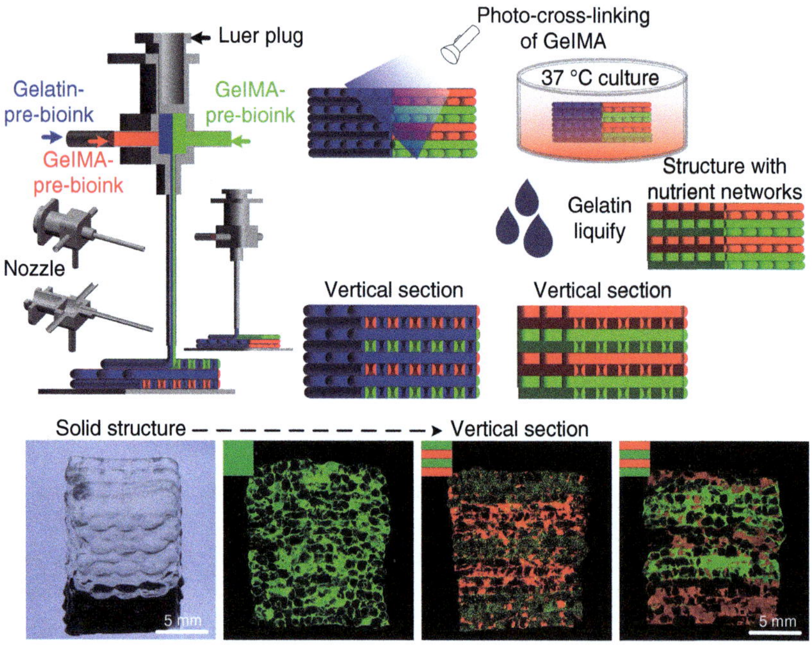

Figure 11.5 Heterogeneous 3D bioprinting of interconnected pore structure through the all-in-one nozzle. Source: Shao et al. [43]. Reproduced with permission of WILEY-VCH.

to the above method, heterogeneous structures with nutrient interconnected pores could be printed, as shown in Figure 11.5. And the vertical section views demonstrated effective nutrient networks. 3D constructs with interconnected pores showed that combining the all-in-one nozzle with 3D bioprinting is a useful technique for the creation of complex and heterogeneous constructs with nutrient networks on-demand. However, as the neighboring different cell types usually show more synergy for cell proliferation and differentiation, to construct more accurate multicellular structures, the printing accuracy of the multicellular structure needs to be further improved by increasing the response speed of switching materials.

11.2.4 Long-term Perfusion Culture on a Chip

Once the pore structure is well constructed, it would be overkill to use only its ability to promote nutrient diffusion, and for large tissue structures, static nutrient diffusion alone is not sufficient. Accordingly, it is necessary to conduct long-term perfusion culture of the whole tissue constructs on perfusion chips. More fascinatingly, multiple different kinds of interactional tissue/organ constructs can be cultured simultaneously with sequential perfusion on a chip. This has implications for studying the interactions between different tissues or organs and even for simulating the circulatory system throughout the organism. It should be known that the system of multi-organ interaction is closer to the real situation in vivo than the single organ, and the relevant application results are more accurate.

Typical perfusion culture chips include import channel, export channel, and tissue constructs storage chambers. For interconnected pore structure, during perfusion culture, the culture medium will gradually flow through each pore to promote nutrient/oxygen exchange and metabolite elimination; however, for independent pore structure, the exchange of matter between the pores is completely dependent on diffusion, and the efficiency of nutrient/oxygen transport will be much lower. Nevertheless, the study on the development and maturation process of large tissue constructs under dynamic perfusion culture on a chip is expected in the future.

11.2.5 Discussions (Properties, Pros, Cons, etc.)

Porous structures, either independent pore structure or interconnected pore structure, can effectively promote nutrition delivery but cannot mimic blood vessels and cannot simulate a blood flow environment. Perhaps, endothelial cells (ECs) can be mixed into the sacrificial template for pore formation, and ECs will adhere to the pore surface during the removal of the sacrificial ink template. With further perfusion culture, angiogenesis may be triggered, thus enriching vascularization of the whole tissue. Additionally, the porous structure may induce other behaviors of cells. These all need to be studied further. Certainly, to simulate blood vessels, vascular channels should be constructed.

In addition, it is worth mentioning that except for promoting nutrient flow, porous structure may play a role in the regulation of stem cell behavior, but it needs to be further verified. Meanwhile, the repair effect of porous structures in vivo may be better than expected.

11.3 Large Scale Cell-laden Structures with Vascular Channel Printing

Apparently, to better simulate the vascular system in vivo, the fabrication of artificial vascular channel structures is particularly important. Meanwhile, large cell-laden structures with vascular channels are closer to the real tissues in the organisms. With the development of 3D bioprinting, specific printing strategies were developed to create vascular channels, such as sacrificial bioprinting and coaxial bioprinting. The vascular channel has a biomimetic vascular structure, which is conducive to the flow of nutrients. Once the vessel channels were constructed, direct long-term perfusion culture can be performed.

11.3.1 Sacrificial Bioprinting

Sacrificial bioprinting involves using sacrificial ink with reversible sol–gel properties to construct structures, which are then removed to construct vascular channels [44, 45]. Accordingly, the key to sacrificial printing is the design and use of sacrificial ink, which should have good printability, such as F127, gelatin, agarose, sugar, etc. Generally, as shown in Figure 11.6, sacrificial bioprinting is divided into four steps: (i) the printing of a sacrificial bioink first; (ii) the embedding of printed sacrificial bioink structures in a hydrogel matrix via a molding process; (iii) dissolving the sacrificial bioink to form hollow vascular channels; (iv) the seeding of ECs on the surface of hollow vascular channels by perfusing a suspension of ECs. The

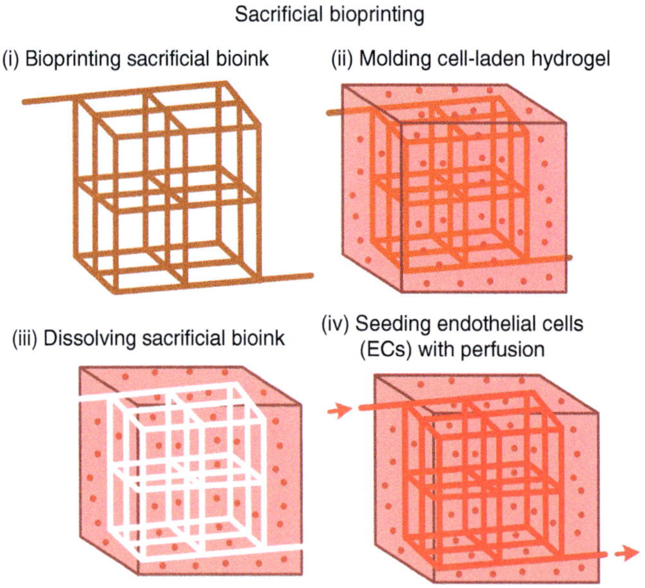

Figure 11.6 Sacrificial bioprinting of vascular channels. Source: Shao et al. [56]. Reproduced with permission of IOP Publishing.

vascular channels constructed through sacrificial printing can be directly perfused for long-term culture.

For sacrificial bioprinting, it is clever and simple to create the vascular channel. The sacrificial bioprinting strategy has achieved some success in the manufacture of vascularized tissues, but due to the molding process for hydrogel matrix, the fabrication of a large-scale complex cell-laden construct with vascular networks is not easily achieved. Also, due to the requirement to seed later ECs individually, it is difficult to seed ECs on complex vessel-like structures within a complex construct.

11.3.2 Coaxial Bioprinting

With the development of bioprinting, coaxial 3D bioprinting, a new method of manufacturing vascular channels, is proposed, which can print bulk cell-laden structures with hollow channels for nutrient transport inside large printed tissues [46]. Specifically, as shown in Figure 11.7, the cross-linker/hydrogel are simultaneously pumped into the inner/outer needles to generate hydrogel tubes, then deposit continuous hydrogel tubes to form 3D constructs. Subsequently, ECs were seeded on the surface of hollow vascular channels by perfusing ECs suspension. Finally, direct long-term perfusion culture can be conducted to promote tissue development and maturation. For coaxial bioprinting, because of their fast ionic cross-linking property, alginate-based hydrogels are commonly used bioinks.

Coaxial printing is a very simple and promising method for creating vascular channels. However, due to the poor biological performance of alginate-based hydrogels, the presence of alginate in this method inevitably affects the normal growth and development of encapsulated cells, although it could be gradually released. Also,

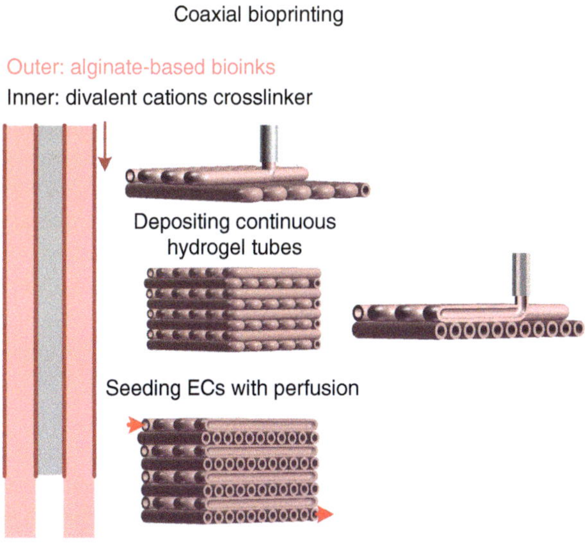

Figure 11.7 Coaxial bioprinting of vascular channels. Source: Shao et al. [56]. Reproduced with permission of IOP Publishing.

due to the continuity of the hydrogel tubes, the generation of complex 3D constructs with vascular networks is not easily achieved. Moreover, as the hydrogel tubes do not have sufficient mechanical strength to support the weight of the constructs, the structures will easily collapse, causing vascular channel failure. Additionally, this method also requires the seeding ECs on vascular networks by perfusing ECs suspension, so it is difficult to effectively vascularize complex structures. Accordingly, biomaterials with better biological properties and more efficient methods of seeding endothelium cells are needed to improve coaxial printing methods.

11.4 One-step Coaxial/Sacrificial Printing of Large Scale Vascularized Tissue Constructs

Accordingly, a complex construct with vascular channels is not easily achievable to date. Moreover, due to the dead zones of endothelialization, seeding ECs by perfusion is not practical for complex cell-laden construct with vascular channels. The strong demand for printing endothelialized vascular networks will lead to the development of relevant manufacturing methods. Thus, a strategy capable of synchronously bioprinting complex tissue constructs and vascular networks with self-seeding ECs without perfusion is proposed (see the details below) [56].

11.4.1 Mechanism

To bioprint complex tissue constructs with endothelialized vascular networks, we designed a one-step coaxial/sacrificial printing strategy of large-scale vascularized tissue constructs, which can self-seed ECs without perfusion, as shown in Figure 11.8 [56]. Cleverly taking advantage of the reversible thermo-cross-linking property of gelatin and GelMA and the irreversible photo-cross-linking property of GelMA, this coaxial 3D bioprinting strategy can directly print solid and self-support structure, which consists of solid sheath-core (GelMA-gelatin) fibers and has good mechanical properties to ensure fidelity of the printed structure the during printing process. Concretely, as shown in Figure 11.8a, first, the ECs-laden pre-cooled gelatin bioink and the other tissue cells-laden pre-cooled GelMA bioink were pumped into the inner and outer channels of the coaxial nozzle, respectively, and the temporary self-supporting solid structure was printed on the thermostatic refrigerating plate. Then, to obtain a stable structure, GelMA permanent photo-cross-linking was performed on the entire structure. Finally, in the subsequent culture process, the core fibers (gelatin) are dissolved to generate an effective hollow channel network, which facilitates nutrient/oxygen transport and improves the bioactivity of the entire structure. Through this direct coaxial 3D bioprinting method, simple and complex large-scale constructs with highly organized vascular networks can be easily printed, as shown in Figure 11.8b. In particular, ECs can automatically deposit and adhere to the inner wall of the vascular networks forming a vascular structure, as shown in Figure 11.8c. Furthermore, the cells (mouse osteoblast cell line [MC3T3-E1]) encapsulated in the bioprinted bulk constructs can stretch, migrate, and connect within a period of culture, as shown in Figure 11.8c. In addition, macroscale tissues

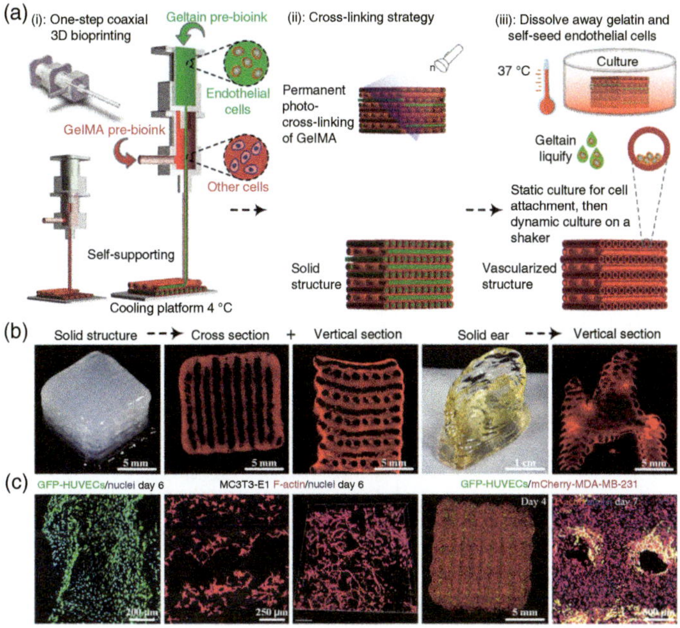

Figure 11.8 (a) Mechanism of one-step coaxial/sacrificial printing strategy of large-scale vascularized tissue constructs. Source: Shao et al. [56]. Reproduced with permission of IOP Publishing. (b) Through this direct coaxial 3D bioprinting method, simple and complex large-scale constructs with highly organized vascular networks can be easily printed. Source: Shao et al. [56]. Reproduced with permission of IOP Publishing. (c) ECs can automatically deposit and adhere to the inner wall of the vascular networks forming a vascular structure; MC3T3-E1 encapsulated in the bioprinted bulk constructs can stretch, migrate, and connect within a period of culture; tumor tissue with vascular networks can be fabricated. Source: Shao et al. [56]. Reproduced with permission of IOP Publishing.

(e.g. tumor tissue) with vascular networks were fabricated to demonstrate the potential functional versatility of the 3D bioprinting strategy for tissue engineering, as shown in Figure 11.8c.

Summarily, by ingeniously utilizing the GelMA (tissue cells-laden bioink) and gelatin (sacrificial ECs-laden bioink), a simple and practical one-step coaxial 3D bioprinting strategy was proposed, which (i) can directly bioprint cell-laden large-scale freeform constructs with vascular networks; (ii) can self-seed ECs on entire vascular networks without perfusion; (iii) enable the biological functionalization of encapsulated cells in the large-scale bioprinted. These characteristics make this strategy superior to current methods for the fabrication of large-scale vascularized tissue constructs, indicating its great potential for tissue engineering, regenerative medicine, organ printing, and drug delivery applications.

11.4.2 Freeform Structure with Vascular Channels Printing

As the structure of organism tissues/organs is mostly complex and irregular, the capability of bioprinting freeform constructs with vascular channels is important

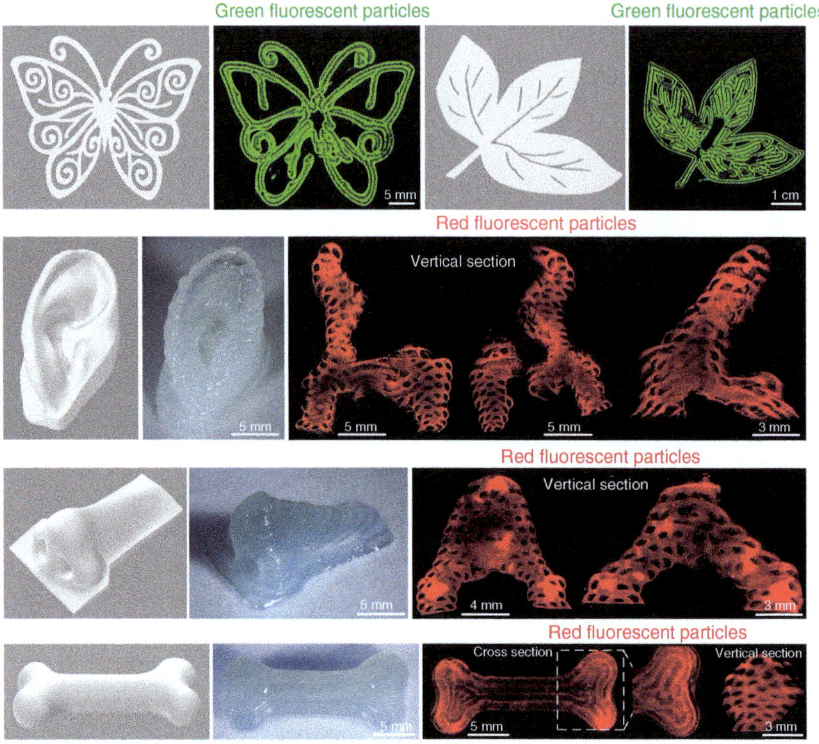

Figure 11.9 Based on the above one-step coaxial/sacrificial printing strategy, the complex desired 2D/3D constructs with high fidelity could be easily created, such as butterfly, leave, ear, nose, and bone. The vertical section or cross-section views showed the vascular channels, which closely mimic the abundant vascular distribution of tissues or organs in vivo. Source: Shao et al. [56]. Reproduced with permission of IOP Publishing.

to mimic the complex structures of natural tissues. Based on the above one-step coaxial/sacrificial printing strategy, as shown in Figure 11.9, the complex desired 2D/3D constructs with high fidelity could be easily created, such as butterfly, leave, ear, nose, and bone [56]. Meanwhile, the bioprinted high-fidelity constructs maintained their vascular channels embedded in the whole constructs. The vertical section or cross-section views showed the vascular channels, which closely mimic the abundant vascular distribution of tissues or organs in vivo. The easy fabrication of complex structures indicates the strong manufacturing capability of the strategy. Accordingly, this advanced bioprinting strategy has significant potentials in building large-scale or organ-level vascularized tissue toward applications in tissue engineering, ultimately, perhaps in organ transplantation and repair.

11.4.3 Heterogeneous Structure with Vascular Channels Printing

The tissues/organs of organisms are generally composed of a variety of cells and extracellular matrices. In addition to building vascular channels to ensure cell survival and development, heterogeneity of tissue structure should also be simulated.

Accordingly, heterogeneous constructs with tunable compositions are highly needed to mimic the physical properties of native tissues comprising multiple cell types and the extracellular matrix (ECM). To bioprint multicomponent/multicellular tissue constructs, a practical all-in-one coaxial nozzle was designed, which allows printing multiple components without changing the nozzle to mimic the heterogeneity of native tissues [56]. Specifically, as shown in Figure 11.10, different cell-laden GelMA bioinks or GelMA inks with different properties were selectively pumped into the outer channels of the all-in-one coaxial nozzle, while ECs-laden gelatin bioink was pumped into the only inner channel of the all-in-one coaxial nozzle, then the pre-designed multicomponent construct is printed on a thermostatic cooling platform. Subsequently, after permanent photocross-linking GelMA, the whole structure was cultured at 37 °C. During the culture process, GelMA would dissolve

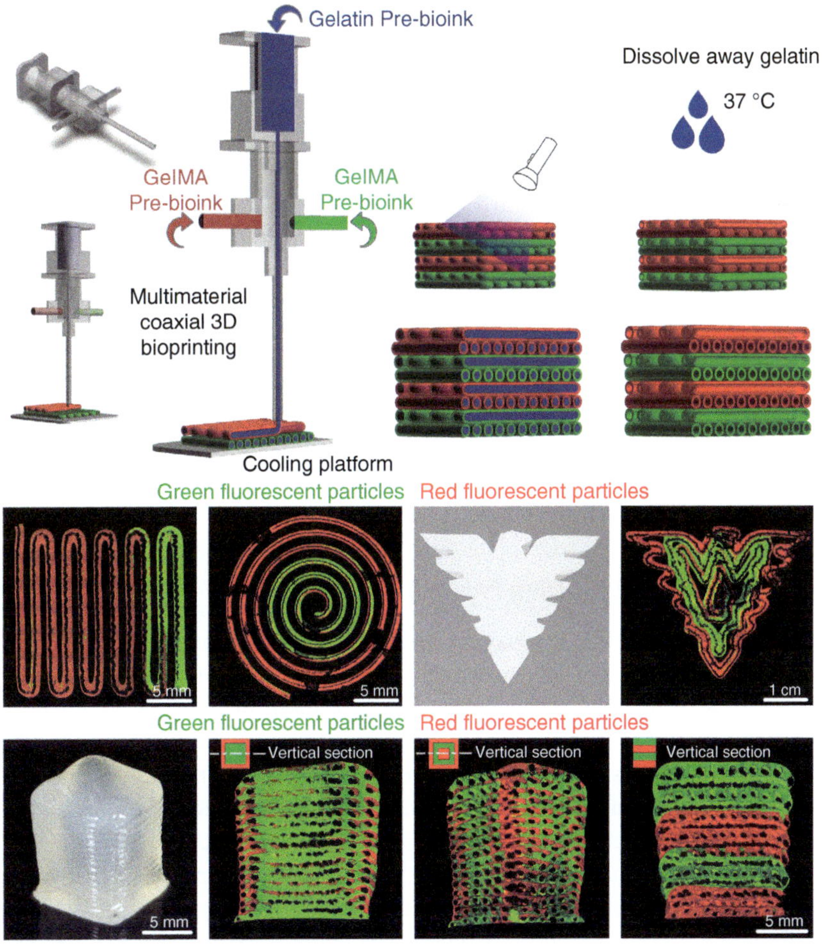

Figure 11.10 Bioprinting multicomponent/multicellular tissue constructs through a practical all-in-one coaxial nozzle. Source: Shao et al. [56]. Reproduced with permission of IOP Publishing.

and sacrifice quickly to form vascular channels, and ECs would automatically deposit and adhere to the inner wall of the channel to form endothelialized vascular channels. As an example, based on a two-in-one coaxial nozzle, two-component structures with vascular networks were printed, as shown in Figure 11.10. The multicomponent 2D patterns showed that this multi-component printing strategy can quickly switch between different components in a programmable way. The multicomponent 3D structures showed that the multi-component printing strategy using an all-in-one coaxial nozzle might provide a simple and effective method for the creation of complex and heterogeneous constructs on demand. Meanwhile, the vertical- or cross-sectional views showed good and rich vascular channels. It is worth mentioning that the internal and external channels of an all-in-one coaxial nozzle can be extended to as many as seven, indicating that this system has a strong multicomponent manufacturing capacity. This unique feature of this bioprinting system could provide unprecedented convenience to fabricate freeform vascularized constructs consisting of multicomponents when necessary, without the need for replacing the nozzle, closely mimicking the real tissues.

11.4.4 Long-term Perfusion Culture on a Chip

Similarly, structures containing vascular channels require long-term perfusion culture to improve nutrient/oxygen exchange and metabolite elimination, thereby facilitating cell growth, development, and tissue maturation. Similar to the porous structure, due to the openness of the vascular channels, the whole tissue constructs need long-term perfusion culture on a chip. However, the nutrient/oxygen flow efficiency of the vascular channel is higher than that of the porous structure. Moreover, with the one-step coaxial/sacrificial bioprinting strategy, the endothelialized vascular channel can be constructed, which might be able to perform some biological functions of blood vessels after a period of perfusion culture. Of course, these need to be studied further.

11.4.5 Discussion (Properties, Pros and Cons, etc.)

Technically and prospectively speaking, 3D bioprinting of endothelialized vascular channel networks is a great advancement in the field of large-scale vascularized tissue manufacturing. Yet, to promote vascularization and biological function, long-term perfusion culture needs to be further performed on a chip. Additionally, nerve cells can be easily introduced into the above-mentioned one-step coaxial/sacrificial bioprinting strategy to fabricate vascularized tissue constructs with nerve conduction. For in vivo applications, such as tissue repair, the constructs with vascular networks is possible to promote repair effect and regeneration rate. It has to be mentioned that although the vascular channel created by the above strategy can simulate the vascular tubular structure, it cannot construct the real vascular morphology in vivo. To create a vascular structure that approximates the real morphology of the organisms, it is necessary to develop high-precision bio-printing technology.

11.5 Advanced Bioprinting Technique-Enabled Printing Highly Biomimetic Tissues

Due to the strong demand for whole tissue or organ manufacturing, the development of more advanced bio-manufacturing technology has always been a hot and difficult issue. Of course, as the latest bio-manufacturing technology, 3D bioprinting is also in rapid development. With the development of bioprinting, some high-precision bioprinting technologies have appeared, such as support bath-assisted bioprinting [35, 57–66] and light-based bioprinting [67–79], which can print high biomimetic tissue structures containing vascular systems.

11.5.1 Support Bath-Assisted Bioprinting

As shown in Figure 11.11, the support bath-assisted bioprinting is deposition and embedding of the cell-laden hydrogels (bioinks) within a support bath that could maintain the desired structure during the print process and significantly improves print fidelity [64]. For support bath-assisted bioprinting, the key lies in the development of support bath materials. The support bath acts like a Bingham plastic during the print process, behaving as a rigid body at low shear stresses but flowing as a viscous fluid at higher shear stresses. Using the support bath mechanism, when a needle-like nozzle moves through the bath, there is little mechanical resistance, yet the cell-laden hydrogel being extruded out of the nozzle and deposited within the bath is embedded in situ. Since support bath acts as a rigid body at low shear stresses, it can well maintain the printed structure. This means that soft materials that would collapse if printed in air are easily maintained in the support bath-assisted bioprinted constructs. Significantly, this strategy greatly reduces the requirement for printability of the bioinks, and bioinks with superior biological performance and weak mechanical properties can be easily printed. It greatly alleviates the contradiction between the low mechanical strength requirement of biological performance and the high mechanical performance requirement of printability for bioinks. With the support bath-assisted bioprinting technology, after removing the support bath material, high biomimetic tissue structures containing vascular systems can be printed. Accordingly, support bath-assisted bioprinting is a promising manufacturing strategy and has great application potential in tissue engineering.

11.5.2 Light-Based Bioprinting

Another novel high-precision bio-printing strategy is light-based bioprinting photocross-link bioinks. Light-based bioprinting can project millions of voxels per time to form light image projection, initiating photocross-linking to cure bioink and form a two-dimensional plane printing unit. Then, 3D constructs can be printed through layer-by-layer projection. The advantages of light-based bioprinting are that it can print with high precision without support, and printing speed is high. Among the many 3D bioprinting technologies, light-based bioprinting is recognized

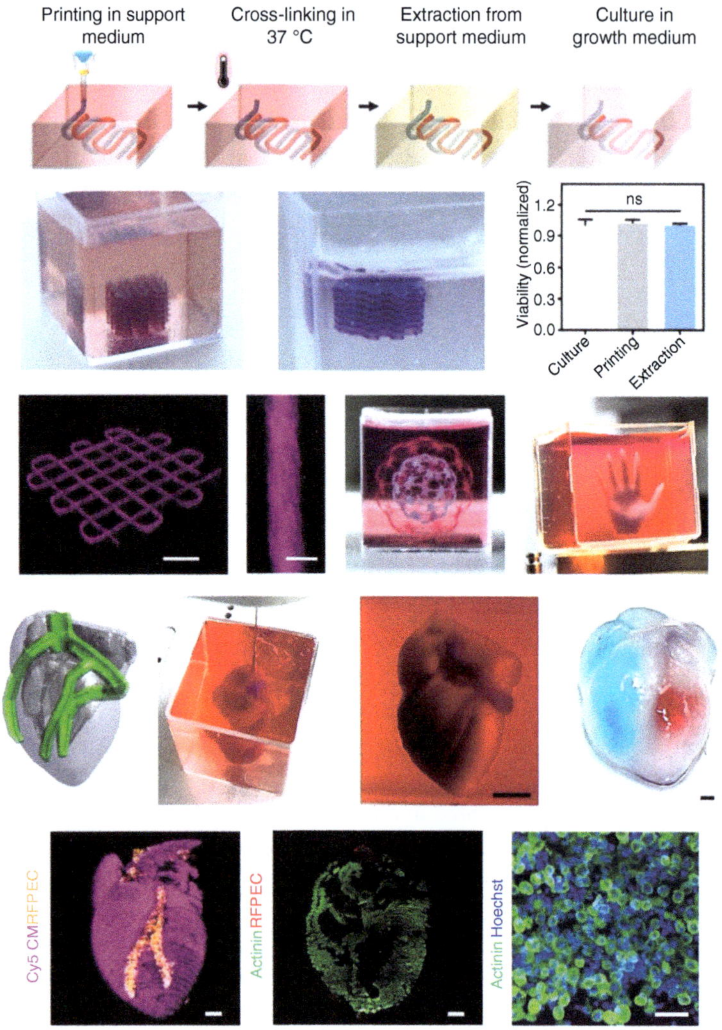

Figure 11.11 Support bath-assisted bioprinting of high biomimetic tissue structures containing vascular systems. Source: Noor et al. [64]. https://onlinelibrary.wiley.com/doi/full/10.1002/advs.201900344 CC-BY 4.0.

as one of the technologies that can meet relatively high standards in terms of model complexity and precision. However, the bioinks used in light-based bioprinting must have the characteristics of photocross-linking, which is quite limited. Moreover, at present, some photocurable bioinks generally have a low viscosity when printing; the cells are easy to deposit, resulting in the uneven distribution in the printed structure. But in general, with light-based bioprinting technology, the high precision and biomimetic tissue/organ constructs containing the vascular system can be printed, as shown in Figure 11.12 [79]. Light-based bioprinting is a promising technology that has great potential and is still in development.

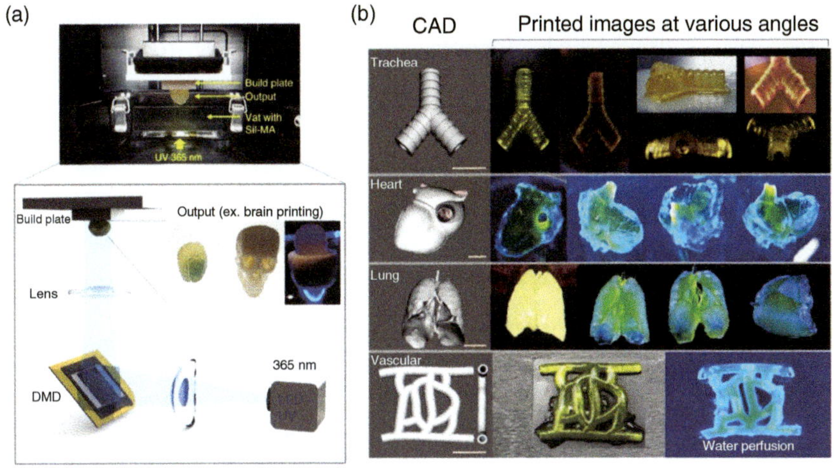

Figure 11.12 Light-based bioprinting of high biomimetic tissue structures containing vascular systems. Source: Kim et al. [79]. https://www.nature.com/articles/s41467-018-03759-y CC-BY 4.0.

11.5.3 Discussion (Properties, Pros and Cons, etc.)

In the field of engineering large-scale tissues/organs, support bath-assisted bioprinting and light-based bioprinting have great potential. These technologies can print highly biomimetic tissue structures containing the vascular system, which can be long-term cultured by direct perfusion. The fabrication of high-accuracy vascular systems can simulate blood circulation by direct perfusion and facilitate vascular anastomosis during tissue or organ transplantation. Although highly biomimetic tissue constructs can be achieved, there is still a long way to go to fabricate functional organ-level substitutes due to the complex physical environment inside the organisms.

11.6 Representative Biomedical Applications

The bioprinted large-scale tissue constructs can lay the foundation for the ambitious goal of organ repair/replacement in vivo and can also serve as an in vitro pathological model for drug screening, organ development, and lesions.

For tissue/organ models in vitro, in the last decade, bioprinting has made considerable progress, and a large number of studies have shown attractive achievements in different applications, such as tissue/organ development, drug discovery, and disease modeling. Moreover, the in vitro application of bioprinting model is still developing rapidly.

For the ultimate goal of bioprinting, due to the clinical problems of biomaterials and cells, the tissues/organs repair/replacement in vivo only stays at the stage of animal experiments, which still has a long way to go. With the development of biomaterials/cells and the realization of clinical application, especially hydrogels and stem cells, bioprinted models for human tissues/organs repair/replacement will eventually be available.

References

1 Kolesky, D.B., Truby, R.L., Gladman, A.S. et al. (2014). 3D bioprinting of vascularized, heterogeneous cell-laden tissue constructs. *Advanced Materials* 26 (19): 3124–3130.
2 Murphy, S.V. and Atala, A. (2014). 3D bioprinting of tissues and organs. *Nature Biotechnology* 32 (8): 773–785.
3 Mandrycky, C., Wang, Z., Kim, K. et al. (2016). 3D bioprinting for engineering complex tissues. *Biotechnology Advances* 34 (4): 422–434.
4 Kang, H.W., Lee, S.J., Ko, I.K. et al. (2016). A 3D bioprinting system to produce human-scale tissue constructs with structural integrity. *Nature Biotechnology* 34 (3): 312–319.
5 Ozbolat, I.T. and Yu, Y. (2013). Bioprinting toward organ fabrication: challenges and future trends. *IEEE Transactions on Biomedical Engineering* 60 (3): 691–699.
6 Mironov, V., Trusk, T., Kasyanov, V. et al. (2009). Biofabrication: a 21st century manufacturing paradigm. *Biofabrication* 1 (2): 022001.
7 Carmeliet, P. and Jain, R.K. (2000). Angiogenesis in cancer and other diseases. *Nature* 407 (6801): 249.
8 Jain, R.K., Au, P., Tam, J. et al. (2005). Engineering vascularized tissue[J]. *Nature Biotechnology* 23 (7): 821.
9 Rouwkema, J. and Khademhosseini, A. (2016). Vascularization and angiogenesis in tissue engineering: beyond creating static networks. *Trends in Biotechnology* 34 (9): 733–745.
10 Rouwkema, J., Rivron, N.C., and van Blitterswijk, C.A. (2008). Vascularization in tissue engineering. *Trends in Biotechnology* 26 (8): 434–441.
11 Mironov, V., Boland, T., Trusk, T. et al. (2003). Organ printing: computer-aided jet-based 3D tissue engineering. *TRENDS in Biotechnology* 21 (4): 157–161.
12 Melchels, F.P.W., Domingos, M.A.N., Klein, T.J. et al. (2012). Additive manufacturing of tissues and organs. *Progress in Polymer Science* 37 (8): 1079–1104.
13 Schubert, C., Van Langeveld, M.C., and Donoso, L.A. (2014). Innovations in 3D printing: a 3D overview from optics to organs. *British Journal of Ophthalmology* 98 (2): 159–161.
14 Gross, B.C., Erkal, J.L., Lockwood, S.Y. et al. (2014). Evaluation of 3D printing and its potential impact on biotechnology and the chemical sciences. *Analytical Chemistry* 86 (7): 3240–3253.
15 Ventola, C.L. (2014). Medical applications for 3D printing: current and projected uses. *Pharmacy and Therapeutics* 39 (10): 704.
16 Xu, T., Jin, J., Gregory, C. et al. (2005). Inkjet printing of viable mammalian cells. *Biomaterials* 26 (1): 93–99.
17 Calvert, P. (2007). Printing cells. *Science* 318 (5848): 208–209.

18 Cui, X., Breitenkamp, K., Finn, M.G. et al. (2012). Direct human cartilage repair using three-dimensional bioprinting technology. *Tissue Engineering Part A* 18 (11-12): 1304–1312.

19 Norotte, C., Marga, F.S., Niklason, L.E. et al. (2009). Scaffold-free vascular tissue engineering using bioprinting. *Biomaterials* 30 (30): 5910–5917.

20 Ferris, C.J., Gilmore, K.G., and Wallace, G.G. (2013). Biofabrication: an overview of the approaches used for printing of living cells. *Applied Microbiology and Biotechnology* 97 (10): 4243–4258.

21 Hölzl, K., Lin, S., Tytgat, L. et al. (2016). Bioink properties before, during and after 3D bioprinting. *Biofabrication* 8 (3): 032002.

22 He, Y., Yang, F.F., Zhao, H.M. et al. (2016). Research on the printability of hydrogels in 3D bioprinting. *Scientific Reports* 6: 29977.

23 Hospodiuk, M., Dey, M., Sosnoski, D. et al. (2017). The bioink: a comprehensive review on bioprintable materials. *Biotechnology Advances* 35 (2): 217–239.

24 Chung, J., H.Y., Naficy, S., Yue, Z. et al. (2013). Bio-ink properties and printability for extrusion printing living cells. *Biomaterials Science* 1 (7): 763–773.

25 Ji, S. and Guvendiren, M. (2017). Recent advances in bioink design for 3D bioprinting of tissues and organs. *Frontiers in Bioengineering and Biotechnology* 5: 23.

26 Ferris, C.J., Gilmore, K.J., Beirne, S. et al. (2013). Bio-ink for on-demand printing of living cells. *Biomaterials Science* 1 (2): 224–230.

27 Gungor-Ozkerim, P.S., Inci, I., Zhang, Y.S. et al. (2018). Bioinks for 3D bioprinting: an overview. *Biomaterials Science* 6 (5): 915–946.

28 Panwar, A. and Tan, L.P. (2016). Current status of bioinks for micro-extrusion-based 3D bioprinting. *Molecules* 21 (6): 685.

29 Stanton, M.M., Samitier, J., and Sanchez, S. (2015). Bioprinting of 3D hydrogels. *Lab on a Chip* 15 (15): 3111–3115.

30 Forget, A., Blaeser, A., Miessmer, F. et al. (2017). Mechanically tunable bioink for 3D bioprinting of human cells. *Advanced Healthcare Materials* 6 (20): 1700255.

31 Parak, A., Pradeep, P., du Toit, L.C. et al. (2019). Functionalizing bioinks for 3D bioprinting applications. *Drug Discovery Today* 24 (1): 198–205.

32 Tarassoli, S.P., Jessop, Z.M., Kyle, S. et al. (2018). *Candidate Bioinks for 3D Bioprinting Soft Tissue[M] 3D Bioprinting for Reconstructive Surgery*, 145–172. Woodhead Publishing.

33 Zhu, W., Ma, X., Gou, M. et al. (2016). 3D printing of functional biomaterials for tissue engineering. *Current Opinion in Biotechnology* 40: 103–112.

34 Zadpoor, A.A. and Malda, J. (2017). Additive manufacturing of biomaterials, tissues, and organs. *Annals of Biomedical Engineering* 45: 1–11.

35 Lee, J.M. and Yeong, W.Y. (2016). Design and printing strategies in 3D bioprinting of cell-hydrogels: A review. *Advanced Healthcare Materials* 5 (22): 2856–2865.

36 Guvendiren, M., Molde, J., Soares, R.M.D. et al. (2016). Designing biomaterials for 3D printing. *ACS Biomaterials Science & Engineering* 2 (10): 1679–1693.

37 Melchels, F.P.W., Blokzijl, M.M., Levato, R. et al. (2016). Hydrogel-based reinforcement of 3D bioprinted constructs. *Biofabrication* 8 (3): 035004.

38 Zhang, Y.S. and Khademhosseini, A. (2017). Advances in engineering hydrogels. *Science* 356 (6337): eaaf3627.

39 Broguiere, N., Husch, A., Palazzolo, G. et al. (2019). Macroporous hydrogels derived from aqueous dynamic phase separation. *Biomaterials* 200: 56–65.

40 Ying, G.L., Jiang, N., Maharjan, S. et al. (2018). Aqueous two-phase emulsion bioink-enabled 3D bioprinting of porous hydrogels. *Advanced Materials* 30 (50): 1805460.

41 Huebsch, N., Lippens, E., Lee, K. et al. (2015). Matrix elasticity of void-forming hydrogels controls transplanted-stem-cell-mediated bone formation. *Nature Materials* 14 (12): 1269–1277.

42 Shao, L., Gao, Q., Xie, C. et al. (2020). Sacrificial microgel-laden bioink-enabled 3D bioprinting of mesoscale pore networks. *Bio-Design and Manufacturing* 3 (1): 30–39.

43 Shao, L., Gao, Q., Xie, C. et al. (2020). Synchronous 3D bioprinting of large-scale cell-laden constructs with nutrient networks. *Advanced Healthcare Materials* 9 (15): 1901142.

44 Kolesky, D.B., Homan, K.A., Skylar-Scott, M.A. et al. (2016). Three-dimensional bioprinting of thick vascularized tissues. *Proceedings of the National Academy of Sciences* 113 (12): 3179–3184.

45 Miller, J.S., Stevens, K.R., Yang, M.T. et al. (2012). Rapid casting of patterned vascular networks for perfusable engineered three-dimensional tissues. *Nature Materials* 11 (9): 768–774.

46 Gao, Q., He, Y., Fu, J. et al. (2015). Coaxial nozzle-assisted 3D bioprinting with built-in microchannels for nutrients delivery. *Biomaterials* 61: 203–215.

47 Yu, Y., Zhang, Y., and Ozbolat, I.T. (2014). A hybrid bioprinting approach for scale-up tissue fabrication. *Journal of Manufacturing Science and Engineering* 136 (6): 061013.

48 Liu, W., Zhong, Z., Hu, N. et al. (2018). Coaxial extrusion bioprinting of 3D microfibrous constructs with cell-favorable gelatin methacryloyl microenvironments. *Biofabrication* 10 (2): 024102.

49 Mistry, P., Aied, A., Alexander, M. et al. (2017). Bioprinting using mechanically robust core–shell cell-laden hydrogel strands. *Macromolecular Bioscience* 17 (6): 1600472.

50 Jia, W., Gungor-Ozkerim, P.S., Zhang, Y.S. et al. (2016). Direct 3D bioprinting of perfusable vascular constructs using a blend bioink. *Biomaterials* 106: 58–68.

51 Gao, Q., Liu, Z., Lin, Z. et al. (2017). 3D bioprinting of vessel-like structures with multilevel fluidic channels. *ACS Biomaterials Science & Engineering* 3 (3): 399–408.

52 Liang, H., He, J., Chang, J. et al. (2018). Coaxial nozzle-assisted electrohydrodynamic printing for microscale 3D cell-laden constructs. *International Journal of Bioprinting* 4 (1): 1–8.

53 Pi, Q., Maharjan, S., Yan, X. et al. (2018). Digitally tunable microfluidic bioprinting of multilayered cannular tissues. *Advanced Materials* 30 (43): 1706913.

54 Zhu, K., Chen, N., Liu, X. et al. (2018). A general strategy for extrusion bio-printing of bio-macromolecular bioinks through alginate-templated dual-stage crosslinking. *Macromolecular Bioscience* 18 (9): 1800127.

55 Liu, X., Carter, S.S.D., Renes, M.J. et al. (2019). Development of a coaxial 3D printing platform for biofabrication of implantable islet-containing constructs. *Advanced Healthcare Materials* 8 (7): 1801181.

56 Shao, L., Gao, Q., Xie, C. et al. (2020). Directly coaxial 3D bioprinting of large-scale vascularized tissue constructs. *Biofabrication* 12 (3): 035014.

57 Hinton, T.J., Jallerat, Q., Palchesko, R.N. et al. (2015). Three-dimensional printing of complex biological structures by freeform reversible embedding of suspended hydrogels. *Science Advances* 1 (9): e1500758.

58 Suntornnond, R., An, J., and Chua, C.K. (2017). Roles of support materials in 3D bioprinting-Present and future. *International Journal of Bioprinting* 3 (1).

59 Jin, Y., Compaan, A., Bhattacharjee, T. et al. (2016). Granular gel support-enabled extrusion of three-dimensional alginate and cellular structures. *Biofabrication* 8 (2): 025016.

60 Jin, Y., Chai, W., and Huang, Y. (2017). Printability study of hydrogel solution extrusion in nanoclay yield-stress bath during printing-then-gelation biofabrication. *Materials Science and Engineering: C* 80: 313–325.

61 Rocca, M., Fragasso, A., Liu, W. et al. (2018). Embedded multimaterial extrusion bioprinting. *SLAS TECHNOLOGY: Translating Life Sciences Innovation* 23 (2): 154–163.

62 Corbett, D.C., Olszewski, E., and Stevens, K. (2019). A FRESH take on resolution in 3D bioprinting. *Trends in Biotechnology* 37 (11): 1153–1155.

63 Mirdamadi, E., Muselimyan, N., Koti, P. et al. (2019). Agarose slurry as a support medium for bioprinting and culturing freestanding cell-laden hydrogel constructs. *3D Printing and Additive Manufacturing* 6 (3): 158–164.

64 Noor, N., Shapira, A., Edri, R. et al. (2019). 3D printing of personalized thick and perfusable cardiac patches and hearts. *Advanced Science* 6 (11): 1900344.

65 Lee, A., Hudson, A.R., Shiwarski, D.J. et al. (2019). 3D bioprinting of collagen to rebuild components of the human heart. *Science* 365 (6452): 482–487.

66 Compaan, A.M., Song, K., and Huang, Y. (2019). Gellan fluid gel as a versatile support bath material for fluid extrusion bioprinting. *ACS Applied Materials & Interfaces* 11 (6): 5714–5726.

67 Nakamura, M., Iwanaga, S., Henmi, C. et al. (2010). Biomatrices and biomaterials for future developments of bioprinting and biofabrication. *Biofabrication* 2 (1): 014110.

68 Zhang, A.P., Qu, X., Soman, P. et al. (2012). Rapid fabrication of complex 3D extracellular microenvironments by dynamic optical projection stereolithography. *Advanced Materials* 24 (31): 4266–4270.

69 Wang, Z., Abdulla, R., Parker, B. et al. (2015). A simple and high-resolution stereolithography-based 3D bioprinting system using visible light crosslinkable bioinks. *Biofabrication* 7 (4): 045009.

70 Miri, A.K., Nieto, D., Iglesias, L. et al. (2018). Microfluidics-enabled multimaterial maskless stereolithographic bioprinting. *Advanced Materials* 30 (27): 1800242.

71 Shanjani, Y., Pan, C.C., Elomaa, L. et al. (2015). A novel bioprinting method and system for forming hybrid tissue engineering constructs. *Biofabrication* 7 (4): 045008.

72 Raman, R. and Bashir, R. (2015). *Stereolithographic 3D bioprinting for biomedical applications[M] Essentials of 3D Biofabrication and Translation*, 89–121. Academic Press.

73 Derakhshanfar, S., Mbeleck, R., Xu, K. et al. (2018). 3D bioprinting for biomedical devices and tissue engineering: A review of recent trends and advances. *Bioactive Materials* 3 (2): 144–156.

74 Morris, V.B., Nimbalkar, S., Younesi, M. et al. (2017). Mechanical properties, cytocompatibility and manufacturability of chitosan: PEGDA hybrid-gel scaffolds by stereolithography. *Annals of Biomedical Engineering* 45 (1): 286–296.

75 Wang, Z., Kumar, H., Tian, Z. et al. (2018). Visible light photoinitiation of cell-adhesive gelatin methacryloyl hydrogels for stereolithography 3D bioprinting. *ACS Applied Materials & Interfaces* 10 (32): 26859–26869.

76 Zhu, W., Qu, X., Zhu, J. et al. (2017). Direct 3D bioprinting of prevascularized tissue constructs with complex microarchitecture. *Biomaterials* 124: 106–115.

77 Koffler, J., Zhu, W., Qu, X. et al. (2019). Biomimetic 3D-printed scaffolds for spinal cord injury repair. *Nature Medicine* 25 (2): 263.

78 Grigoryan, B., Paulsen, S.J., Corbett, D.C. et al. (2019). Multivascular networks and functional intravascular topologies within biocompatible hydrogels. *Science* 364 (6439): 458–464.

79 Kim, S.H., Yeon, Y.K., Lee, J.M. et al. (2018). Precisely printable and biocompatible silk fibroin bioink for digital light processing 3D printing. *Nature Communications* 9 (1): 1–14.

12

3D Printing of Vascular Chips

12.1 Introduction

Vascular network is crucial to the survival and functionality of most tissues/organs, while vascularization represents one of the critical challenges in tissue engineering and organ-on-a-chip fields [1]. The vascular system, containing both large blood vessels and capillaries, is the fundamental element of the human body. It supplies oxygen and nutrients to tissues/organs and also intently interacts with multiple biological occurrences of many tissues/organs. As a consequence, the reconstruction of the integrated blood supply and circulation system is of vast significance for the establishment of *in vitro* physiological-relevant models as well as the simulation of multiple *in vivo* physiological behaviors of the human body [2–4].

Plenty of strategies have been proposed for the building of large blood vessels, including the deformation of cell sheets into tubular structures [5], preparation of fibers based on cell-laden hydrogels [6], thin wire-based molding [7], application of sacrificial templates [8], coaxial 3D bio-printing [9], light-assisted curing [10], and layer-by-layer lamination [11].

As a robust advanced technology, 3D printing could be involved in the building of vascular chips in both direct and indirect ways. Hollow vascular constructs could be easily fabricated through coaxial 3D printing. On the other hand, fiber templates could also be printed, serving as a sacrificial layer of vascular chips.

However, the major difficulty lies in the construction of the bifurcation structures and the capillaries at the end [12]. Until recently, few studies have been reported about the engineering of capillaries, which are the necessary components of the vascular system. To remodel the organized and stable capillaries, a self-assembly strategy is proposed by extending new capillaries from the pre-existing vessels toward angiogenic signals [13]. However, the wide introduction of growth factors is cost-prohibitive and inefficient [14]. An alternative approach, referred to as a physical structure-based strategy, is to engineer ultrafine channels in combination with endothelial cell seeding [15]. This strategy brings convenience to regulate the distribution of the capillaries in both direction and range.

In spite of the considerable advancements, demands still exist for a practical protocol to construct large-scale tissue/organ models with the complete vascular system containing both large blood vessels and capillaries.

Cell Assembly with 3D Bioprinting, First Edition. Yong He, Qing Gao, and Yifei Jin.

At present, the mainstream micro-manufacturing process is mostly developed from the silicon-based micro processing technology in the semiconductor industry, which is not appropriate for handling the "fragile and brittle" hydrogel materials. In addition, hydrogel materials with poor mechanical strength and waterborne interface are incompatible with the conventional bonding strategies of traditional microfluidic chips. Furthermore, due to the inherent limitations of the traditional process, the construction of the vascular chips is mostly restricted to a certain scale range, while it is almost impossible to build the complete vascular system containing both large blood vessels and capillaries.

In this chapter, a damage-free demolding approach is proposed based on a novel soft fiber template, altering the surface contact of the traditional rigid mold to line contact so as to decrease the demolding stress while improving the precision and quality of the micro/nanostructures on the hydrogel surface. A twice cross-linking strategy is designed to realize the bonding of independent hydrogels between layers and the encapsulation of open microstructures. The critical rationale for this method is to prepare hydrogel sheets with microgroove structures based on a casting process and then layer-by-layer assembly of the prepared hydrogel sheets to encapsulate the channels based on the bonding process. The whole process is dependent on the inherent cross-linking properties of hydrogel materials. In addition, a multi-scale 3D printing technology is developed to achieve the manufacturing of soft fiber templates scaling from hundreds of nanometers to hundreds of microns through a shift of the printing mode between fused deposition modeling (FDM) printing and electro-hydraulic fluid dynamics (EHD) printing.

Finally, a robust and novel process is presented to construct hydrogel-based vascular chips with multi-scale and biomimicking structures. The whole process includes: (i) template design (vascular system), (ii) template printing (soft fiber network), (iii) hydrogel material casting (formation of partially-cross-linked hydrogel sheets), (iv) template peeling off (creation of microgrooves on the surfaces of hydrogel sheets), (v) hydrogel sheet bonding (formation of a closed channel network), and (vi) cell loading (specific cells distributed on specific positions mimicking *in vivo* situations). Based on these processes, hydrogel-based vascular chips could be constructed containing the entire blood supply and circulation network (from arteries, capillaries to veins). It is further validated that it is feasible to construct the ubiquitous architectures of *in vivo* vascular system with bio-mimicking vascular morphology based on this method. In addition, the engineered endothelium channel network could model the native vascular functionality on different physiological and pathological conditions.

12.2 Construction Process of Hydrogel-Based Vascular Chips

12.2.1 Damage-Free Demolding Process Based on Soft Fiber Template

Due to the brittle and fragile nature of hydrogel materials, the demolding stress introduced by the rigid mold during the traditional demolding process is frequently

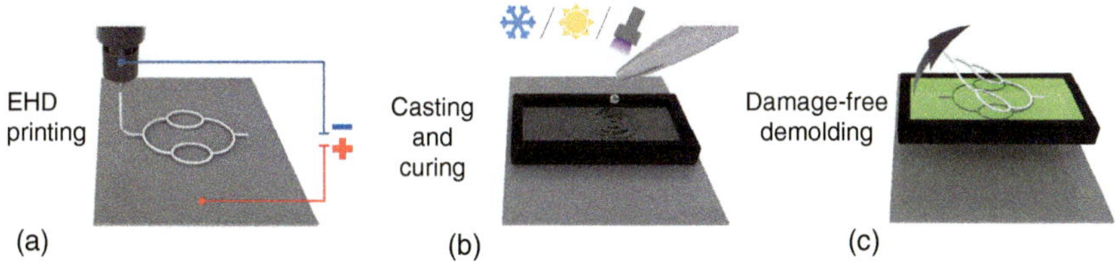

Figure 12.1 Fabrication of the hydrogel micro structure: (a) 3D printing of fiber template. (b) casting and curing the hydrogel precursor. (c) fiber peeling off. Source: Lv et al. [16]. Reproduced with permission of IOP Publishing.

causing defects and cracks within hydrogel structures, hindering the precise generation of the hydrogel microstructure, which is particularly unsuitable in cases of structures with high aspect ratio and fine complex geometry. The main cause for this destructive demolding stress is the large contact area between the hydrogel material and the rigid integrated mold, while stress concentration occurs in the corner of the micro–nanostructures, which will further cause the generation of huge friction and adhesion force.

12.2.1.1 Damage-Free Demolding Process

It could be envisaged that through the decrease of the contact area between hydrogels and molds, the demolding stress could be decreased, while the demolding quality could be improved. A novel soft fiber template is proposed to replace the traditional rigid mold, thereby changing the mold contacting forms. Based on this, a damage-free demolding method is proposed for the formation of microstructures based on the brittle hydrogel materials, as shown in Figure 12.1. The entire damage-free demolding process is presented as follows:

First, template materials are deposited on the substrate according to the pre-defined printing path through extrusion 3D printing to obtain a soft fiber template with a specific geometry; next, an appropriately-sized frame is placed on the substrate surrounding the entire fiber template. Then, the hydrogel precursor solution is poured onto the template within this frame, and the corresponding cross-linking conditions are applied according to the hydrogel type so as to realize the pre-forming of the hydrogel. Finally, the cured integrity (soft fiber template, outer frame, and hydrogel structure) is detached from the substrate, and the fiber template is little by little peeled off the surface of the hydrogel to obtain the microstructure on the hydrogel surface.

12.2.1.2 Comparative Analysis of Damage-Free and Conventional Demolding Processes

The body material of the traditional mold is usually rigid material; this kind of rigid mold is generally with an inseparably integrated structure. The casted structure is closely fitting with the mold and is constrained by the mold from all directions.

The contact area between the material and the mold covers the entire bottom surface, while the cured structure has to be completely detached from this surface. This process will bring about large friction and adhesion force, which are positively

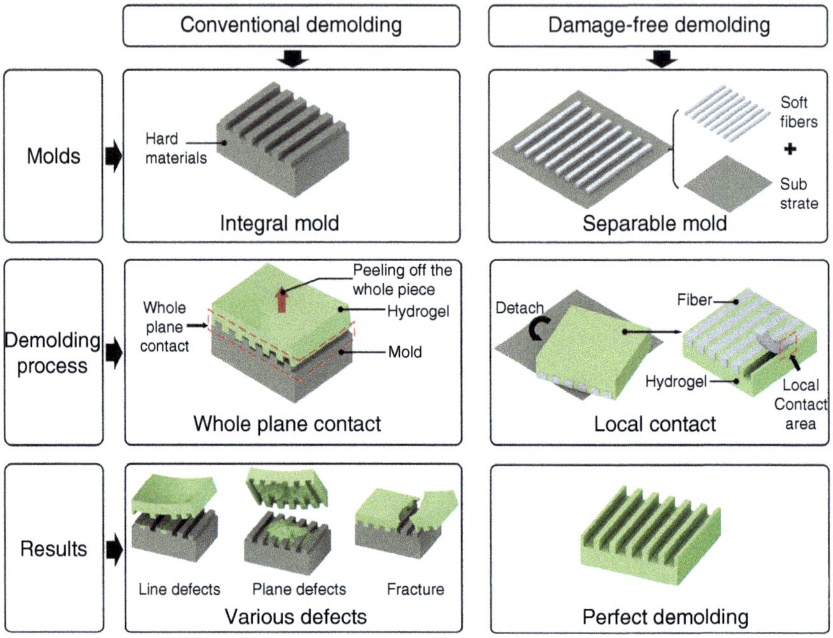

Figure 12.2 Preliminary comparison of damage-free demolding and the conventional one with respect to molds, demolding process and quality of demolding. Source: Lv et al. [16]. Reproduced with permission of IOP Publishing.

related to the contacting area. As a result, although this method is suitable for high-strength and high-elasticity materials, it is not compatible with brittle materials such as hydrogels, especially not applicable for structures with high aspect ratios and micro-scale sizes.

Unlike traditional molds, the soft fiber frame template is composed of detachable soft fiber, a peripheral frame, and a substrate. After the cross-linking is completed, the soft fiber template is detached from the substrate along with the hydrogel structure. Further, the point-by-point peeling of the soft fiber template off the hydrogel is conducted to replace the integrated detachment of the hydrogel structure off the rigid mold, making the demolding process more flexible and easier to operate, and the contacting area is converted into a local micro-scale line-contact, which could greatly reduce the demolding stress, effectively avoid structural damage and defects, and improve the integrity and precision of the structure (Figure 12.2).

Different from the overall surface contact between the rigid mold and the hydrogel material during the traditional demolding process, there exists only local linear contact between the soft fiber template and the hydrogel material in the damage-free demolding process, which greatly decreases the friction and adhesion forces. Moreover, during the traditional demolding process, the hydrogel structure has to undergo bending deformation to detach from the rigid mold, leading to bending stress which influences the structure precision. As for the damage-free demolding method, the soft fiber template itself could realize bending and deforming, so that it could be peeled off the surface of the hydrogel material little by little, avoiding the excessive stress caused by the deformation of the overall hydrogel structure. In a word, the damage-free demolding method reduces the demolding stress and improves the demolding quality by changing the template form and the demolding means.

During the demolding process, there are four key factors affecting the stress condition, including interface adhesion, mechanical friction, bending stress, and additional bending moment.

First of all, since the crosslinked hydrogel material is closely fitted with the mold surface, intermolecular interactions inevitably exist on the contacting surface, including electrostatic force, van der Waals force, and capillary force, further leading to the generation of interfacial adhesion. Second, during the demolding process, there is a relative movement trend between the hydrogel material and the mold. Thus, the two independent contacting layers pull each other, which makes the two parts simultaneously withstand mechanical friction, resulting in resistance deformation. Due to the rigidity of the mold, the elasticity of the hydrogel material itself, and the uneven detachment process, deformation will inevitably happen to the hydrogel structure, resulting in additional bending stress. In addition, the asymmetry of the demolding process prevents the continuous maintenance of its vertical direction, and the inclination angle formed during the process causes the hydrogel to resist the extrusion of the mold side wall, thereby introducing an additional moment.

The above four factors are firstly analyzed from a microscopic perspective: the contact area of a single fiber with the hydrogel is significantly smaller than the overall contact between the bottom surface of the mold and the hydrogel in the traditional way, leading to smaller interface adhesion. During the damage-free demolding process, the existing friction force is unidirectional, which is half of the friction force in the traditional demolding method. As the fibers are independent of each other, there is no need to peel them off at the same time, and only a single soft fiber is pulled at a single time point to detach from the hydrogel surface. Only the soft fibers are bending during the peeling process, while the overall hydrogel structure is not deforming, and thus, there is no excess bending stress.

The specific implementation of the demolding process is more likely to be related to the scale and pattern of the overall structure. Thus, further macroscopic mechanical analysis is conducted. In the traditional demolding method, the entire hydrogel structure has to be detached from the mold; thus, the entire surface of the hydrogel is contacted and interacted with the mold. The macroscopic overall resistance is the sum of the forces and moments of each tiny unit. The larger contact surface and the more complicated geometry of the mold will lead to a greater overall adhesion force. On the contrary, during the damage-free demolding process, the entire detaching is replaced by the peeling of a single fiber, and there is only a stable adhesion between a single fiber and the local area of the hydrogel during the whole process, which has nothing to do with the size and structure. In addition, as the thickness of the hydrogel is increased, the bending deformation is more likely to occur, which in turn will generate greater bending stress; whereas, bending deformation of the hydrogel

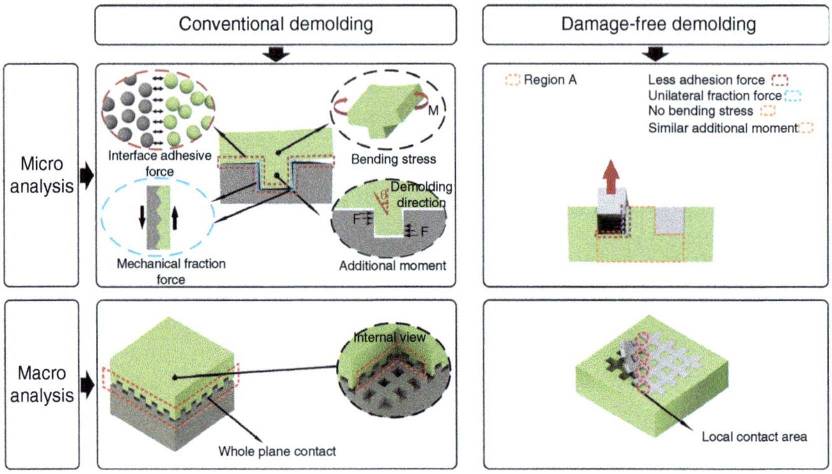

Figure 12.3 Mechanical analysis of two demolding processes including the micro and macro analysis of the conventional demolding and damage-free demolding. Source: Lv et al. [16]. Reproduced with permission of IOP Publishing.

structure is not involved in the damage-free demolding process, and the thickness of the overall structure has no influence on the demolding effect. In a word, for the traditional demolding method, the increase in contacting area, structure thickness, and geometry complexity of the mold will lead to an increase in demolding stress as well as structural defects. As for the damage-free demolding process, there is only one local contact in real-time during the peeling of fiber template; thus, the stress condition from the macroscopic perspective is almost the same as that from the micro aspect, while there is no size or structure effect. Obviously, in the cases where the pattern structures are complex, the aspect ratio is high, and the overall structure size and thickness are large, the superiorities of damage-free demolding over traditional demolding are more apparent (Figure 12.3).

The comprehensive mechanical analysis results indicate that the soft fiber template has certain advantages over the traditional rigid mold, and this advantage is becoming more and more obvious as the contacting area between the hydrogel and the mold increases.

12.2.2 Hydrogel Bonding Strategy Based on Twice-Cross-linking Mechanism

Targeting at the poor strength and water interface of the hydrogel material, it is envisaged to realize the bonding process based on the unique cross-linking reaction properties of the hydrogel material itself. A twice-cross-linking strategy is designed and proposed to achieve efficient bonding between hydrogel layers and the encapsulation of hydrogel microstructures, so as to obtain a hydrogel-based microfluidic chip with reliable perfusion capability. Based on the inherent cross-linking properties of the hydrogel material, the specific scheme is to obtain a composite hydrogel by mixing two hydrogel materials with different cross-linking systems, one of which is applied for preliminary molding, while the other is applied for subsequent bonding.

12.2.2.1 Manufacturing Process of Hydrogel-Based Microfluidic Chips

The entire construction process of the hydrogel-based microfluidic chip includes casting, demolding, and bonding. Hydrogel sheets with microstructure on the surface are first obtained through the injection molding process, and then a hydrogel-based microfluidic chip with internally closed channels are realized through aligning and bonding two such hydrogel sheets.

The specific process covers the obtaining of partially-cross-linked hydrogel sheets with a surface micro-groove structure based on the injection demolding process and first cross-linking condition, as well as the obtaining of a fully-cross-linked hydrogel-based microfluidic chip based on the bonding process and second cross-linking condition (Figure 12.4).

12.2.2.2 Mechanism Study

The hydrogel prepolymer solution exhibits good fluidity to automatically fill the entire mold. Based on the independent cross-linking system of different hydrogel compositions, the step-by-step cross-linking could be realized during this process. The internal microscopic changes are shown in Figure 12.5:

During the casting process, the composite hydrogel prepolymer solution is in a liquid state. No cross-linking condition is involved, and neither of the two components is crosslinked (Uncross-linked).

During the demolding process, the composite hydrogel is in a reversible solid state. One of the cross-linking conditions is applied, and one component achieves a reversible cross-linking (Partially-cross-linked).

During the bonding process, the composite hydrogel gradually takes on an irreversible solid state. Another cross-linking condition is applied, and the other component is cross-linked, while the whole structure is irreversibly cross-linked (Fully-cross-linked).

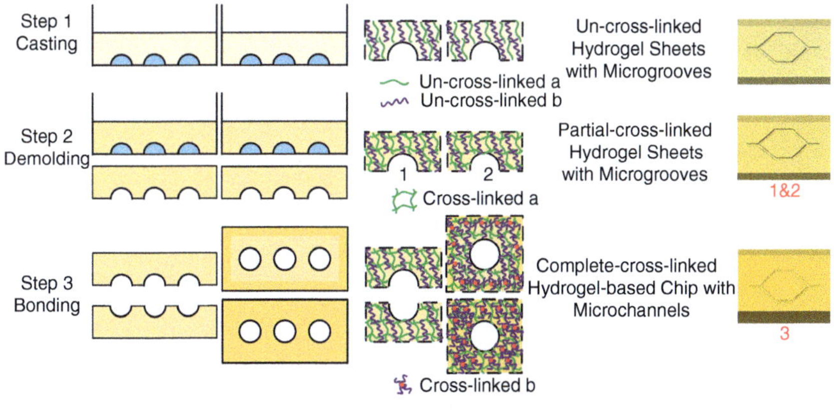

Figure 12.4 Schematic of the fabrication process for the hydrogel-based chips and its mechanism. Source: Nie et al. [17]. Reproduced with permission of Wiley-VCH.

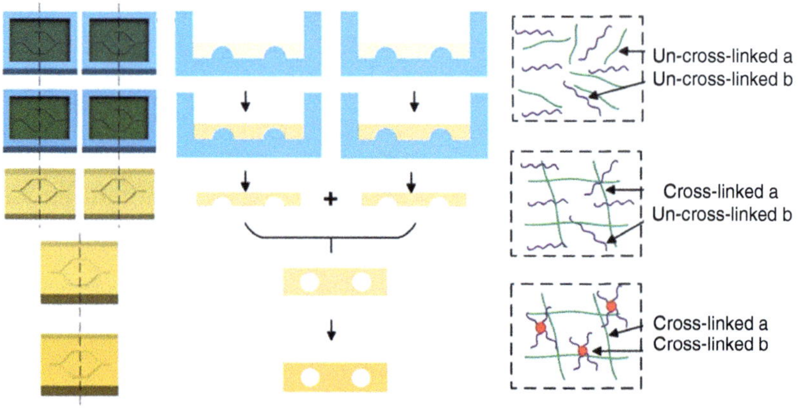

Figure 12.5 Mechanism of twice-cross-linking bonding strategy. Source: Nie et al. [17]. Reproduced with permission of Wiley-VCH.

Throughout the molding process and the obtained structure, the interactions between the two hydrogel components may be van der Waals forces, hydrogen bonds, electrostatic forces, etc.

12.2.2.3 Material Selection

As commonly applied biomaterials, sodium alginate, gelatin, and GelMA each own theirs unique cross-linking conditions. By combining any two of these three hydrogel materials, three different composite hydrogel materials are obtained: gelatin–sodium alginate, sodium alginate–GelMA, and gelatin–GelMA. The three composite hydrogels are compared in terms of biocompatibility to optimize the material selection.

The biocompatibility of the three composite hydrogel materials is explored through cell seeding. As displayed in Figure 12.6, the cells seeded on the surface of

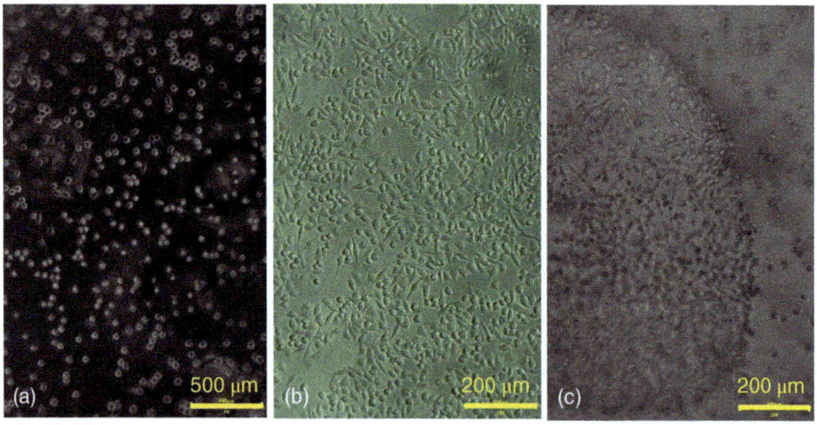

Figure 12.6 Images of cell attachment and spreading for the three composite hydrogels: gelatine-sodium alginate (a), gelatine-GelMA (b), and sodium alginate-GelMA (c). Source: Nie et al. [17]. Reproduced with permission of Wiley-VCH.

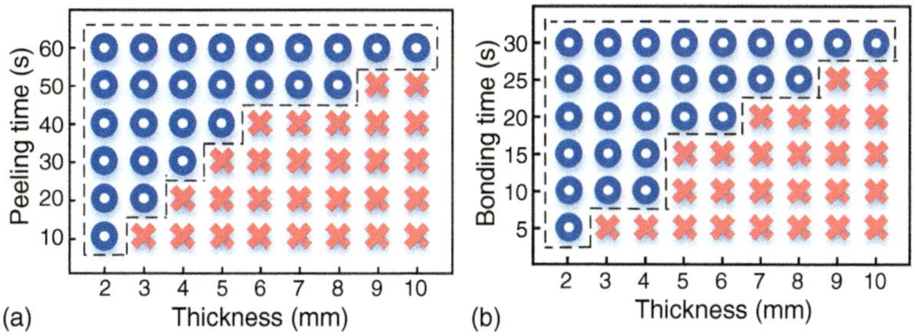

Figure 12.7 The feasible domain for stripping (a) and bonding (b) processes. Source: Nie et al. [18]. Reproduced with permission of The Royal Society of Chemistry.

the gelatin–sodium alginate composite hydrogel are showing a sphere shape and could not be attached or stretched on the surface of the hydrogel material. The cells seeded on the surface of the sodium alginate–GelMA material could achieve attachment and a certain degree of extension. However, the cells inoculated on the surface of gelatin–GelMA material are showing attachment and full extension with normal cell morphology, confirming the good biocompatibility of this material combination. So far, the gelatin–GelMA composite hydrogel material system is selected as the final material combination for further research.

12.2.2.4 Feasible Domain

Based on the gelatin–GelMA composite hydrogel material system, to optimize the process and improve the reliability and controllability of the manufacturing process, a series of relevant processing parameters are explored and studied. First of all, to improve the manufacturing efficiency and reduce the subsequent cell damage as much as possible, the lithium phenyl-2, 4, 6-trimethylbenzoylphosphinate (LAP) photoinitiator with high cross-linking efficiency and little cell damage is selected. Under this reaction system, targeting at various structural thickness settings, demolding, and bonding processes are explored separately about the specific feasible parameters and domains.

Specific setting of the experimental parameters of the demolding and bonding process is displayed in Figure 12.7. The results indicate that the minimal time for successful achievement of demolding and bonding is positively related to the thickness of the hydrogel structures. As a result, the corresponding minimum time could be feasibly selected according to the size of the hydrogel structure in a specific experimental process so as to improve efficiency and decrease the damage of temperature and light conditions to the biologically active components such as cells during the molding process, thereby improving the biocompatibility of the process while extending relevant biomedical applications.

12.2.2.5 Bonding Results

The reliable bonding of two hydrogel sheets, as well as the integrity and strength of the channel, are essential for cell seeding on the inner wall surface of the channel and subsequent simulation of physiological flow conditions. To validate the reliability of this bonding strategy, comparative analysis has been conducted in aspects of bonding interface channel morphology and structural tensile strength.

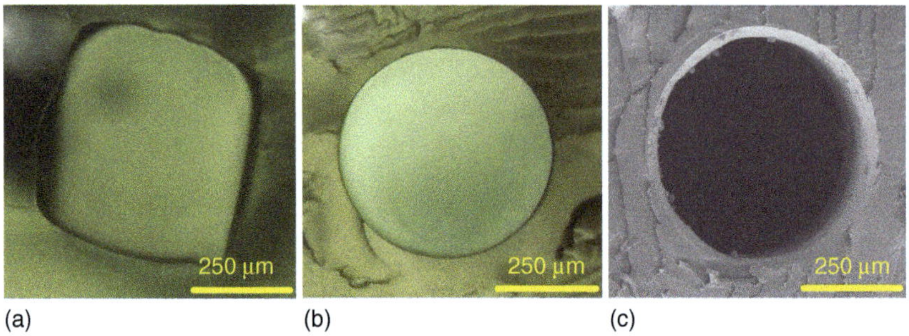

Figure 12.8 Cross-section images of channels: rectangular channel (a) and circular channel (b) under optical microscopy, and circular channel under SEM (c). Source: Nie et al. [17]. Reproduced with permission of Wiley-VCH.

First, the cross section of the internal channel is observed through the optical microscope and scanning electron microscope. As shown in Figure 12.8, the channel presents a completely closed structure where there is no visual boundary between the interface of the two independent hydrogel sheets, and they successfully merge into a whole structure. In particular, compared with the square cross-section channel, the round cross-section channel provides an ideal experimental platform for the subsequent construction of the blood vessel model, as well as the simulation of vascular morphology and perfusion condition.

To carry out the tensile experiment to determine the bonding strength between hydrogel sheets, bonding samples are first prepared and cut into a standard size (width: 5 mm, height: 2 mm, length: 10 mm), while integral casting samples are prepared and applied as a control group. Before the tensile test, the samples are tightly fixed on a specific fixture, and then assembled on the gel sample tensile mechanical tester. Then, loading is applied at a rate of 4 mm/min, and the stress–strain curve is automatically recorded until the sample is broken. The collected pressure and tensile displacement data are normalized into stress and strain data based on the sample width, thickness, and length, respectively.

After undergoing a complete tensile process, the bonded hydrogel structures are broken at random positions outside the bonding interface, indicating that it is similar to the integral casting hydrogel structure, and the bonding process does not cause weakness of interface strength.

Figure 12.9 displays the tensile stress–strain curve of hydrogel structures constructed in two different ways. It could be found that several sets of integral casting samples fracture at about 240% strain, while the bonded samples fracture at the strain range of 230–240%, which is not much different from the integral samples.

The final stress when the bonded samples and integral samples are broken is not much different, which is 115.3 ± 2.1 and 117.6 ± 1.7 kPa, respectively. It is further known through a calculation that the bonded hydrogel structures own elastic modulus similar to the overall hydrogel structures. These data verify that partially cross-linked hydrogel sheets could be completely fused together to form an integral structure while there is no bonding interface between them, validating the reliability of this bonding strategy.

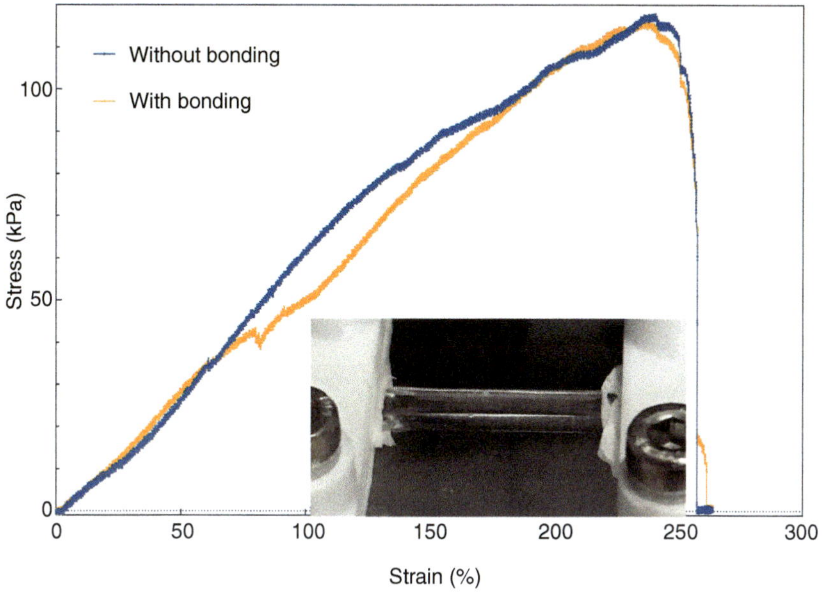

Figure 12.9 The tensile curves for both casting and bonding hydrogels. Source: Nie et al. [17]. Reproduced with permission of Wiley-VCH.

12.2.3 Multi-Scale 3D Printing Process

To break through the scale limitation of traditional mold manufacturing, a multi-scale 3D printing technology is developed to flexibly fabricate multi-scale diversified fiber template covering both large blood vessels and capillaries.

12.2.3.1 Mechanism of Multi-Scale 3D Printing Process

By introducing a high-voltage electric field, the printing material is stretched a second time, thereby improving the resolution of conventional 3D printing technology. The multi-scale 3D printing system is composed of FDM printing and EHD printing. The real-time shift between FDM and EHD printing modes is achieved by controlling the on and off of the high-voltage power supply, thereby obtaining multi-scale fiber templates. The specific mechanism is displayed in Figure 12.10: when the high-voltage power supply is turned off, the FDM printing process is conducted, and the thick filament is extruded through air pressure; when the high-voltage power supply is turned on, the EHD printing process is conducted, and the thin filament is extruded through electric field force.

Based on the multi-scale 3D printing process, through the flexible regulation of printing materials and printing modes, the printing of fibers scaling from a few millimeters to a few microns could be achieved. Based on the obtained soft fiber template, the flexible construction of the entire vascular network model containing vessels and capillaries, as well as a variety of personalized models with complex topological structures, could be realized, thereby promoting the personalized customization of hydrogel-based microfluidic chips and the simulation and reproduction of vascular biomimicking morphology.

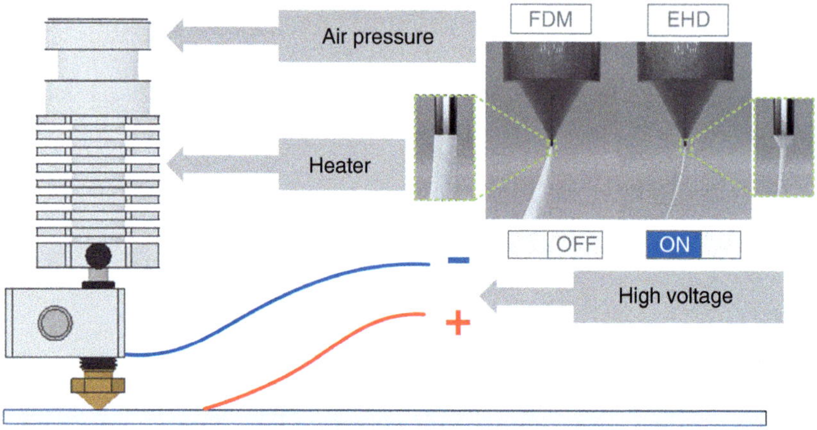

Figure 12.10 Printing principle of the multi-scale template. Source: Nie et al. [18]. Reproduced with permission of The Royal Society of Chemistry.

12.2.3.2 Printing Parameters

To achieve precise control over the channel scale, the effects of nozzle distance, printing speed, and extrusion pressure on the fiber template diameter in different scales as well as the effects of the swelling phenomenon during the subsequent cultivation process on the hydrogel channel diameter are systematically explored, targeting at the two printing modes of FDM and EHD.

First, the extrusion pressure and printing speed are selected for exploration, and the changing trend of fiber diameter is studied by setting a series of discrete specific parameters. The results show that the fiber diameter increases with the increase of extrusion pressure and decreases with the increase of printing speed. The reason is that the increase in extrusion pressure will directly increase the volume of the extruded material, which will naturally lead to an increase in fiber diameter at a set printing speed. In addition, as the printing speed increases, the extruded fiber is stretched and becomes thinner. During the printing process, the two parameters of printing speed and extrusion pressure could be comprehensively adjusted to achieve real-time regulation of the diameter of the extruded fiber within a specific scale range. In this study, by adjusting the printing parameters (i.e. printing speed and extrusion pressure), the thick fibers in the range of 253–732 μm could be obtained by the FDM printing process, as shown in Figure 12.11; while the thin fibers in the range of 3.16–94.19 μm could be obtained by the EHD printing process, as shown in Figure 12.11. Finally, a multi-scale hydrogel channel network containing both macro channels in 233–589 μm and micro channels in 4.17–102.36 μm could be obtained.

It is worth noticing that during the demolding and bonding processes, the different cross-linking degrees of the hydrogel components is inevitably causing changes in the channel diameters, showing good repeatability. Furthermore, the subsequent swelling of the hydrogel structure when immersed in the medium or buffer could cause a certain reduction in the channel diameter. Sufficient experiments have been

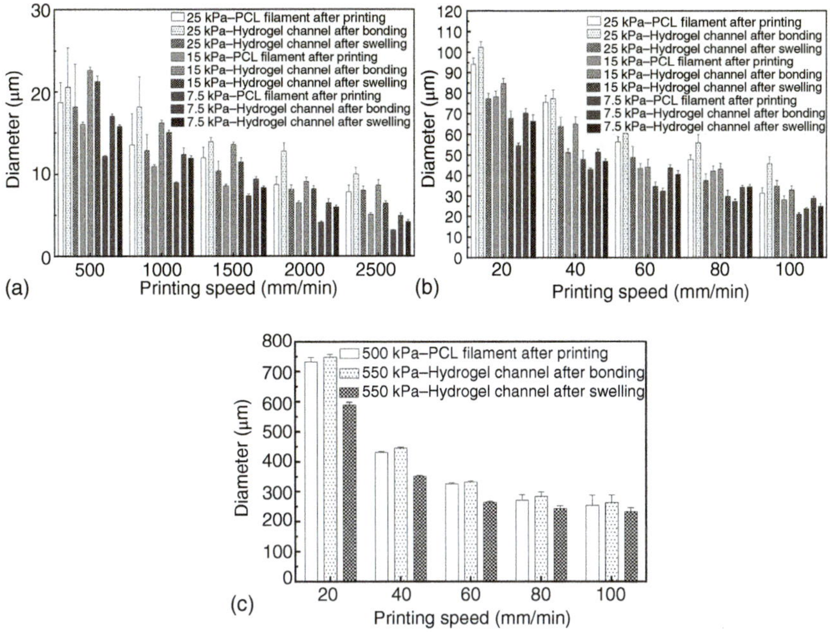

Figure 12.11 Statistical analysis of the processing parameters during the EHD (a–b) and FDM (c) printing processes. Source: [18]. Reproduced with permission of The Royal Society of Chemistry.

Figure 12.12 Quantitative analysis of the various channel diameters during the swelling process. Source: Nie et al. [18]. Reproduced with permission of The Royal Society of Chemistry.

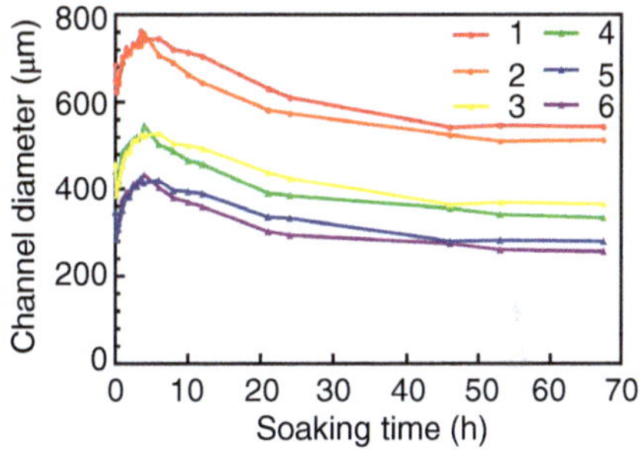

carried out to confirm that this change could reach a stable equilibrium point after 53 hours, as shown in Figure 12.12.

12.2.4 Construction Process of Hydrogel-Based Vascular Chips

By combining multi-scale 3D printing technology, damage-free demolding process, and twice-cross-linking strategy, a novel method for constructing multi-scale channel network and personalized channel structures within hydrogel materials has been designed and proposed, and a new method is further developed to flexibly construct multi-scale blood supply circulatory system and bionic vascular models with personalized morphology.

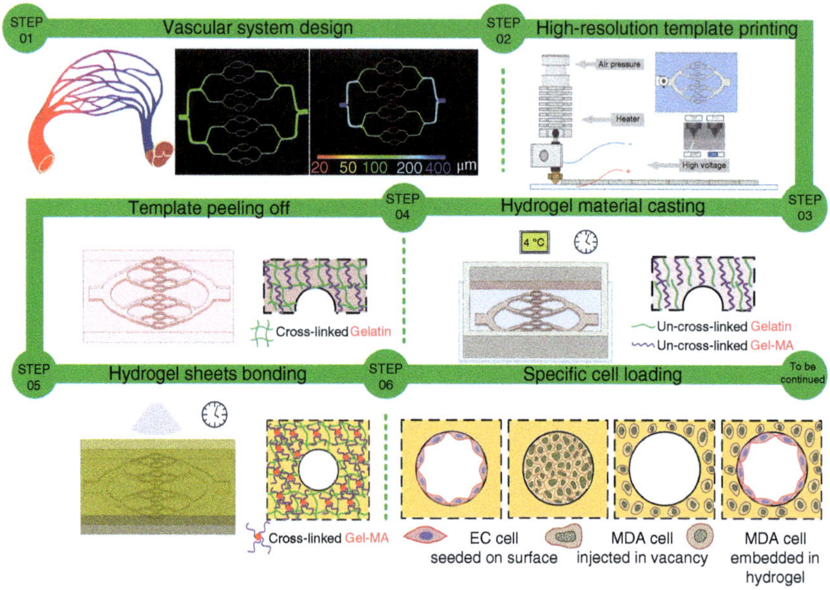

Figure 12.13 Schematic for the construction of a multi-scale vascular chip. Source: Nie et al. [18]. Reproduced with permission of The Royal Society of Chemistry.

The process is displayed in Figure 12.13. First, the soft fiber template is manufactured based on multi-scale 3D printing technology. Then, the multi-scale surface microstructure is constructed based on the damage-free demolding process. Next, the channel encapsulation is realized based on the mechanism of twice-cross-linking. Finally, through the seeding of vascular endothelial (VE) cells on the channels, the *in vitro* reconstruction of a complete vascular network system (artery–capillary–vein) and other biomimicking vascular models such as spiral vessels and stenosis has been successfully achieved.

The specific scheme of this method is as follows: a multi-level channel network model is designed inspired by the *in vivo* blood supply circulation system, and a multi-scale template is manufactured by multi-scale 3D printing technology, while the subsequent steps are based on the above-mentioned hydrogel-based microfluidic chip construction process system. Finally, the endothelial cells could be seeded on the channel network to form a hydrogel-based vascular chip. The whole process includes: (i) design of vascular network template (from artery, capillary to vein); (ii) printing of vascular template; (iii) casting of gelatin–GelMA composite hydrogel precursor solution; (iv) peeling off of vascular template; (v) hydrogel sheets bonding (formation of closed vascular network); (vi) seeding of vascular endothelial cells, flips for uniform attachment, and static cultivation at a constant temperature.

12.3 Characterization of Vascular Chips

12.3.1 Fundamental Characterization of Vascular Chips

12.3.1.1 Characterization of Endothelium Function of Channels

Figure 12.14 displays the microscopic morphology of vascular endothelial cells on the channels in the diameter of 300 and 500 μm after two and seven days of culture, respectively. It could be observed that the endothelial cells on the channels are showing good adhesion while gradually stretching and proliferating on the surface of the channel until a uniform and complete endothelial cell monolayer is formed. After seven days of cultivation, the attachment of vascular endothelial cells covers almost the entire length of the channel.

After the seeding of endothelial cells, the growth and proliferation of endothelial cells in the channel and the endothelium process of the channel are observed and recorded in different magnifications. Figure 12.15 shows the optical microscopic figures of the endothelium channel at 1, 4, 7, 14, and 21 days. It could be observed that the number of cells and the area of cell coverage are gradually increasing, indicating that the endothelial cells on the channel could maintain the proliferation and migration ability and then gradually cover the entire channel to form a dense endothelial cell monolayer.

12.3.1.2 Characterization of Endothelial Cells Viability

Cell viability is detected using the Cell Live/Dead Staining Kit. The calcein addictive manufacturing (AM) in the kit could label the cell membrane of live cells and produce green fluorescence, while propidium iodide (PI) could penetrate the

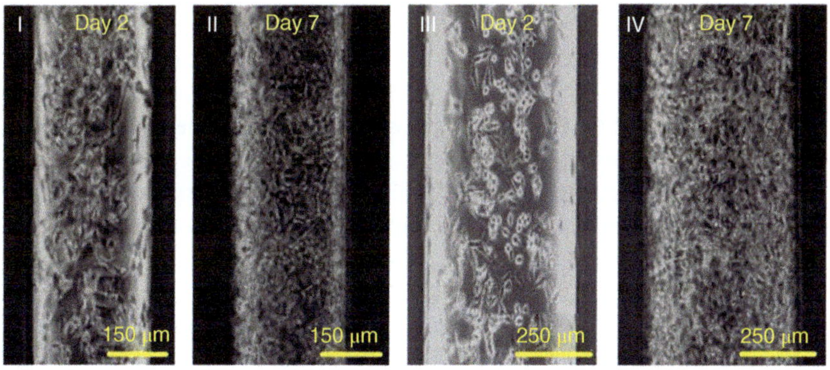

Figure 12.14 Photographs of 300 and 500 μm microchannels with human umbilical vein endothelial cells (HUVECs) covering the entire length of the channels after two and seven days of static culture. Source: Nie et al. [17]. Reproduced with permission of Wiley-VCH.

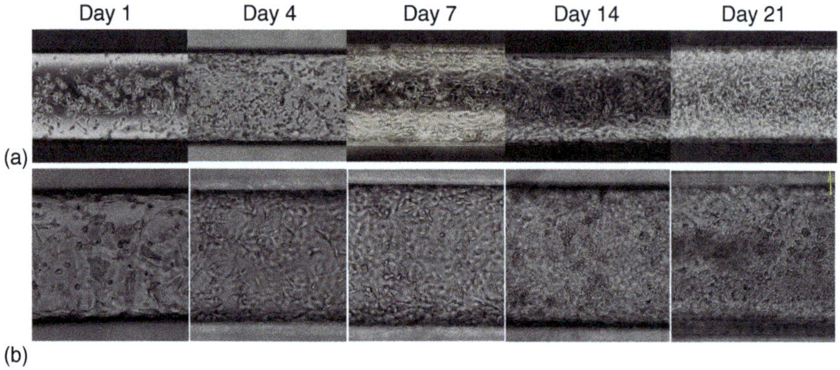

Figure 12.15 Schematic for the endothelialization process of HUVECs along with time (a) and under higher magnification (b). Source: Nie et al. [17]. Reproduced with permission of Wiley-VCH.

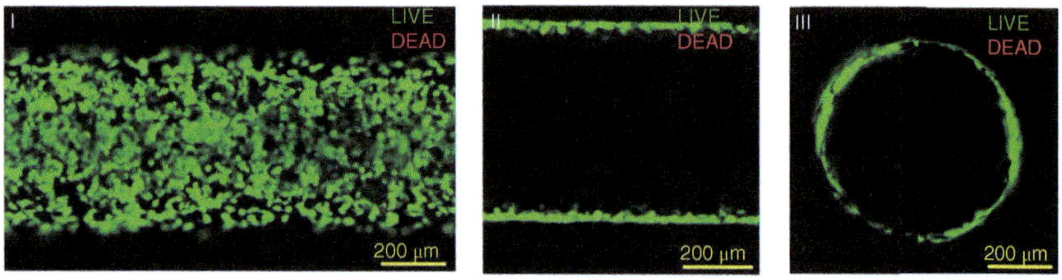

Figure 12.16 Confocal images of LIVE/DEAD staining of the endothelialized channels. Source: Nie et al. [17]. Reproduced with permission of Wiley-VCH.

cell membrane of dead cells and combine with the cell nucleus, producing red fluorescence. The endothelium channel has been stained with the Cell Live/Dead Staining Kit and observed under a confocal fluorescence microscope. As shown in Figure 12.16, the results indicate that most cells in the channel survive (green), while almost no cells are dead (red).

This structure is first cut in the radial direction for observation, and it could be seen that the seeded endothelial cells form a complete circle of adhesion along the channel. Then, the structure is cut in the axial direction and observed at different focal planes, results showing that the vascular endothelial cells are evenly distributed and completely covering the entire channel surface to form a complete endothelium monolayer. The cell survival rates at different time points during the culture process are further calculated, showing a high survival rate of over 98%.

12.3.1.3 Characterization of Endothelial Cells Morphology

Cytoskeleton staining (F-actin and nucleus) is performed using phalloidin and 2-(4-amidinophenyl)-6-indolecarbamidinedihydrochloride (DAPI) staining solution to observe cell morphology. The experimental results of cytoskeleton staining further confirm the even distribution, attachment and stretching, and maintenance of cell morphology of endothelial cells on the channel. It could be observed from

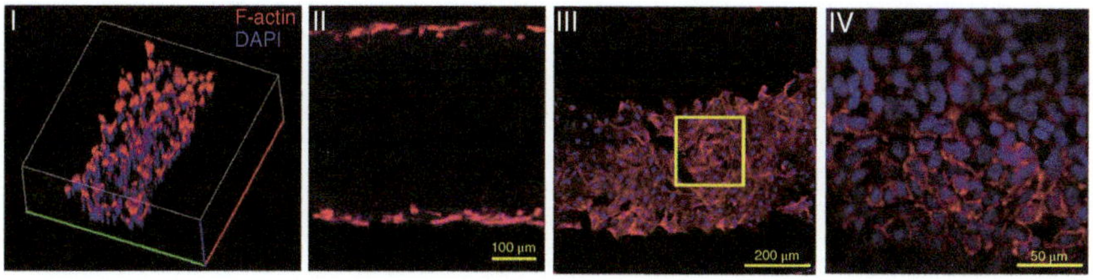

Figure 12.17 Confocal images of cytoskeleton staining of the endothelialized channels. Source: Nie et al. [17]. Reproduced with permission of Wiley-VCH.

the figure that the endothelial cells seeded on the inner wall surface of the channel are forming a confluent monolayer while maintaining the normal cell phenotype and the expression of F-actin and nuclei, as shown in Figure 12.17.

12.3.1.4 Characterization of Endothelium Channel

As shown in Figure 12.18, the three-dimensional stacked confocal images show the complete distribution of endothelial cells along the entire channel.

High-magnification scanning electron microscopy images display the microscopic morphology of endothelial cells seeded on the channel from the radial and axial directions, as shown in Figure 12.19. The results show that the vascular endothelial cells seeded on the inner wall of the channel could achieve good adhesion and extension, display a healthy growth state while forming a dense endothelium monolayer, further confirming the good biocompatibility of this system.

The vascular endothelial cells in specific positions are further observed with the help of a scanning electron microscope, and the results show that the connections between individual cells are forming, as shown in Figure 12.20. At the same time, the cross section of the endothelium channel is observed through a scanning electron microscope. The results indicate that a complete circle of endothelial cells is attaching along the channel. In particular, the formation of intercellular connections is clearly observed.

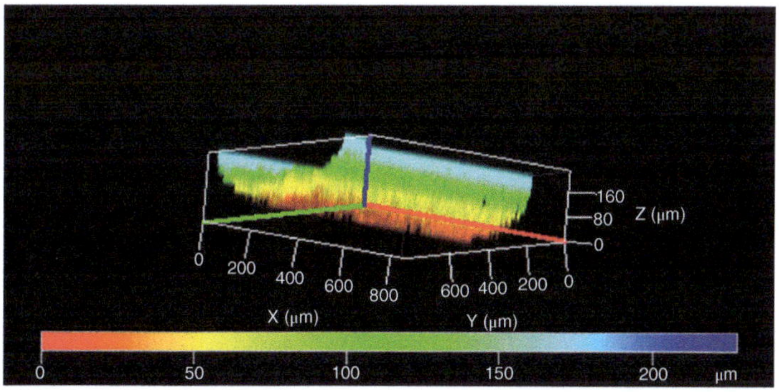

Figure 12.18 3D stacking confocal images of the endothelial cell (EC) channels. Source: Nie et al. [18]. Reproduced with permission of The Royal Society of Chemistry.

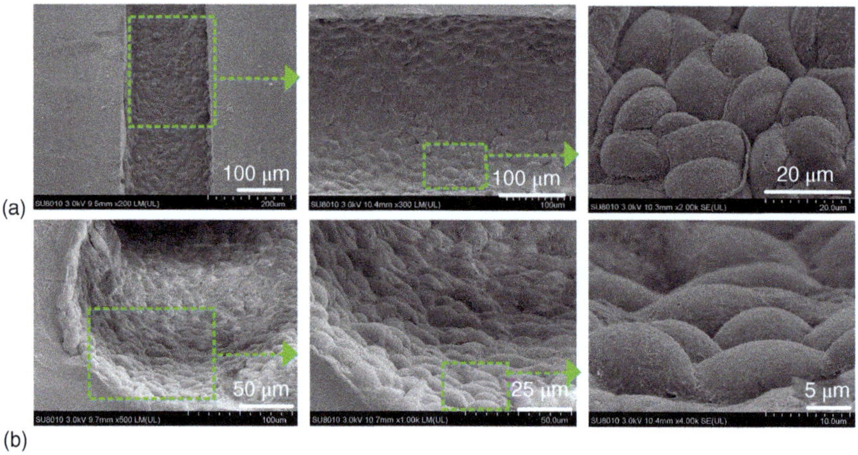

Figure 12.19 (a) Microscopy morphology of the longitudinal section of the EC channels. (b) Microscopy morphology of the cross-section of the EC channels. Source: Nie et al. [18]. Reproduced with permission of The Royal Society of Chemistry.

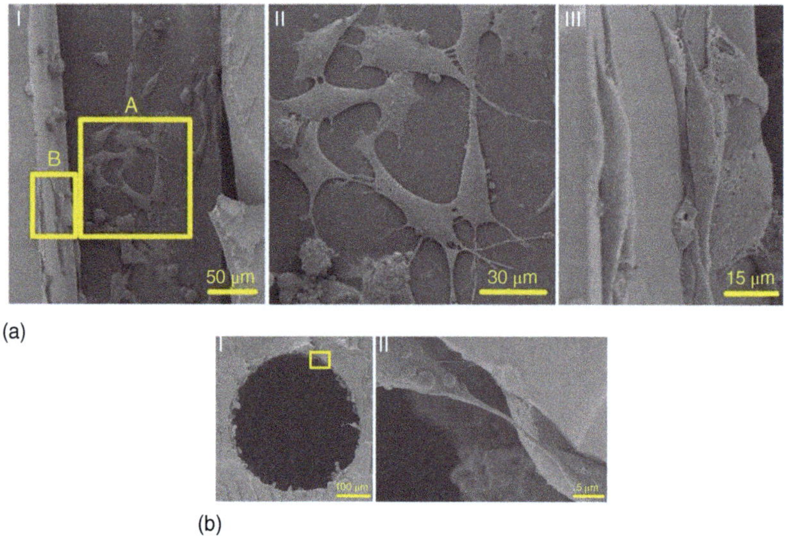

Figure 12.20 (a) Scanning electron microscope (SEM) image of the inner wall of the endothelialized channels. (b) SEM image of cross-section of the endothelialized channel. Source: Nie et al. [17]. Reproduced with permission of Wiley-VCH.

12.3.2 Morphology Characterization of Hydrogel-Based Vascular Chips

12.3.2.1 Multi-Level Bifurcated Channel Network Structure

Based on the above-mentioned hydrogel-based microfluidic chip construction process system and related results of processing exploration experiments, a nine-level bifurcated channel network is constructed by selecting specific combinations of processing parameters, as shown in Figure 12.21.

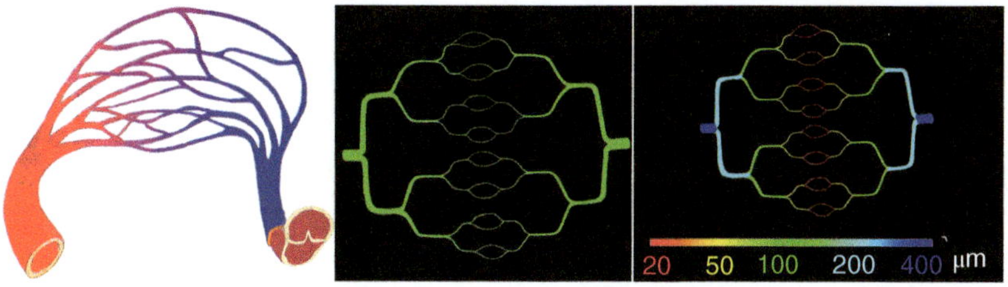

Figure 12.21 *In vivo* vascular system. Source: Nie et al. [18]. Reproduced with permission of The Royal Society of Chemistry.

This channel network is sliced along the axial direction of the channel to expose the inner wall of the channel, treated with gradient dehydration and critical point drying, and then microscopically characterized targeting at each bifurcation point of the entire network by means of a scanning electron microscope.

It is further envisioned that this process could be applied to construct multiple complicated and precise internal channels within hydrogel materials, such as multi-scale structures, ultrafine patterned structures, and varying-scale structures, which are expected to be applied to simulate the blood supply circulation system, capillary network, as well as vascular-related diseases.

12.3.2.2 Multi-Scale Vascular Model

The nine-level bifurcated hydrogel channel network is further seeded with vascular endothelial cells and cultured under specific conditions, and a multi-scale vascular model containing both large blood vessels and capillaries is constructed to simulate the entire vascular network. The *in vitro* reconstruction of a complete blood supply circulation system has been successfully achieved, and the concept of a multi-scale vascular network chip has been put forward. Based on a series of experimental data and related phenomena, the feasibility of the method for constructing a multi-scale hydrogel vascular chip (artery–capillary–vein) is validated, while displaying the scale effect of the channel diameter on cell growth and stretch, as well as the good biocompatibility of hydrogel-based microfluidic system.

By selecting specific combinations of processing parameters, endothelium channels with diameters of 10, 20, 40, 100, 150, 250, and 500 µm have been obtained, which initially confirms the ability of this method to construct a complete vascular system model. Figure 12.22 displays the microscopic cytoskeletal morphology of vascular endothelial cells seeded on channels in different diameters.

Lentiviral green fluorescent protein (GFP) constructs are used for transduction, and endothelial cells successfully transfected with GFP are further selected through puromycin to obtain a batch of high-purity green fluorescent-labeled endothelial cell populations.

12.3.2.3 Biomimicking Vascular Model

The morphology of *in vivo* vascular is not restricted to a single linear structure, while its specific morphology is closely related to concrete physiological and pathological

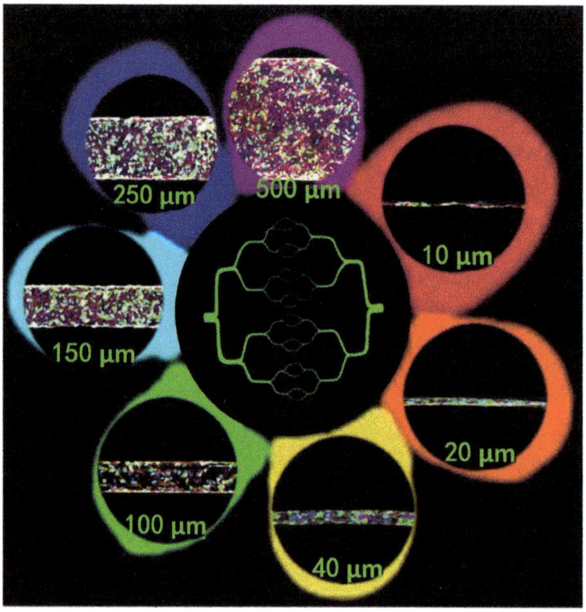

Figure 12.22 Display of the multi-scale channel network and EC channels with various diameters. Source: Nie et al. [18]. Reproduced with permission of The Royal Society of Chemistry.

conditions. The simulation and reconstruction of common structures in the biological vascular system is very important for the construction of *in vitro* vascular models and the realization of vascularization. However, traditional methods cannot meet the needs of personalized customization and are limited by the complicated micro-machining process, the introduction of toxic substances, and the consumption of disposable molds.

On the contrary, in the above method, according to the specific demand of a single unique structure, only a small amount of polycaprolactone (PCL) material is needed to realize the rapid manufacture of the corresponding vascular template. As proof of extending applications, a variety of physiologically meaningful vascular models have been constructed to simulate the *in vivo* vascular morphology based on this flexible method of constructing personalized hydrogel channel structures, including bifurcated blood vessels, spiral blood vessels, and blood vessel stenosis.

Inspired by the *in vivo* blood circulation system, a multi-level and multi-scale hydrogel channel model has been designed and constructed to simulate the vascular network hierarchy. Figure 12.23 displays a three-level bifurcated endothelium vascular channel network, within which it could be observed that the vascular endothelial cells spread to a large extent, covering the entire channel network while showing normal cytoskeletal morphology.

Inspired by the spiral artery existing in the endometrium, an endothelium channel with spiral geometry has been designed and constructed, as shown in Figure 12.24.

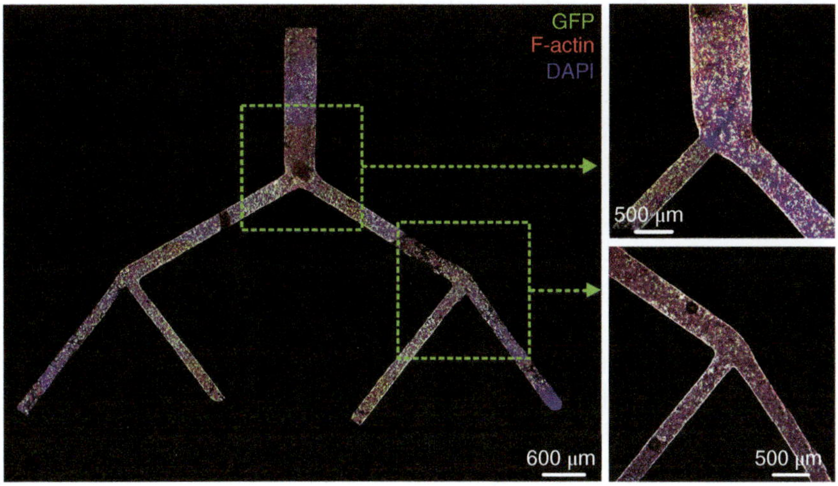

Figure 12.23 Fluorescence micrograph of a three-level EC network. Source: Nie et al. [18]. Reproduced with permission of The Royal Society of Chemistry.

Figure 12.24 Fluorescence micrograph of a spiral EC channel. Source: Nie et al. [18]. Reproduced with permission of The Royal Society of Chemistry.

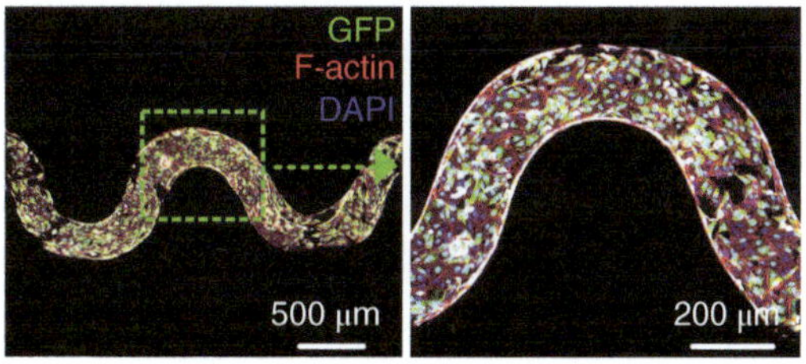

Figure 12.25 Fluorescence micrograph of a stenosis-mimicking EC channel. Source: Nie et al. [18]. Reproduced with permission of The Royal Society of Chemistry.

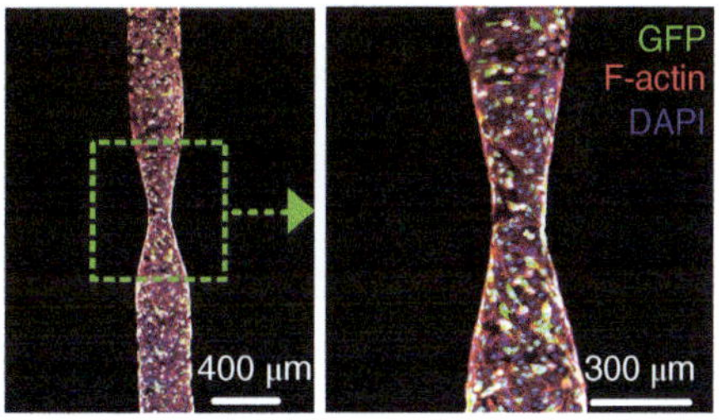

As the main cause and risk factor of cardiovascular disease, stenosis is directly causing blood flow reduction and ischemia, which will lead to symptoms such as blood vessel inflammation and dysfunction. To simulate the phenomenon of vascular stenosis, a continuously-varying endothelium channel is further constructed as a disease model for disease mechanism research and related drug development, as shown in Figure 12.25.

12.3.3 Characterization of Vascular Function

12.3.3.1 Nutrition Supply Function

By applying the construction processing system of the vascular chip to cell-laden hydrogel materials, the *in vitro* construction of a vascularized artificial tissue model could be achieved, and the realization of the basic functions of its internal vascular structures has been further verified. A vascularized breast tumor tissue model is constructed, the nutritional supply function of the internal vascular network is confirmed through the survival and proliferation data of tumor tissue cells, and the expression of key vascular function proteins is displayed through immunofluorescence staining. In addition, the ability of the simulating vascular pathological state is displayed through the application of vascular inflammation factors, and the realization of vascular barrier function is proved through molecule diffusion experiments.

Targeting at the above-mentioned tumor tissue model, the cell viability was tested, and the number of live and dead cells are counted at different time points during the culture period to calculate the cell survival rate. The results indicate high survival rates of breast tumor cells at different time points during culture.

Cell proliferation is further detected using the Cell Counting Kit-8 (CCK-8). The 2-(2-Methoxy-4-nitrophenyl)-3-(4-nitrophenyl)-5-(2,4-disulfophenyl)-2H-tetrazolium sodium salt (WST)-8 in the kit could be oxidized by an intracellular dehydrogenase to produce orange-yellow formazan dyes, which are dissolved in the culture medium. The amount of formazan produced is proportional to the number of living cells. The optical density is collected and recorded by the microplate reader to display the changes in cell numbers. The results indicate that the tissue cells are maintaining normal activity and proliferation levels, as shown in Figure 12.26, further confirming that the existence of an internal vascular network could provide nutrition for the growth of tissue cells.

12.3.3.2 Expression of Key Functional Proteins in Endothelial Cells

A series of immunofluorescent staining is conducted targeting at vinculin protein antibody, CD31 protein antibody, and VE-cadherin antibody to study the adhesion of endothelial cells on the channel, the connection between cells, and

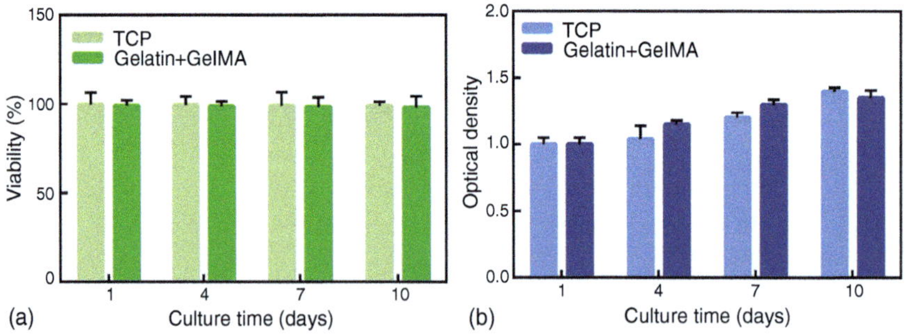

Figure 12.26 (a) Viability of the tumor cells in the model. (b) Cell proliferation rate of the tumor cells in the model. Source: Nie et al. [18]. Reproduced with permission of The Royal Society of Chemistry.

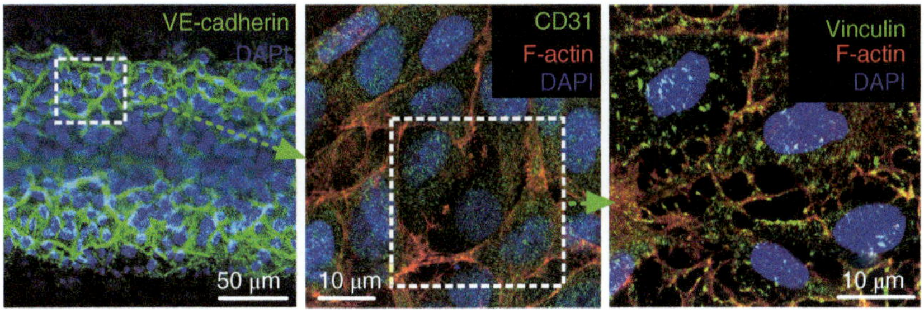

Figure 12.27 Fluorescence micrograph of VE-cadherin/CD31/vinculin markers of the ECs. Source: Nie et al. [18]. Reproduced with permission of The Royal Society of Chemistry.

the vascularization function of the endothelium channel. The expression of key function proteins and the maintenance of normal cell morphology are displayed in Figure 12.27.

First, immunofluorescence staining is performed on vascular endothelial cadherin protein (VE-cadherin), which is an important protein marker of intercellular connection. The results are shown in Figure 12.27. The high-level expression of the corresponding protein proves that the cells are forming tight interconnection, and obvious information exchange could be achieved between the cells, which is essential for the stable generation of fusion endothelial monolayer and the integrity of blood vessels. This dense endothelial cell monolayer is the basis for the subsequent simulation and realization of vascular barrier function.

Figure 12.27 displays an optical microscopic comparison of the loose endothelium channel immediately after the seeding of endothelial cells with the dense endothelium channel formed after cell proliferation and migration.

Then, immunofluorescence staining of CD31 protein is carried out, and the expression of CD31 marker protein in endothelial cells confirms the maintenance and realization of its key endothelial function, further verifying the realization of vascular functionalization, as shown in Figure 12.27.

Furthermore, immunofluorescence staining of vinculin protein is performed on the sample, which is a key adhesion protein between cell-ECM (extracellular matrix) and cell scaffold.

The emergence of adhesion point is observed by fluorescence confocal microscopy (Figure 12.27, marked with a yellow arrow), thus confirming the formation of tight adhesion between endothelial cells and hydrogel channels.

12.3.3.3 Simulation of Vascular Inflammation Reaction

In order to simulate the inflammation reaction of vascular under pathological conditions such as atherosclerosis, pathological conditions are simulated based on this vascular model through the application of inflammation-inducing factors while reconstructing a vascular inflammation model. Endothelial cells are cultured for 24 hours on standard 2D Petri dishes and hydrogel channels using standard medium and medium containing tumor necrosis factor (TNF)-α (10 ng/ml), respectively. After that, real-time polymerase chain reaction (RT-PCR) is conducted to detect the

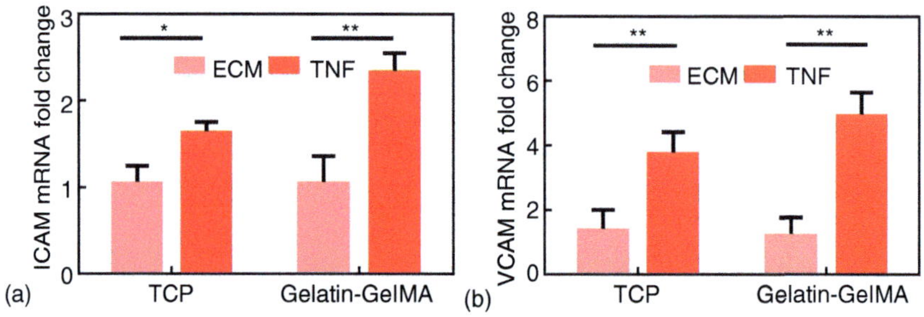

Figure 12.28 (a) The messenger RNA (mRNA) levels of ICAM-1 in ECs. (b) The mRNA levels of VCAM-1 in ECs. Source: Nie et al. [18]. Reproduced with permission of The Royal Society of Chemistry.

adhesion molecules of two endothelial activation markers: intracellular adhesion molecule-1 (ICAM-1) and vascular cell adhesion molecule-1 (VCAM-1).

The experimental results are displayed in Figure 12.28: in both the 2D control group and the channel experimental group, the expression of ICAM-1 in the TNF-α treated group is 1.5 and 2.3 times higher than that in the control group, while the expression level of ICAM-1 molecule is not much different between the 2D control group and the channel experimental group. Moreover, the detection on VCAM-1 molecule is showing similar experimental results. These results together indicate that the hydrogel endothelium channel could be used as an effective vascular inflammation model, which could further be applied in vascular pathology research.

12.3.3.4 Characterization of Vascular Barrier Function

As a polysaccharide molecule with multiple molecule weights, dextran could be transported between individual endothelial cells through cell bypass. To characterize the diffusion permeability of the applied hydrogel materials and the barrier function of the endothelium channel, fluorescein isothiocyanate (FITC)-labeled dextran solution is injected into the blank channel and endothelium channel incubated for 48 hours under static condition. On this basis, to better simulate the sustained and slow-releasing process of the molecule, the GelMA prepolymer solution containing FITC-dextran is first injected into the blank channel and the endothelium channel, and then subjected to photo-curing operation. The GelMA hydrogel material is introduced as a slow-releasing carrier of the molecule. The experimental schematic diagram of the above four situations is displayed in Figure 12.29.

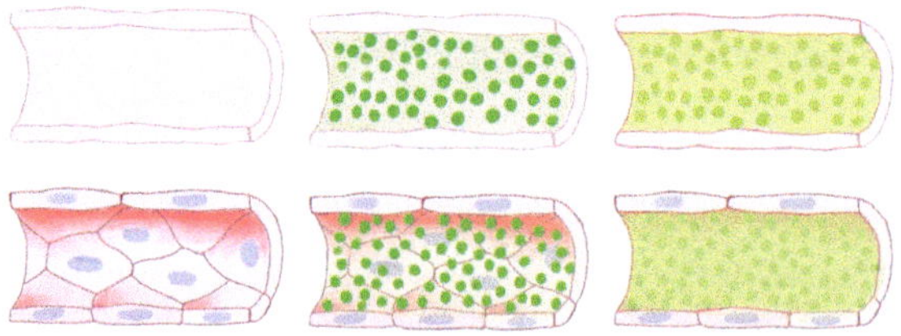

Figure 12.29 Presentation of the empty channels and the EC channels for both diffusion and release cases. Source: Nie et al. [18]. Reproduced with permission of The Royal Society of Chemistry.

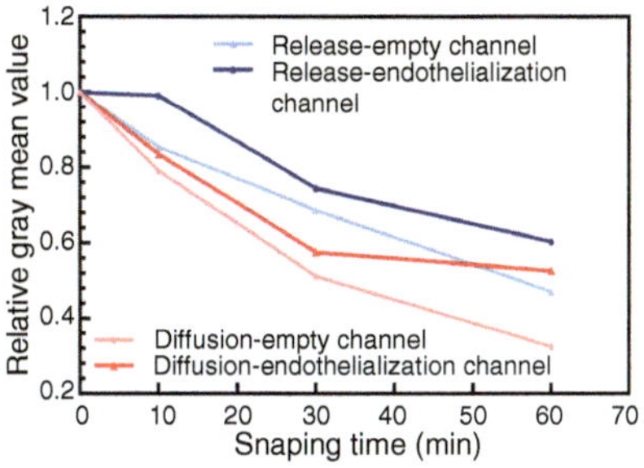

Figure 12.30 Schematic for the endothelialization process of HUVECs along with time. Source: Nie et al. [18]. Reproduced with permission of The Royal Society of Chemistry.

After that, a laser confocal microscope is used to capture and record the diffusion pattern of dextran. The fluorescence images are extracted at different time points. The Image J software is applied to convert the fluorescence intensity into a grayscale value at a specific location to quantitatively characterize the penetration rate of fluorescence. In other words, the penetration rate of molecules in the case of diffusion and release is calculated by the rate at which the fluorescence intensity of the cavity decreases with time. As shown in Figure 12.30, the statistical results indicate that the penetration rate in the blank channel is significantly higher than that in the endothelium channel in both diffusion and release conditions.

As shown in Figure 12.31, the grayscale value of fluorescence intensity at a specific time point qualitatively displays the molecule penetration rate under diffusion and release conditions. It could be observed that FITC-dextran gradually penetrates into the surrounding hydrogel bulk from the cavity of the channel, and more FITC-dextran penetrates from the cavity into the bulk in the blank channel group than that in the endothelium channel group. This indicates that the permeability of the blank channel is better than the endothelium channel. Furthermore, several

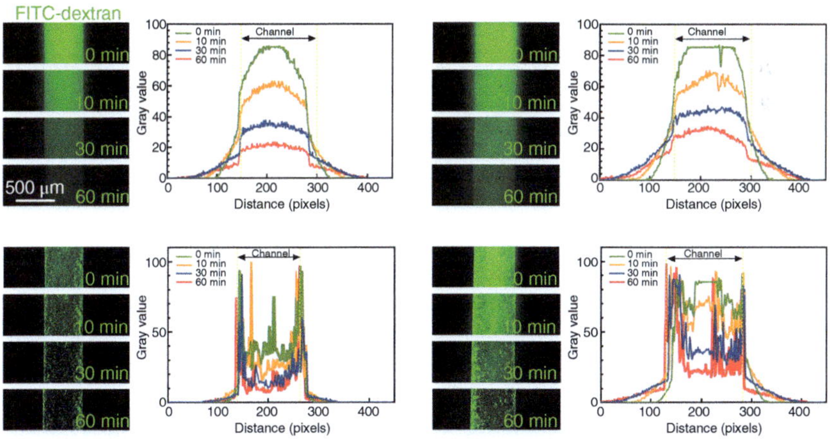

Figure 12.31 Time-lapse confocal images of the diffusional permeability patterns of FITC-dextran through the channels. Statistical data of the fluorescence intensity across the channel at different time points. Source: Nie et al. [18]. Reproduced with permission of The Royal Society of Chemistry.

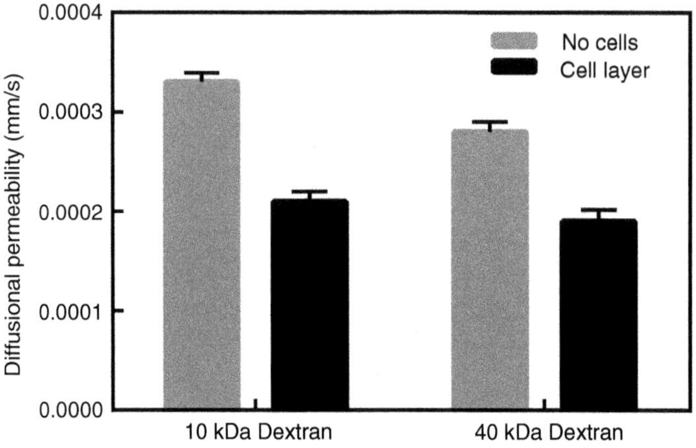

Figure 12.32 Diffusional permeability calculation of dextran. Source: Nie et al. [17]. Reproduced with permission of Wiley-VCH.

sets of fluorescence images are applied to display the distribution of fluorescence intensity along the diameter of the channel at different time points. The results indicate that the overall fluorescence intensity distribution pattern is presenting a sinusoidal curve in all experimental groups except for the singular points caused by the distribution of GFP-transfected endothelial cells.

On this basis, the barrier function of endothelium channels for molecules with different molecular weights is further characterized in a comparative way. Fluorescence-labeled 10 and 40 kDa dextran solutions are injected into the blank channel and endothelium channel, respectively.

The diffusion rate is calculated by moving distance of the fluorescence particles within a certain time, as shown in Figure 12.32. The results indicate that for glucans with two molecular weights of 10 and 40 kDa, the diffusion rate of molecules in the endothelium channel is about three-fifths and two-thirds of that in the blank channel, respectively.

In addition, fluorescence images are collected after 5, 10, 20, and 30 minutes of diffusion. It could be observed from Figure 12.33 that as the diffusion time increases, the fluorescent dextran gradually diffuses from the cavity of the channel into the surrounding hydrogel bulk (the channel is indicated by a dashed red line in the figure). Through comparative observation, it could be found that the fluorescent dextran permeating into the surrounding hydrogel from the endothelium channel is less than that from the blank channel. As a result, the molecule diffusion rate in the endothelium channel is slower than that in the blank channel, which further verifies the barrier function of the endothelial cell monolayer on molecules of different sizes.

The molecule diffusion is demonstrated by collecting the fluorescence intensity at specific locations in the radial direction. Figure 12.34 displays the fluorescence intensity distribution curve along the radial direction of the channel after the fluorescent particles of two molecule weights are injected into the channel and diffused for 30 minutes (the channel is indicated by a yellow dotted line). It could be observed

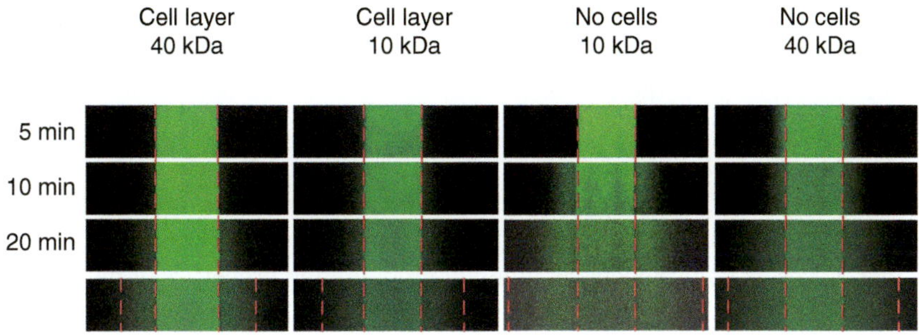

Figure 12.33 Time-lapse fluorescence images of diffusion patterns of 10 and 40 kDa FITC-dextran into surrounding hydrogel bulk from the channel. Source: Nie et al. [17]. Reproduced with permission of Wiley-VCH.

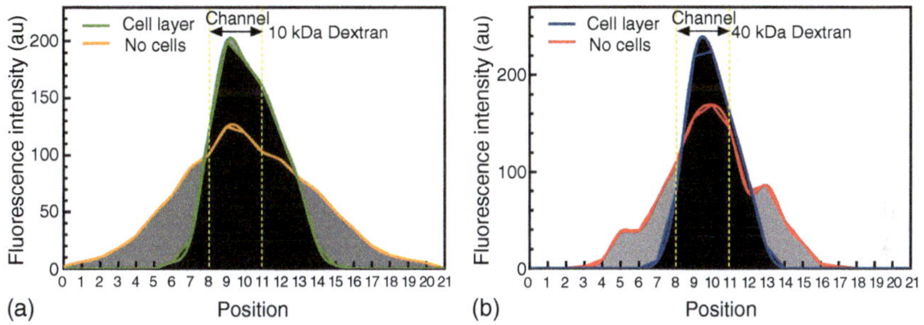

Figure 12.34 Line plots of fluorescence across the channel after 30 minutes of perfusion with 10 kDa dextran (a) and 40 kDa dextran (b). The position of channel is indicated with yellow dotted line.. Source: Nie et al. [17]. Reproduced with permission of Wiley-VCH.

that in both channels, the cylindrical structure of the channel exhibits a diffusion pattern of the sinusoidal profile, and the higher fluorescence intensity distribution outside the cavity represents a higher diffusion rate. It is proved again that the engineered endothelial cell channel owns a powerful barrier function, rendering it a reliable model for further study.

12.4 Conclusion

This chapter designs and proposes corresponding schemes for the creation of microstructures and bonding strategies targeting hydrogel materials. A flexible method of constructing hydrogel-based vascular chips based on a lamination process is presented. The multi-scale 3D printing process is further applied for the manufacture of soft fiber templates, thereby overcoming the channel scale limitations of the hydrogel-based microfluidic chips to realize *in vitro* reconstruction of the complete blood circulation system. A multi-scale 3D printing technology is realized through the real-time shift between FDM and EHD printing modes based on the control over a high voltage power. A damage-free demolding method based on a soft fiber template is proposed to fabricate micro/nanostructures on hydrogels. A twice cross-linking strategy is designed to obtain a bonding interface that has the same strength with the hydrogel bulk, which could be applied to arbitrary combinations of hydrogels.

There is no need of a complicated experimental setup or configuration, and the whole process is easy to conduct. Different forms of channels can be easily achieved through the design and fabrication of different molds for casting. The demolding and bonding processes both rely on the inherent cross-linking properties of the hydrogel material, without the introduction of any other materials or physical–chemical reactions. The channel accuracy completely depends on the molds for casting, which can be easily guaranteed by the process of mold manufacture. Most importantly, the bonded hydrogel sheets eventually become one complete bulk without any dividing line between the bonding interfaces.

This study pioneers the combination of multi-scale 3D printing and a twice-cross-linking strategy to construct multi-scale structures within hydrogel materials. This method has the ability to construct the whole vascular hierarchy (from arteries to veins) and to emulate the ubiquitous structures of biological vascular systems. Based on these, a complete vascular system with vascular function in both physiological and pathological situations is established, providing a promising model for further investigations such as vascularization, vascular inflammation, tissue engineering, and drug development.

References

1 Bae, H., Puranik, A.S., Gauvin, R. et al. (2012). *Science Translational Medicine* 4 (160): 160ps23–160ps23.
2 Lanza, R., Langer, R., Vacanti, J.P., and Atala, A. (2020). *Principles of Tissue Engineering*. Academic Press.
3 Jain, R. (2005). *Nature Biotechnology* 23: 821.
4 Kaully, T., Kaufman-Francis, K., Lesman, A., and Levenberg, S. (2009). *Tissue Engineering Part B: Reviews* 15 (2): 159–169.
5 Yuan, B., Jin, Y., Sun, Y. et al. (2012). *Advanced Materials* 24 (7): 890–896.
6 Yu, Y., Shang, L., Guo, J. et al. (2018). *Nature Protocols* 13 (11): 2557–2579.
7 Trappmann, B., Baker, B.M., Polacheck, W.J. et al. (2017). *Nature Communications* 8 (1): 1–8.
8 Song, K.H., Highley, C.B., Rouff, A., and Burdick, J.A. (2018). *Advanced Functional Materials* 28 (31): 1801331.
9 Gao, Q., He, Y., Fu, J.-Z. et al. (2015). *Biomaterials* 61: 203–215.
10 Zhang, R. and Larsen, N.B. (2017). *Lab on a Chip* 17 (24): 4273–4282.
11 Cuchiara, M.P., Allen, A.C., Chen, T.M. et al. (2010). *Biomaterials* 31 (21): 5491–5497.
12 Paulsen, S. and Miller, J. (2015). *Developmental Dynamics* 244 (5): 629–640.
13 Lee, K.Y., Peters, M.C., Anderson, K.W., and Mooney, D.J. (2000). *Nature* 408 (6815): 998–1000.

14 Rafii, S. and Lyden, D. (2003). *Nature Medicine* 9 (6): 702–712.
15 Xue, D., Wang, Y., Zhang, J. et al. (2018). *ACS Applied Materials & Interfaces* 10 (23): 19428–19435.
16 Lv, S., Nie, J., Gao, Q. et al. (2020). *Biofabrication* 12 (2): 025015.
17 Nie, J., Gao, Q., Wang, Y. et al. (2018). *Small* 14 (45): 1802368.
18 Nie, J., Gao, Q., Xie, C. et al. (2020). *Materials Horizons* 7 (1): 82–92.

13

3D Printing of In Vitro Models

13.1 Introduction

The promotion of new drugs and the analysis of the pathogenesis involve high investment and high risk, mainly associated with preclinical testing and clinical trials.

Conventional cell-monolayer planar culture model with easy manipulation is accessible to high-throughput operation. However, it is commonly known that the 2D growth condition largely varies from the in vivo microenvironment, and the drawback of this model is increasingly evident. Specifically, such models lack the necessary cell–cell and cell–extracellular matrix (ECM) interaction, they are unable to recreate the complicated in vivo condition or to mimic heterogeneous diseases, and cells lose specific morphology, function, as well as metabolic activity. The limitations of this model impede its ability to accurately predict drug performance and toxic effects.

Animal experiments are an alternative during the process of drug screening, which are costly and time-consuming. Primates in high evolution degree cannot be widely applied due to their limited sources and difficulty in maintenance. On the other hand, most rodent animal models are not valid enough for recapitulating in vivo situations or predicting clinical reactions. The major concerns with animal models applied in biomedical fields are the clinical limitation and poor performance. Due to the different behaviors of animals and humans, about 50% of drug candidates that have passed preclinical animal tests are verified to be toxic to humans, whereas some drugs may be effective in clinical trials even if they fail in early animal testing. These results indicate that drugs would have to be screened based on a sufficiently large number of multiple experiments to ensure that no active drugs are missed, which is rather a high-cost proposition.

Considering the inevitable restrictions of 2D cell-culture models and animal models, better predictive models representing the complexity of human tissues/organs, in vivo physiological/pathological conditions, as well as inner mechanism must be developed to complement the existing models for more precise drug screening in the safety and efficacy of drug candidates.

Based on the principle of additive manufacturing, 3D bioprinting technology owns the capability of precisely distributing specific types of cells or biomaterial in the target area, offering exciting prospects for the construction of precise and

Cell Assembly with 3D Bioprinting, First Edition. Yong He, Qing Gao, and Yifei Jin.

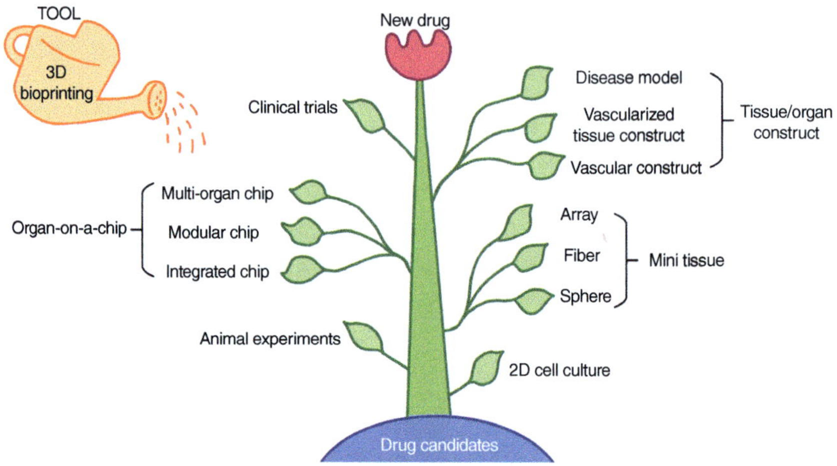

Figure 13.1 Schematic diagram displaying the routing of drug development. Drug candidates undergo a series of tests before being put into use: 2D cell culture, animal experiment, in vitro validation, and clinical trial. 3D bioprinting technology may be grafted to the in vitro drug screening. The in vitro methods for drug screening are generalized into three types: mini-tissue, organ-on-a-chip, and tissue/organ construct. Source: Nie et al. [1]. Reproduced with permission of Wiley-VCH.

individualized in vitro models. These in vitro models can provide an alternative to conventional animal experiments and clinical trials, which can promote drug development of high efficiency while saving time and cost.

This chapter focuses on the existing and potential involvement of 3D bioprinting technology in the construction of in vitro models, which are classified into three types: mini-tissue, organ-on-a-chip, and tissue/organ construct, as shown in Figure 13.1. The developing process of the different kinds of models and the relationship between different models are introduced in detail, with corresponding researches listed as proof of concept. The current limitations and challenges, as well as the future directions of the grafting of 3D printing technology to the construction of in vitro models, are also discussed.

13.2 Typical 3D Bioprinting Technologies and Common Target Tissue/Organ Demand

Specific physiological phenomenon is targeting at specific tissue/organ, while in some cases, several different tissues/organs are involved in a single in vivo activity. In order to obtain more instructive results, it is necessary to construct relevant models as precisely as possible. Multiple tissue/organ modeling demands different sources, materials, and methods. Researchers are devoted to constructing more reliable and predictive in vitro models for precision medicine and biological study. The accurate detection and efficient prediction of efficacy and safety rely on robust assay models which could mimic in vivo situations specifically targeting the individual patient.

3D bioprinting technologies are being involved in building such models in a versatile and individualized way. In order to acquire vigorous models for trustworthy results, cell types and sources, printing bioinks, as well as printing strategies have to be seriously decided and improved. Bioprinted individualized in vitro models own the potential of facing the present challenges of physiological, pathological, and pharmacological study. Multiple organ types varies a lot in the component, architecture, as well as functionality, putting corresponding demands on cell sources, matrix biomaterials, constructing methods, and culture conditions. Currently, there are several classical printing strategies with correspondence to different organ types. The presentation and analysis of three broadly applied 3D printing methods and relevant suitable tissue/organ models, as well as the concrete application in specific tissue/organ models, are roughly shown in the following part.

13.2.1 Inkjet-Based Bioprinting

Inkjet-based bioprinting refers to distributing individual droplets of bioink on the substrate layer. Inkjet printing can be realized through thermal, piezoelectric, and electro-hydrodynamic ways, as shown in Figure 13.2. Discrete droplets are deposited as a basic block, with accurate control over the bio-parameters. In particular, this printing strategy is showing potential in small-scale tissue/organ model construction with strict demand in resolution. Further cooperated with diverse nozzles, the printing of multiple cells, biomaterials, and biological factors could be flexibly realized at one time.

However, the operation mode and equipment configuration are confining the size and holistic of the printed models, blocking its further application in the building of large-scale tissue/organ structures. In addition, the printing ink is limited to biomaterials of low viscosity and concentration, hindering its application in the construction of tissue/organ models with high mechanical strength, cell density, or metabolism activity.

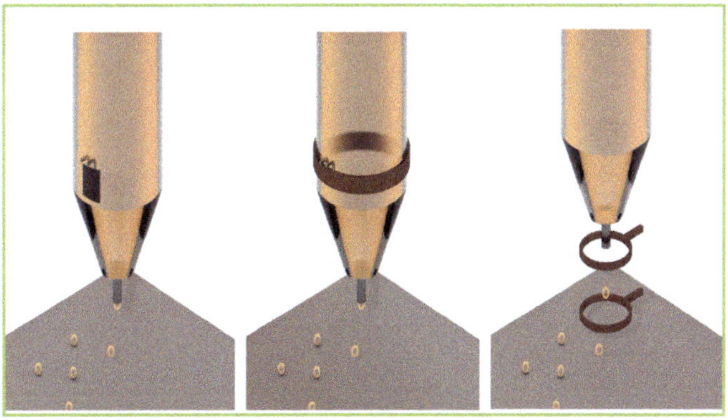

Figure 13.2 Inkjet-based bioprinting systems. Source: Gu et al. [2]. https://www.sciencedirect.com/science/article/pii/S1818087619311869 CC-BY 4.0.

On the other hand, this technology is better appropriate for cases of cell-free tissue/organ models. The inkjet-based printing strategy is involved in the establishment of a skin model based on keratinocytes and fibroblasts to represent the epidermis and dermis, respectively, using collagen to stimulate the dermal matrix of the skin. The inkjet-based bioprinting has also been used to print bone models based on acrylated peptides, acrylated poly(ethylene glycol), and marrow-derived human mesenchymal stem cells (hMSCs). Using a droplet-based bioprinting method, Graham et al. generated a renal-like tissue using human embryonic kidney cells (HEK cells) and mesenchymal stem cells.

13.2.2 Extrusion-Based Bioprinting

Extrusion-based bioprinting refers to the deposition and stacking of continuous fibers instead of discrete droplets. Extrusion nozzles of various forms could be assembled into the printing system to fabricate fibers of multiple architectures, approving for the realization of multiple extrusion modes, including pneumatic, piston, and screw schemes, as shown in Figure 13.3. Extrusion-based printing has displayed the potential of patterning wide materials in a "line-by-line" operation mode. Extrusion-based bioprinting displays excellent printability of a rather broad range of biological-relevant components. In particular, multi-material and coaxial extrusion modes obtain definite dominance in the generation of complicated heterogeneous tissue/organ models and those models abundant with internal perfusion channel network for circulation and metabolic demands, respectively.

On the other hand, this strategy is time-consuming when applied to fabricate large-scale models with complex architecture. Relevant factors such as the extrusion nozzle, substrate platform, and extrusion force are limiting its resolution, flexibility, and reliability. In addition, the extrusion and stacking process put strict demands on the mechanical properties of the applied biomaterials.

An extrusion-based bioprinting technology was applied to print alginate-encapsulated HepG2 cells to construct the liver model. Extrusion-based bioprinting

Figure 13.3 Extrusion-based bioprinting systems. Source: Gu et al. [2]. https://www.sciencedirect.com/science/article/pii/S1818087619311869 CC-BY 4.0.

was also smoothly applied in one research to simulate intestinal villi based on the collagen bioink carried with Caco-2 cells. Johnson et al. built a nervous system using the extrusion-based 3D printing method by applying cervical ganglia and hippocampal neurons. In combination with patient-specific cells, hydrogel is applied to build thick, vascularized, and perfusable cardiac patches based on the extrusion-based 3D bioprinting system. A well printable ink composed of alginate and nanocellulose was prepared and supplemented with chondrocytes and stem cells to build cartilage models.

13.2.3 Light-Assisted Bioprinting

Light-assisted bioprinting refers to the local cross-linking of photo-sensitive bioink on the printing stage, as shown in Figure 13.4. With its major superiority in printing resolution and efficiency, more and more light-assisted printing technologies have been introduced for the construction of in vitro models to address the urgent demands of precise models with defined features. Large-scale tissue/organ models with multiplex surface patterns can be attained in a shorter time. Nevertheless, the contact-free operation mode supports the activity and function of the printed constructs. Based on its nozzle-free configuration, this printing strategy could be adapted to various materials with a broader range of properties, without disturbance of mechanical damage to cells, including plugging and shear stress.

Among them, projection-based 3D printing, referring to a continuous layer-by-layer process, is based on a digital light processing (DLP) system to modulate light. With the advantages of high efficiency, precision, and consistency, projection-based printing is potential for the establishment of stereotype in vitro models equipped with personalized biological-relevant properties in a standardized way. This curing mechanism demands photosensitive ink to realize the photo-cross-linking process. The printing results are affected by the process and materials at the same time. Based on a layer-by-layer operation mode, DLP can significantly modify the printing

Figure 13.4 Light-assisted bioprinting systems. Source: Gu et al. [2]. https://www.sciencedirect.com/science/article/pii/S1818087619311869 CC-BY 4.0.

efficiency and effect, pushing toward its application in the fabrication of large-size physiological models with improved mechanical strength and structural integrity.

Specifically, gelatin methacryloyl (GelMA) – a photo-cross-linkable hydrogel material with superior biological performance, can be readily cured under specific light to generate 3D porous structure approving cell proliferation and migration. Up to recent, it has been widely applied in the fields of tissue engineering and regenerative medicine.

In one study, a laser-assisted bioprinting method was applied to generate the skin substitute model based on fibroblasts and keratinocytes mixed with a stabilizing matrix (Matriderm). In one video article published in the journal of visualized experiments, the detailed protocol for the construction of complicated and intricate 3D constructs using a DLP bioprinter is presented, which could be further adapted to fabricate predefined 3D GelMA structures of more complicated forms in a high-throughput and high-resolution way.

13.3 Developing Process of In Vitro Models

The present and potential involvement of 3D bioprinting technology in the construction of in vitro model, as well as its developing process, is analyzed in detail. The in vitro models have been classified into three categories, which are presented as follows in relevance to the involvement of 3D bioprinting respectively.

13.3.1 Mini-Tissue in 3D Growth State

It has been indicated that the 3D culture condition can better mimic the in vivo physiological microenvironment, and cells growing in a 3D state can preserve a better metabolism activity, better reporting its in vivo situation. Based on this, the mini-tissue models with 3D structures have been broadly adopted in biomedical study, facilitating high-throughput experiments with low reagents assumption. Through the adoption of multiple cell sources and matrix materials, a set of 3D mini-tissue models with specialized structures and functions could be easily built using general 3D bioprinting technologies, as shown in Figure 13.2. Targeting some independently-operated tissue/organ instead of the complicated organ system, the mini-tissue model is adequate for the simulation of in vivo function.

13.3.1.1 Sphere Mini-Tissue Model

The mini-tissue model in spherical form is becoming a promising model of complex cell-loaded constructs due to its outstanding structural properties, processing convenience, and low consumption, which has been widely applied in tissue engineering and drug screening. Hydrogel microspheres have been widely applied, among which heterogeneous microspheres are drawing attention as a promising model to include multiple cell types or biological factors in independent phases.

In an initial attempt to generate micro-droplets, an electro-assisted inkjet printing approach was proposed to construct micro-droplets based on bone mesenchymal

(a)

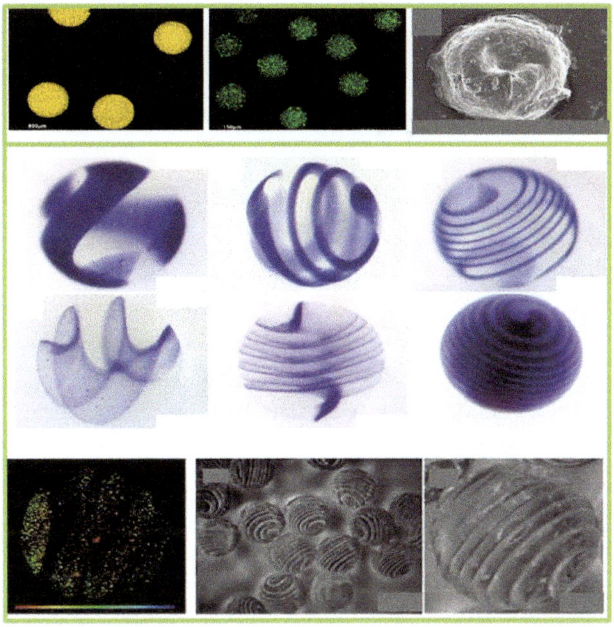

(b)

Figure 13.5 Sphere-shaped mini-tissue: (a) GelMA microspheres. Source: [3]. Reproduced with permission of Wiley-VCH. (b) Microspheres with spiral-compartmental geometries. Source: Zhao et al. [4]. Reproduced with permission of Wiley-VCH.

stem cell (BMSC)-loaded GelMA hydrogel (Figure 13.5). In another research, an airflow-assisted 3D bioprinting method was designed to generate microspheres with various spiral microarchitectures for precise cell distribution. Zhao et al. reconstructed a human multicellular organoid of spirally vascularized ossification through this strategy, revealing its capability of building mini-tissues in micro-spheroids (Figure 13.5).

13.3.1.2 Fiber Mini-Tissue Model

Bio-active fiber-shaped mini-tissues have drawn wide attention for the recapitulation of specific human tissue/organ because of their bio-mimicking structures, including muscle strips, nerve bundle, as well as blood vessels. One critical concern is the generation of morphology-adjustable microfibers with desired characteristics for the simulation of the relevant physiological functions.

Among them, cell-encapsulated GelMA-based microfibers with controllable structures and concrete functions have been obtained through the coaxial extrusion-based 3D bioprinting technique. In one research, GelMA hydrogel was applied as the printing ink, whereas a novel coaxial bioprinting technology was designed to construct morphology-controllable GelMA microfibers enclosed in calcium alginate hydrogel. Endothelial cells (ECs) were encapsulated in the GelMA fibers, further realizing the corresponding function (Figure 13.6). In a separate study, Cheng et al. proposed a versatile method for the generation of microfibers in multiple desired forms through capillary microfluidics. HepG2 cells or NIH

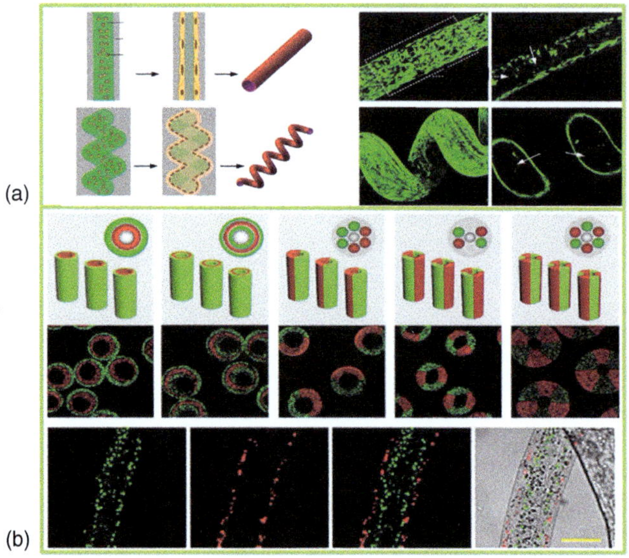

Figure 13.6 Fiber-shaped mini-tissue: (a) Straight and spiral fibers, vessel fibers. Source: Shao et al. [5]. Reproduced with permission of Wiley-VCH. (b) Multicompartmental microfibers and multicell-laden microfibers. Source: Cheng et al. [6]. Reproduced with permission of Wiley-VCH.

3T3 cells were suspended in the Na-alginate pre-polymer solution to prepare the necessary bioink for the following formation of fibers (Figure 13.6).

13.3.1.3 Array Mini-Tissue Model

Micro-array models have emerged due to their unified and stable performance, which could help for high-throughput drug testing.

In a relevant research, a new method was proposed to build an array of composite hydrogel based on the photo-cross-linking mechanism. Human periodontal ligament stem cells embedded in the hydrogel materials were validated in aspects of viability, proliferation, osteogenic differentiation, and in vivo osteogenesis. Individual spheroids were generated based on a high-throughput method. In addition, co-culture models involving ECs with fibroblasts and/or adipose tissue-derived mesenchymal stem cells have been established. A 3D projection-based printing approach of constructing concave hydrogel microstructures was also proposed for sphere growth and long-term conservation. The possibility of sphere culture was illustrated based on two kinds of cells, that is, breast cancer cells and pluripotent stem cells (Figure 13.7). A microarray of multiple forms has been built through an approach named dynamic optical projection stereolithography. A polyethylene glycol diacrylate (PEGDA)-based prepolymer solution was applied as the photo-cross-linking material. Moreover, the versatility of the platform was highlighted through its application in generating multiple shapes and topographic structures that can possibly be adopted in chip-based cell research and the simulation of in vivo extracellular microenvironments (Figure 13.7).

(a)

(b)

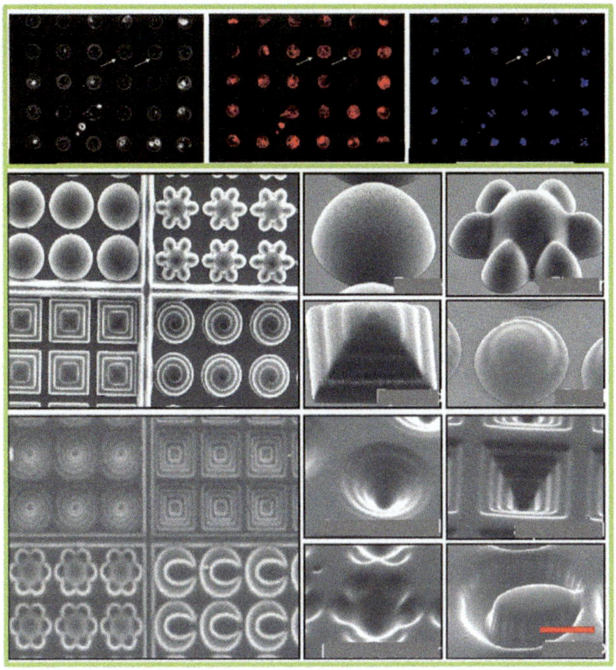

Figure 13.7 Mini-tissue array: (a) Cell-laden mini-tissue array. Source: Hribar et al. [7]. Reproduced with permission of The Royal Society of Chemistry. (b) Mini-tissue array with various microstructures. Source: Zhang et al. [8]. Reproduced with permission of Wiley-VCH.

13.3.1.4 Limitations

While the 3D growth situation can better mimic the in vivo state, the mini-tissue model is lacking the multi-scale structure as well as the tissue interface which are essential for tissue/organ function. Mini-tissue model can only recreate the finite function of a tissue, limiting its expansion for building complicated tissue/organ models. What's more, the transportation of nutrition and the loading of biochemical stimuli mostly depend on the soaking of the whole construct in the liquid, leading to a lack of control over the accurate gradient profile, while impeding the efficient mimicking of the in vivo physiological microenvironment. On the other hand, the cells are generally lacking communication with mechanical clues, which are significant for tissue/organ development.

13.3.2 Organ-on-a-Chip with Multiplex Microenvironment

In order to deal with the challenges of the mini-tissue models, another type of in vitro model is developed based on microfluidic chips. The main difference between the mini-tissue model and the organ-on-a-chip model is the dynamic perfusion condition.

Current advances in microfluidics grant the valid spatiotemporal transport of biological factors and dynamic cell culture conditions for effective metabolic circulation and essential physiological stimulation. Microfluidics has been widely involved in drug testing, in vitro microenvironment simulation, and organ-on-a-chip construction due to their miniaturization, fewer reagents assumption, and capability of in situ analysis. Assisted by the perfusion condition to act as the bloodstreams, the microfluidic chip has gradually been introduced as a robust platform for mimicking the structural layouts and complicated performances of in vivo tissues/organs equipped with sufficient internal vascular network. In the meantime, microfluidics offers a superior choice for the development of in vitro biological-related reaction analysis platform. The involvement of the microfluidic technique can modify the physiological connection of in vitro models and better mimic the in vivo physiological microenvironment in vivo as well as the primary structural properties of targeting tissues/organs. At the same time, it offers an appropriate device for physiological/pathological/pharmacological research and drug discovery, including cell metabolism analysis, intercellular interplay and therapy mechanism study, drug side effects, and dose-response. The perfusion condition could facilitate metabolites exchange while simulating the mechanical shear stress of the in vivo microenvironment. Drug testing could be realized through carrying out real-time, high-throughput, and in situ analysis of cells on the microfluidic chip. The dynamic perfusion based on the microfluidic channels could be applied to act as a circulation system for the adoption of drug candidates or other stimuli in a more flexible and versatile way. Reliable generation and accurate regulation of gradient profiles could be realized, which could further be applied to the study of the gradient effect and dose-reaction of drug candidates.

Based on the microfluidic chips, organ-on-a-chip owns the superior merits of microfluidic technology, with the ability to manipulate trace amounts of liquids as well as steady perfusion condition. In combination with bioactive components, it further owns the capability of precise control over multiplex biological-relevant parameters to simulate specific targeting biological microenvironment, such as spatiotemporal biochemical clues, dynamic physical stimuli, as well as interactions between tissues or organs. With the help of the dynamic perfusion capability, organ-on-a-chip are contributing to the understanding of series physiological phenomena, disease mechanism, as well as drug metabolism and effect. One of the key merits is that it could make use of the microfluidic reaction device to carry out dynamic stimuli loading upon tissue/organ models, such as growth factors and drug candidates, and thus permitting fast testing of the efficacy of physical stress, as well as the influence of drug components and concentrations on organ growth and development.

Since the proposal of the concept of organ-on-a-chip, many honorable pharmaceutical research and discovery corporations have been devoted to fundamental organ-on-a-chip study. The National Transformation Science Promotion Center has carried out some programs to offer funding to the development of an organ-on-a-chip platform to deal with the ethical/time/cost limitations in consideration with the species/individual disparity between animal experiments and clinical trials during the process of new drug discovery. Organ-on-a-chip is referred to as a microfluidic chip loaded with specific cells in order to establish a biomimicking system replicating the in vivo physiological microenvironment and the main functions of specific human tissue/organ. Multiple physical, chemical, and biological stimuli could be applied through microfluidic means, while the in situ analysis of cells and factors could be realized for drug testing.

13.3.2.1 Integrated Organ-on-a-Chip

The classical construction process of organ-on-a-chip, introduced first by Harvard University, covers three steps: manufacturing of the microfluidic chip body, loading of cells through perfusion, and dynamic culture based on the chip. Polydimethylsiloxane (PDMS)-based microfluidic chips have taken the place of initial microfluidic chips based on silicon or glass micromachining with the help of the properties of low toxicity and biochemical inertness.

Several specialized microfluidic chips have been designed and fabricated for precise analysis, testing, as well as gradient concentration generation, potentially offering a versatile platform for chemical and biomedical study. A robust gradient-generation microfluidic chip has been built for the multidrug resistance analysis of single-cell. The obtained results verified the high throughput, flexibility, stability, and low sample waste of the platform (Figure 13.8). In combination with cell loading and induction culture, a microfluidic chip could further act as an organ-on-a-chip with better physiological relevance. Huh et al. established a biomimicking model to reconstruct the essential alveolar-capillary interface

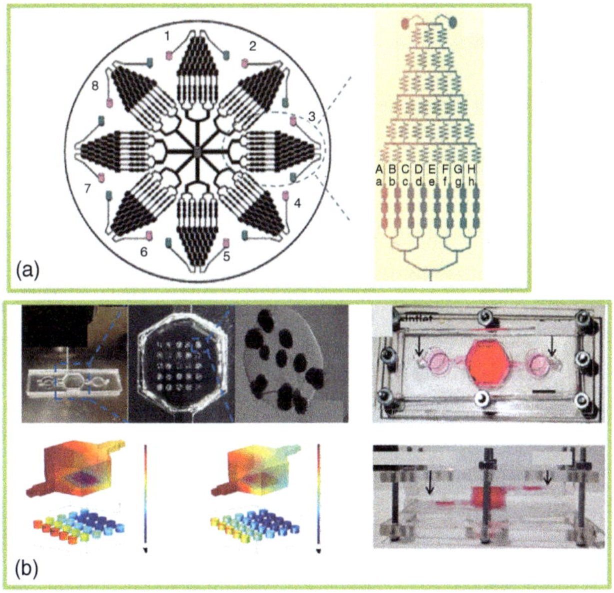

Figure 13.8 Integrated cell/organ-on-a-chip: (a) Schematic diagram of the integrated microfluidic chip for cell-based screening, consisting of eight uniform structural units containing a concentration gradient generator and parallel cell culture chambers. Source: Ye et al. [9]. Reproduced with permission of The Royal Society of Chemistry. (b) A liver-on-a-chip platform. Source: Bhise et al. [10]. Reproduced with permission of The Royal Society of Chemistry.

of the lung, which further reproduced the organ-level reaction to series factors (Figure 13.8). The development of a liver-on-a-chip platform was presented for the long-term culture of bioprinted 3D human HepG2/C3A spheroids for the assessment of drug toxicity (Figure 13.8). The bioreactor design provided direct access to the hepatic construct during the whole experiment. In addition, this engineered bioreactor could be connected with a bioprinter to in situ printing 3D hepatic constructs of spheroids embedded within photo-cross-linkable GelMA hydrogel. Macetaminophen induced toxic response was conducted through this system.

However, common processing methods for PDMS-based microfluidic chip (soft lithography, hot embossing, laser direct writing, microinjection, etc.) are complicated, time-consuming, as well as of high cost. As a result, 3D printing has been progressively adopted in the fabrication of microfluidic chips as an alternative to the classical micromolding methods.

During the construction process of such an organ-on-a-chip device, the 3D bioprinting technology could be potentially involved for both the manufacturing of the microfluidic chip body as a reaction container, as well as the direct deposition of the biomimetic tissue/organ construct model situated on the microfluidic chip as a reagents receiver. The advantages of convenience and customization are translated into a potential technique for the construction of sacrificing templates for the creation of internal microchannels, as well as for the in situ printing of cell-loaded constructs on the platform.

13.3.2.2 Modular Microfluidic System

The closed forms of the channels and chambers of the integrated microfluidic chip are bringing difficulty and uncontrollability to cell loading while preventing the 3D constructs to be deposited in the chamber through in situ printing. In order to face these limitations, a concept of a paper-based microfluidic chip has been introduced to realize liquid delivery through capillary driving force. This concept demands no closed channels for pump driving, and can easily program the flowing rate by adjusting the channel scale. In addition, complete tissues and organs are quite varying and complicated in the structure, composed of individual practical building bricks as the primary units of tissue/organ structure and function, including liver lobules. In order to accomplish the prospective property, various microfluidic chips need to be designed and improved over and over again, which is time-consuming and resource-wasting. As a solution, the idea of modularity is involved in the building of organ-on-a-chip. Fundamental building bricks could be manufactured by units and further gathered into a complete system.

Yuen et al. proposed a truly “plug-and-play” modular microfluidic system for the design and establishment of integrated modular microfluidics for biological and chemical applications. Bhargava et al. presented a microfluidic system design based on easy-to-assemble elements and built a library of standardized components fabricated by stereolithography (Figure 13.9). Qiu et al. designed a modular microfluidic system fabrication process based on sugar printing. The modular design moves the bonding process in the time following cell seeding, fairly simplifying the cell loading step. Also, the modular design of the structure decreases the time and

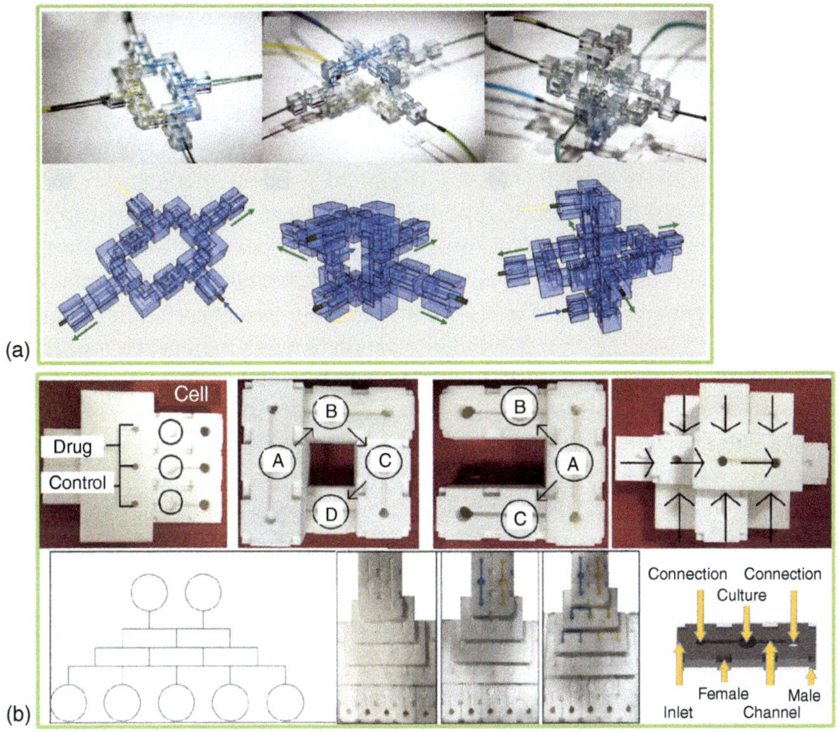

Figure 13.9 Modular microfluidic system: (a) Discrete elements with standardizing interconnect footprints assembled to enable plug-and-play operation and reconfiguration of 3D modular circuits. Source: Bhargava et al. [11]. Reproduced with permission of The National Academy of Science. (b) 3D-printed modules assembled to form various paper-based microfluidic systems with the extensive application. Source: Nie et al. [12]. Reproduced with permission of IOP Publishing.

cost of customized demands. Nie et al. presented a modular microfluidic platform based on the capillary driving force. Fundamental functional modules are designed and manufactured based on 3D printing and further assembled into a larger system through plugs, assembling LEGOs, offering the probability of building a multiple-organ system model (Figure 13.9).

What's more, when combined with 3D bioprinting technique, cells and matrix materials could be constructed directly on the open chamber to form tissue/organ models in situ. Multiple cells on the microfluidic chip are embedded within matrix materials to better simulate the 3D morphology and phenotype of in vivo cells, while multiplex clues could be applied based on capillary action.

13.3.2.3 Multiple-Organ System

The interplay between different tissues/organs directs series of in vivo physiological phenomena, which is particularly decisive for preclinical drug testing. In vivo drug operation is related to multiple tissues and organs. Drug metabolism within liver tissue is influencing the drug clearance time, while producing secondary metabolites in relation with drug efficacy. Drug liver toxicity is one of the major side-effects of the drug candidates. Furthermore, drug absorption and metabolism in the small intestine, drug metabolism in the liver, as well as drug excretion in the kidney are of vital importance to the safety and effectiveness of drug candidates.

Through the application of a continuous perfusion platform, the final purpose is to establish a multi-organ model to reveal the interdependent interaction of various organs and the systematic reaction of drug candidates. The multi-tissue/organ system can represent the mechanism of message transport and interconnection between different tissues and organs, further reproducing in vivo metabolism. Multiple tissues or organs could be related to each other by a microfluidic perfusion platform to reproduce the crosstalk and interaction between various tissues/organs in vivo for interpreting drug mechanism, metabolism, and side-effect.

The effects of drug candidates are intently associated with the crosstalk between different tissues/organs, including the heart–liver, adipose tissue–liver, as well as crosstalk among others (Figure 13.10). The crosstalk between different tissues/organs, covering the brain, gut, muscle, liver, and adipose, directs a series of in vivo physiological and pathological phenomena and is rather important to drug testing and development (Figure 13.10). Earlier, such relevant researches were

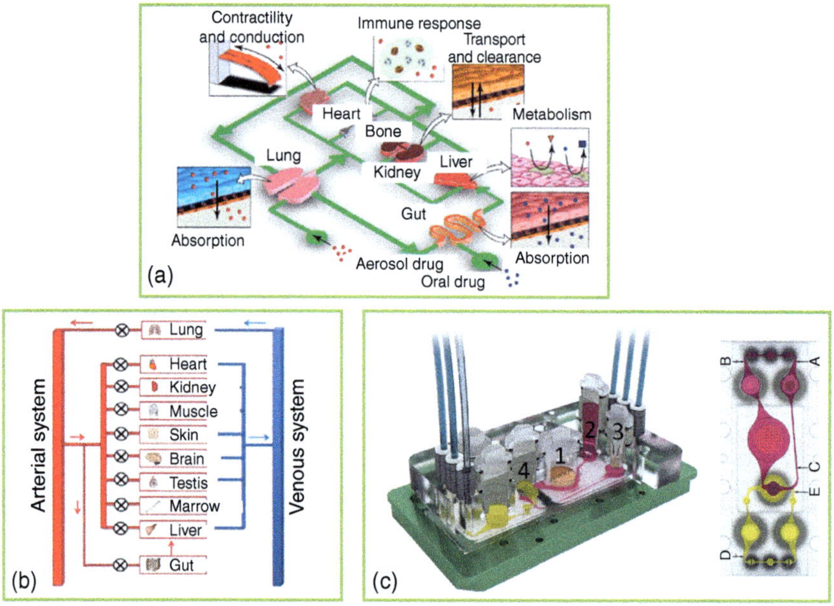

Figure 13.10 Multiple-organ system: (a) Schematic diagram of a human-on-a-chip concept for modeling the process of drug absorption, distribution, metabolism, and excretion, and evaluating drug efficacy and toxicity. Source: Huh et al. [13]. Reproduced with permission of Elsevier. (b) Schematic of an integrated multi-organ system with 10 different organs connected through a microfluidic circulatory system. Source: Huh et al. [14]. Reproduced with permission of The Royal Society of Chemistry. (c) A four-organ system containing intestine, liver, skin, and kidney. Source: Maschmeyer et al. [15]. Reproduced with permission of The Royal Society of Chemistry.

generally carried out based on 2D cell culture models or animal models. As an alternative, 3D bioprinting technology could be involved to model various tissues/organs which are related to each other, modifying the strategy of drug testing influenced by multiple tissues/organs as well as their interactions. A multiple-tissue/organ system could reproduce the information exchange and the interaction between tissues/organs to mimic in vivo metabolism. What's more, the versatile design and miniaturization could help to realize individualized and cost-effective protocols.

A 3D-printed multi-scale model has been designed to reproduce the critical properties of the glial cell–axon interface in the nervous system, representing the organ-level reactions to viral infections within the nervous system. A coculture system maintaining the main functions of four organs for 28 days has also been proposed and has been incorporated with intestine and skin models. A 3D sphere was introduced to mimic the liver function. A barrier separating the medium for the culture of organs from liquids excreted by the kidney was generated by a cell-seeded membrane. The other microfluidic cycle ensured the unloading of the fluid excreted by the kidney cell layer. Metabolism and genetic analysis indicated the establishment of the balance in the process of coculture within two to four days (Figure 13.10).

13.3.2.4 Limitations

Even though the organ-on-a-chip concept could overcome the limitations of mimicking the in vivo complicated microenvironment of the mini-tissue model, its body material constrains its deeper application. The curing of PDMS contains a high-temperature condition, therefore blocking the introduction of bioactive materials in the process of device manufacturing. The PDMS material demands additional surface modification to approve cell adhesion, exhibiting poor permeability and degradability. As a result, PDMS material could only be adapted to building a reaction device, without the capability of simulating ECM for the package of living cells. A modified model demands better materials with improved biological performances.

13.3.3 Tissue/Organ Construct with Biomimicking Property

Faced with the biological restrictions of the PDMS material, multiple natural hydrogels, synthetic hydrogels, and decellular matrix, including collagen, gelatin, GelMA, agarose, alginate, and multiple ECM proteins, have been adopted to the fabrication of in vitro tissue/organ models. Such hydrogel materials could better simulate the ECM in aspects of biocompatibility, permeability, as well as degradability. Cells could be seeded on the material surface to form 2D adhesion while packaged within the material bulk to form 3D growth. Progress in 3D bioprinting technology could help for the precise loading of cells and biomaterials to synchronously establish tissue/organ constructs and the physiological microenvironment, achieving the construction of in vitro 3D biomimicking physiological/pathological models in one step. Apart from building healthy human tissue/organ models, specific disease models could also be constructed for the assessment of drug effects. These models are of significant value in different periods during the whole drug discovery, covering initial drug testing, clinical phase I drug validation, accurate drug adoption, new drug screening, as well as individualized drug development.

13.3.3.1 Vascular Construct

In vivo vascular system, covering the large blood vessels and capillaries, is the fundamental constitution of the human body. It supplied oxygen and nutrition for tissues/organs, while interacting with multiple biological occurrences of various tissues/organs. The vascular network is critical for the viability and function of multiple tissues/organs. The building of the vascular system model is the foundation of the construction of complicated tissues/organs. The vascular system could serve as a blood circulation network, closely relating to the interplay among tissues/organs. Multiple tissues/organs could be bound together by the vascular network, and at the same time, different elements could be conveyed among varied tissues/organs based on the vascular network. The model of the vascular system could act as the platform for related researchers to study the deeper mechanism and novel therapy for vascular-relevant diseases. In addition, it could serve as a barrier to particular factors within ECM.

Massed 3D printing strategies have been designed for engineering vascular constructs embedded in 3D tissue models, such as coaxial printing, sacrificing template printing, as well as DLP-based printing.

In recent work, a 3D vascular construct with multilevel channels was built through extrusion-based 3D bioprinting technology (Figure 13.11a). Gao et al. generated a free-standing vascular model based on the direct printing of vascular equivalents as well as the perfusion system within a single step. These practical models could be involved in the generation of multiple systems for wide biomedical fields (Figure 13.11b).

One microfluidic chip based on hydrogel materials was manufactured by incorporating the casting and bonding steps. In addition, a model of vessel equipped with vascular-relating functions under both the physiological and pathological conditions was built. He et al. designed an approach of engineering the microfluidic channel network within hydrogel materials through a micromolding process as well as an enzymatic-cross-linking mechanism (Figure 13.11c). Song et al. designed an approach in which one kind of hydrogel was directly printed in another kind of hydrogel based on the properties of shear-thinning and self-healing (Figure 13.11d).

Xue et al. proposed a projection-based printing strategy to realize the construction of hydrogel scaffolds equipped with inner complex channels for realizing cell patterning while simulating the multi-scale vascular network (Figure 13.11e). Zhu et al. built the prevascularized tissues with intricate 3D structures through fast bioprinting. Multiple cell sources could be embedded within hydrogels in predefined distribution patterns for mimicking in vivo cell composition (Figure 13.11f).

Such models are limited to sole modeling of vascular structure, while it is challenging for them to simulate more definite physiological/pathological conditions.

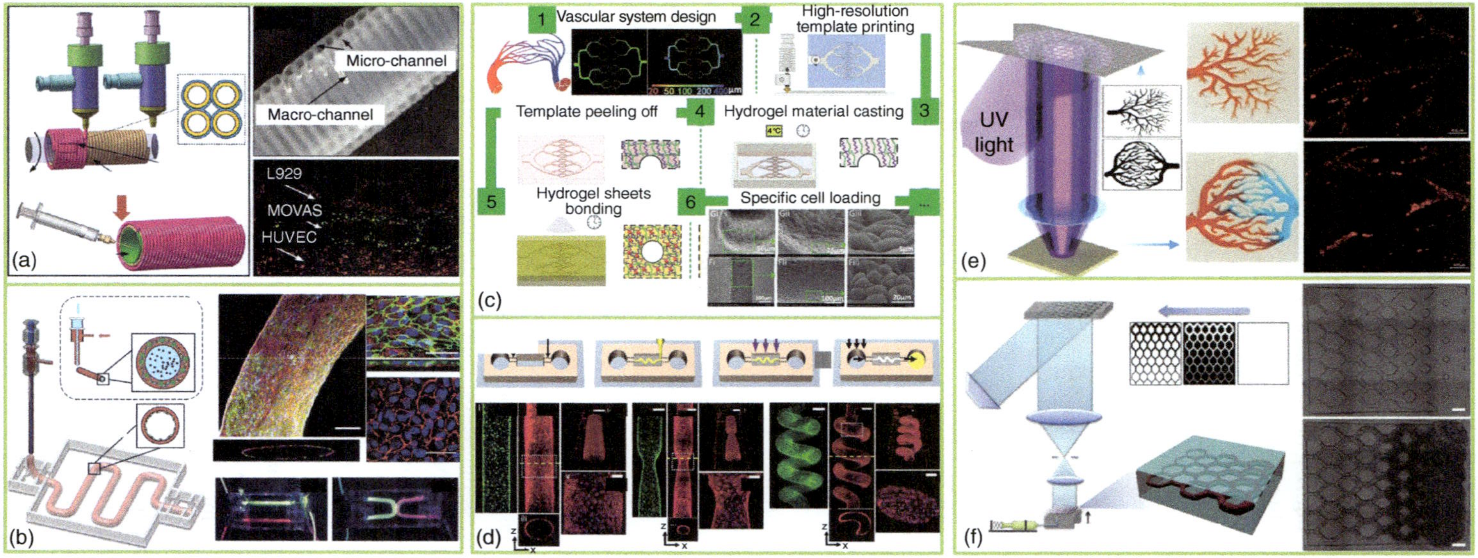

Figure 13.11 Vascular constructs: (a) Schematic diagram of the construction process of vessel structures with multiscale channels, and images of a double-layer structure and a vessel-like structure containing three kinds of vascular cells. Source: Gao et al. [16]. Reproduced with permission of American Chemical Society. (b) Schematic diagram of the construction of freestanding functional vascular models and the maturation of printed vasculatures. Source: Gao et al. [17]. Reproduced with permission of Wiley-VCH. (c) Schematic of the construction of the multiscale vascular construct and microscopic morphology of the endothelial cells. Source: Nie et al. [18]. Reproduced with permission of The Royal Society of Chemistry. (d) Schematic of the microchannel fabrication process and display of the endothelial cell-seeded microchannels in various forms. Source: Song et al. [19]. Reproduced with permission of Wiley-VCH. (e) Schematic of the projection-based 3D printing platform and images showing the capillary action of the channels and cell distribution in the capillary-like scaffold. Source: Xue et al. [20]. Reproduced with permission of American Chemical Society. (f) Schematic of the bioprinting platform for the prevascularized tissue constructs and images displaying the heterogeneous cell-laden tissue constructs. Source: Zhu et al. [21]. Reproduced with permission of Elsevier.

13.3.3.2 Vascularized Tissue Construct

Vascularization is the prior condition for the realization of tissue/organ function, while vascularization remains the critical challenge for tissue engineering and organ-on-a-chip fields. Just as the in vivo organs covering an intricate vascular network, a valid tissue/organ substitute includes the establishment of an effective vascular system. The in vitro modeling of the vascularization exhibits an improved physiological connection with the native tissues/organs, which could more precisely predict relevant responses to environment clues. Multiple tissue/organs could be integrated into the single system based on the vascular network, in order to realize a higher level of mimicking of the in vivo organ system of the human body.

Lately, 3D bioprinting technology has been involved in the construction of large-size 3D tissue models equipped with the inner vascular system. The vascularized tissue/organ models promote the process of biomedical study due to the improved in vivo relevance and higher prediction precision. The vascular system supplies the required nutrient and oxygen delivery for engineering large-scale in vitro tissue/organ models. In addition, the imitation of the vascular barrier function further lessens the disparity within the in vitro models and in vivo physiological situations. Referred to as the three-point elements, cells, matrix materials, and the vascular network are distributed in predefined patterns through 3D bioprinting technology to simulate in vivo tissues/organs.

Gao et al. proposed a 3D bioprinting approach based on hollow alginate fibers through a coaxial nozzle. Cell-laden hydrogel constructs equipped with microchannels could be obtained based on the fusion of adjacent hollow fibers (Figure 13.12a).

Miller et al. used the carbohydrate glass networks as the sacrificing templates to generate cylindrical networks within hydrogel materials, compatible with various cell sources, matrix materials, as well as specific cross-linking mechanisms (Figure 13.12b). Kolesky et al. designed a new 3D bioprinting method to construct well-defined tissue/organ constructs supplied with vascular networks, cells, as well as ECMs (Figure 13.12c). Bertassoni et al. proposed a 3D micromolding method through a bioprinted agarose template to generate channel networks within hydrogel material (Figure 13.12d). Nie et al. constructed a breast cancer model containing tumor stroma, functional vascular system, and tumor cells at the same time (Figure 13.12e).

Ma et al. reported a 3D hydrogel-based coculture model containing human induced pluripotent stem cell-derived hepatic progenitor cells (hiPSC-HPCs), ECs, as well as adipose-derived stem cells. Enhancements in both phenotypic and functional aspects have been validated through morphological organization, gene expression, metabolic product secretion, and cytochrome P450 induction (Figure 13.12f). A stereolithography-based bioprinting system for the on-demand generation of heterogeneous hydrogel constructs has further been designed. A digital micro-mirror device and a moving stage, synchronized with a microfluidic chip, were involved in the generation of 3D constructs covering multiple materials through a layer-by-layer operation (Figure 13.12g).

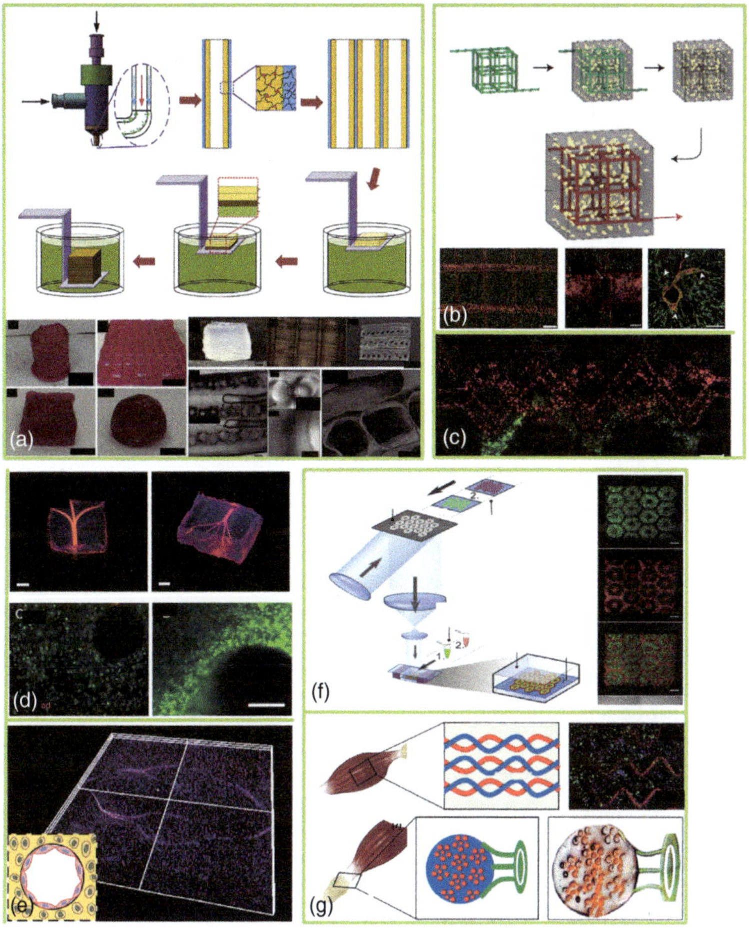

Figure 13.12 Vascularized tissue constructs: (a) Schematic of the fabrication process and photographs of printed 3D alginate structures with built-in microchannels. Source: Gao et al. [22]. Reproduced with permission of Elsevier. (b) Schematic showing the process for the establishment of a tissue construct containing vascular structures and cells, and images of the three key components of vascularized tissues, including the vascular lumen, ECs lining the vascular wall, and the interstitial zone containing the matrix and encapsulated cells. Source: Miller et al. [23]. Reproduced with permission of Springer Nature. (c) 3D bioprinted vascularized, heterogeneous cell-laden tissue constructs containing three cell types. Source: Kolesky et al. [24]. Reproduced with permission of Wiley-VCH. (d) Photographs of the microchannels perfused with a fluorescent microbead suspension and images showing the viability of cells encapsulated in hydrogel constructs with fabricated microchannels. Source: Bertassoni et al. [25]. Reproduced with permission of The Royal Society of Chemistry. (e) Display of a vascularized tumor model. Source: Nie et al. [18]. Reproduced with permission of The Royal Society of Chemistry. (f) Schematic of the 3D bioprinting of hydrogel-based hepatic construct and images showing patterns of cells within the hepatic model. Source: Ma et al. [26]. Reproduced with permission of The National Academy of Science. (g) Schematic of the skeletal muscle tissue and the tendon-to-bone insertion site, and images of the bioprinted skeletal muscle model and tendon-to-bone insertion model. Source: Miri et al. [27]. Reproduced with permission of Wiley-VCH.

13.3.3.3 Limitations

The exceeding fragile virtue of the matrix materials causes challenges in terms of both manipulation and preservation. There is an inevitable contradiction between the mechanical characteristic and biological performance of the matrix materials. The pattern resolution and size precision are limited by the deformation during the swelling process.

13.4 3D Printing of In Vitro Tumor Models

In this part, the in vitro tumor model is introduced as an instance to illustrate the potential grafting of 3D bioprinting technology in the construction of in vitro models.

Cancer is inherently complicated, covering multiple cell sources with a specific microenvironment. Cell–cell interplay (tumor cells, vascular cells, and immune cells) and cell–ECM interaction are of vital importance to the origin, developing, and propagation of the tumors. Cells are changed by a specific clue to make up a primary tumor, while the tumor progressively develops through the interplay with the microenvironment. The tumor microenvironment is extremely complicated, containing the vascular network, matrix, as well as a wide range of biological, mechanical, and chemical clues. Tumor cells prompt tumor angiogenesis through the secretion of angiogenic growth factors. Newly-formed capillaries later provide nutrition and oxygen for tumor development, while serving as the essential pathway for facilitating tumor metastasis, as shown in Figure 13.5(a). The rather complex tumor microenvironment is also connected with the resistance of tumor cells to anti-cancer drugs. On the other hand, it is impossible for the widely applied 2D monolayer cell-culture model to rebuild the intricacy of tumor tissues or represent its distinguishing microenvironment. As for the animal experiments, the mechanism of drug resistance is not the same as that in vivo, while cells display distinct drug responses.

In order to overcome these challenges, it is crucial to building an in vitro tumor model to reproduce the physiological microenvironment of the tumor tissues to mimic the tumor developing process in vivo.

Multiple in vitro tumor models have been constructed to modify the translatability of possible anticancer drugs, including cell-laden constructs, multicellular spheroids, as well as tumor metastasis models. 3D bioprinting technology could help realize the concrete structural layouts of cells and matrix materials to generate tumor models as well as its complicated microenvironments.

13.4.1 Tumor Cell-Laden Construct

The solid tumor existing in vivo is a 3D cell aggregate. Some studies have shown that the tumor cells cultured on 3D condition display better biocompatibility and drug resistance, which could better mimic the native behavior of tumors in aspects of phenotype, genotype, as well as drug response in comparison with 2D cultured cells. The 3D growth condition is closer to the practical condition of the microenvironment of the human body.

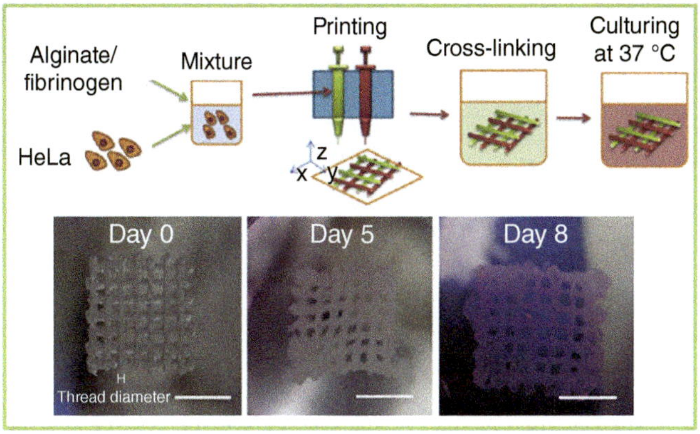

Figure 13.13 Schematic of the printing process and images of 3D HeLa cell-laden hydrogel constructs. Source: Zhao et al. [28]. Reproduced with permission of IOP Publishing.

Different from the 2D models, 3D tumor cell-laden constructs better represent the tumor microenvironment, in aspects of cell–cell and cell–matrix interactions. As a result, in comparison with the 2D culture systems, 3D tumor cell-laden models could potentially lead to more instructive consequences.

Zhao et al. carried out the generation and study of cervical tumor models based on the 3D printing of HeLa cells and hydrogel matrix materials to simulate the ECM properties and cervical cancer microenvironment (Figure 13.13). Dai et al. involved a 3D bioprinting technique to establish the brain tumor model based on glioma stem cell-laden gelatin/alginate/fibrinogen (GAF) hydrogel.

13.4.2 Multi-Cell Tumor Sphere

Cells embedded within 3D hydrogels grow and proliferate in the form of a sphere, which may promote the exchange of factors or message between cells and the surrounding environment, generate a gradient of nutrition and oxygen, and decline the influence of drug candidates on tumor tissues, accordingly enhancing the drug resistance of tumor tissues.

The multicellular tumor sphere model further arises. This spheroid is composed of aggregated cells, including cell–cell interaction, cell–ECM interplay, as well as relevant factors. These spheroids exhibit structural heterogeneity and spontaneously form a gradient from the external parts to the core, showing drug resistance. As a result, multicellular tumor spheres could better forecast the effect of drug application.

Some studies have merged a microfluidic device with multicellular tumor spheroids for the screening of anticancer drug candidates, which is extremely integrated into a more physiologically related case. Chen et al. proposed an original microwell array microfluidic system, intending to achieve the high-throughput culture of tumor spheroids for drug development research (Figure 13.14).

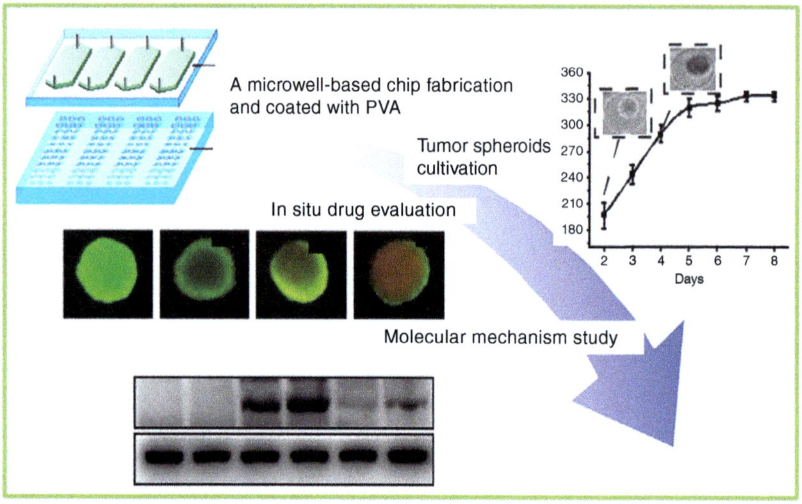

Figure 13.14 Schematic of drug cytotoxicity and signaling pathway analysis with 3D tumor spheres in a microwell-based microfluidic device. Source: Chen et al. [29]. Reproduced with permission of Elsevier.

13.4.3 Tumor Metastasis Model with Angiogenesis

Tumor metastasis is the hallmark of malignant cancer, while it is reliable for almost 90% of the cancer-relevant mortality; however, it remains the most mystical part of the cancer disease mechanism. Tumor angiogenesis is the essential course for tumor development and spread. When the angiogenesis is impeded or invalid, the tumor tissue will not grow or migrate anymore. The adoption of specific drug candidates to inhibit angiogenesis could block the growth and metastasis of tumors. Antiangiogenic-based treatment has been one of the hotspots of the worldwide antitumor therapy study. By means of the precise distribution of tumor cells, vascular cells, as well as matrix materials, a tumor metastasis model covering the tumor microenvironment as well as cell–cell/cell–ECM interactions could be generated, which could act as a robust platform for metastasis-relevant drug testing.

Microfluidic systems have been introduced as the vascular models to build a tumor-on-a-chip for the modeling of the point process of tumor metastasis, through which multiple kinds of research have been carried out to look into metastasis-relating drug effect. Such microfluidic devices could be manufactured in a more convenient and versatile way based on 3D bioprinting technology.

Chen et al. constructed a biomimicking 3D microstructure through DLP-based bioprinting technology for a deeper understanding of cancer spread (Figure 13.15b). Roger Kamm's laboratory has carried out several studies on the involvement of the microfluidic devices in the study of tumor cell intravasation and extravasation. The intravasation model is composed of two microchannels loaded with either tumor cells or ECs. A collagen gel bridge is built between these two channels, paving the

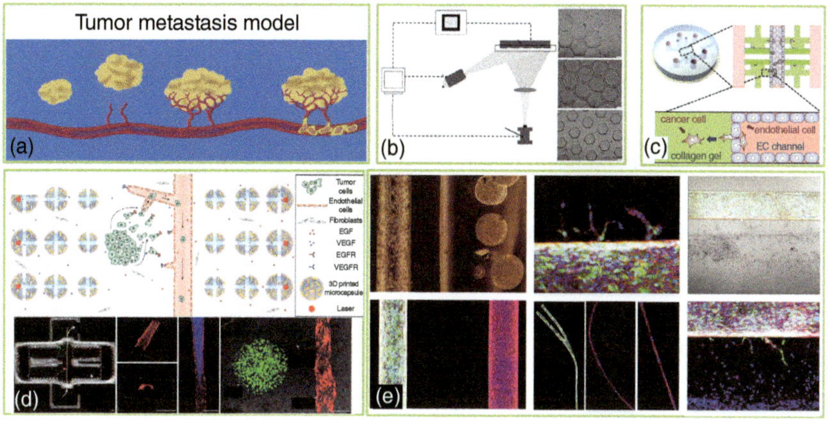

Figure 13.15 Tumor metastasis model: (a) Schematic of the tumor development process. Source: [18]. Reproduced with permission of The Royal Society of Chemistry. (b) Schematic of the bioprinting system for HeLa cell-seeded PEGDA construct as a biomimetic structure for cancer migration. Source: Huang et al. [30]. Reproduced with permission of Springer Nature. (c) General schematic of the in vitro model of tumor cell extravasation containing an endothelial monolayer, cancer cells, ECs, and collagen. Source: Jeon et al. [31]. https://journals.plos.org/plosone/article?id=10.1371/journal.pone.0056910 CC-BY 4.0. (d) Schematic and photograph showing the 3D-printed in vitro tumor models mimicking metastatic dissemination by the integration of tumor cells, vascular conduits, and factor signals within a cell-laden fibrin gel to reconstruct tumor microenvironments. Source: Meng et al. [32]. Reproduced with permission of Wiley-VCH. (e) Images displaying the modeling of multistep tumor development. Source: Nie et al. [18]. Reproduced with permission of The Royal Society of Chemistry.

way for tumor cells to migrate toward the ECs channels for the reproducing of tumor cell intravasation process into the blood vessel. A tumor cell extravasation model was also generated on a microfluidic platform based on two fibrin gels seeded with either ECs or normal lung fibroblasts. After the formation of the vascular network, cancer cells were introduced into the platform, while the extravasation process of tumor cells out of the microvascular network was validated. Jeon et al. used the microfluidic platform to investigate the key process of cancer extravasation (Figure 13.15c).

To better simulate tumor metastasis, a 3D glioma-vascular niche model was established through 3D bioprinting technology to study the cell–cell interaction. Tumor constructs have also been fabricated through 3D bioprinting technologies based on the precise patterning of cells, matrix materials, as well as release capsules. This approach helps for the spatiotemporal regulation of instructive factor gradients for the dynamic control of cell behaviors. In one relating study, vascularized tumor models were generated to reproduce the critical process of cancer dissemination based on the directional metastasis of tumor cells and ECs by stromal cells and growth factors. The key processes during the tumor development were replicated based on a parallel-channel model covering the tumor stroma, chemical environment, functional vascular network, as well as tumor cells (Figure 13.15e).

13.5 Summary and Prospect

13.5.1 Key Virtue and Comparison

The unique superiorities of the 3D bioprinting technology advance the achievement of the modular and flexible building of in vitro models in a high-throughput and customized means. Furthermore, the distinct advantages and limitations of these models have advanced the development of the in vitro models, as shown in Table 13.1.

13.5.2 Outlook

Most of the current in vitro models are exhibiting poor individuality, heterogeneity, intricacy, as well as manipulation difficulty because of the primary processing technologies. The introduction of the 3D bioprinting technique could effectively deal with these concerns. Based on its capability of accurately patterning cells and matrix materials, 3D bioprinting technology has been widely involved in the generation of in vitro models. However, limitations still lie in the printing process, cell source, as well as matrix material, such as challenges to fully reproduce the physiologically relevant cell organization and function, as well as the special microenvironment of native tissues/organs. Technical advancements in imaging means and testing tools would also be an urgent requirement for further testing and validation.

Table 13.1 Advantages and disadvantages of the in vitro methods for drug screening of different types.

	Advantages	Limitations
Mini-tissue	Simple manipulation Approving high-throughput Sample consistency Acting as building bricks Low reagent assumption	Discrepancy with in vivo microenvironment Static growth condition Lack of mechanical factors Lack of tissue/organ interaction
Organ-on-a-chip	Dynamic perfusion Multiplex stimuli Controlled microenvironment	Multi-step process
Tissue/organ construct	Biomimicking One-step construction Simulation of tissue/organ-specific structure Integration with microfluidics Involvement of a vascular network Generation of tissue/organ interface	Poor mechanical strength Manipulation and storage Resolution and precision

13.5.2.1 3D Bioprinting Technology

In particular, challenges with 3D printing cover the printing resolution, printing efficacy, biocompatibility, as well as scaling up. High resolution leads to the construction of complex single-cell structures, including a capillary network. The biocompatibility of 3D printing technology has been restricted in cell survival and fundamental functions, while the influences on gene or protein aspects have yet to be investigated. Additionally, the scaling up of the bioprinted constructs is still limited. To generate large-sized tissue/organ constructs, additional researches are required in both aspects of the printing strategy and printing ink. Standardization and modification of the printing strategy in adaptation with the final demand are crucial for the building of functional models within a shorter period. As a result, the interaction between the printing-relating parameters and the properties of the achieved constructs must be studied, which could be realized through both computer simulation and experimental validation.

13.5.2.2 Individual Differences

The discrepancy among different disease types and within the single kind of disease vary to a great extent. The heterogeneity is leading to rather high intricacy and inconstancy, while demanding personalized treatment for an individual patient. Patient-defined models could facilitate decide the most proper therapy option for different individuals. Further researches that involve patient-derived cells could contribute insights into patient-customized drug discovery. In addition, in vitro disease models could be generated based on the diseased or dysfunctional human cells for disease mechanisms investigation as well as new therapy validation.

13.5.2.3 Systematic Interaction

What's more, in vivo drug operation covers multiple tissues and organs, while the most crucial of which are target tissues/organs and drug metabolism-relating tissues/organs. The metabolism of drugs in liver tissue could influence the period of drug clearance, while producing secondary metabolites that affect the effects of drugs. In addition, the hepatotoxicity of drugs is the major side effect of drugs. As a result, an effective in vitro drug screening model covers targeting tissues, liver tissues, as well as the interaction between them. The absorption and metabolism of drugs in the small intestine, metabolism in the liver, and the excretion within the kidney are critical for the safety and efficacy of drug candidates. Through introducing a dynamic culture platform, the final objective is to construct a multiple-tissue/organ model for the investigation of the interdependent influence of multiple tissues/organs as well as the comprehensive reaction of drug candidates.

13.5.2.4 Industrialization

During the further process of industrialization, sample consistency is of vital importance for standardized drug screening protocol and reliable results. The 3D bioprinted in vitro models are supposed to be highly uniform in each aspect, covering the structure, function, as well as developing level. In a word, developments in both 3D printing technology and biomedical study are essential for addressing present limitations, and thereby to thoroughly achieve the valid application of 3D bioprinting technology in drug discovery.

13.6 Conclusions

To relieve the urgent demand for better predictive in vitro models for biomedical study, 3D bioprinting technology has been involved in the construction of in vitro models. The in vitro models have experienced three developing levels: mini tissue, organ-on-a-chip, and tissue/organ construct. At first, to overcome the limitations of 2D cell culture models, mini tissue models have been proposed, providing a 3D growth condition for cells to maintain their morphology and function. The limitations of mini-tissue models in mimicking the intricate in vivo microenvironment further prompted the integration of an organ-on-a-chip model with increased physiologic relevance. Finally, tissue/organ construct came into emergency, aiming at solving the biological limitations of chip materials by introducing ECM mimicking materials. The experimental protocol for biomedical study must meet both standardization and quantification requirements. 3D bioprinting technology has shown its valuable potential for in vitro model construction in biomedical fields.

References

1 Nie, J., Gao, Q., Fu, J., and He, Y. (2020). *Advanced Healthcare Materials* 9 (7): 1901773.
2 Gu, Z., Fu, J., Lin, H., and He, Y. (2019). *Asian Journal of Pharmaceutical Sciences*.
3 Xie, M., Gao, Q., Zhao, H. et al. (2019). *Small* 15 (4): 1804216.
4 Zhao, H., Chen, Y., Shao, L. et al. (2018). *Small* 14 (39): 1802630.
5 Shao, L., Gao, Q., Zhao, H. et al. (2018). *Small* 14 (44): 1802187.
6 Cheng, Y., Zheng, F., Lu, J. et al. (2014). *Advanced Materials* 26 (30): 5184–5190.
7 Hribar, K.C., Finlay, D., Ma, X. et al. (2015). *Lab on a Chip* 15 (11): 2412–2418.
8 Zhang, A.P., Qu, X., Soman, P. et al. (2012). *Advanced Materials* 24 (31): 4266–4270.
9 Ye, N., Qin, J., Shi, W. et al. (2007). *Lab on a Chip* 7 (12): 1696–1704.
10 Bhise, N.S., Manoharan, V., Massa, S. et al. (2016). *Biofabrication* 8 (1): 014101.
11 Bhargava, K.C., Thompson, B., and Malmstadt, N. (2014). *Proceedings of the National Academy of Sciences of the United States of America* 111 (42): 15013–15018.
12 Nie, J., Gao, Q., Qiu, J.-J. et al. (2018). *Biofabrication* 10 (3): 035001.
13 Huh, D., Hamilton, G.A., and Ingber, D.E. (2011). *Trends in Cell Biology* 21 (12): 745–754.
14 Huh, D., Torisawa, Y.-s., Hamilton, G.A. et al. (2012). *Lab on a Chip* 12 (12): 2156–2164.

15 Maschmeyer, I., Lorenz, A.K., Schimek, K. et al. (2015). *Lab on a Chip* 15 (12): 2688–2699.

16 Gao, Q., Liu, Z., Lin, Z. et al. (2017). *ACS Biomaterials Science & Engineering* 3 (3): 399–408.

17 Gao, G., Park, J.Y., Kim, B.S. et al. (2018). *Advanced Healthcare Materials* 7 (23): 1801102.

18 Nie, J., Gao, Q., Xie, C. et al. (2020). *Materials Horizons* 7 (1): 82–92.

19 Song, K.H., Highley, C.B., Rouff, A., and Burdick, J.A. (2018). *Advanced Functional Materials* 28 (31): 1801331.

20 Xue, D., Wang, Y., Zhang, J. et al. (2018). *ACS Applied Materials & Interfaces* 10 (23): 19428–19435.

21 Zhu, W., Qu, X., Zhu, J. et al. (2017). *Biomaterials* 124: 106–115.

22 Gao, Q., He, Y., Fu, J.-Z. et al. (2015). *Biomaterials* 61: 203–215.

23 Miller, J.S., Stevens, K.R., Yang, M.T. et al. (2012). *Nature Materials* 11 (9): 768–774.

24 Kolesky, D.B., Truby, R.L., Gladman, A.S. et al. (2014). *Advanced Materials* 26 (19): 3124–3130.

25 Bertassoni, L.E., Cecconi, M., Manoharan, V. et al. (2014). *Lab on a Chip* 14 (13): 2202–2211.

26 Ma, X., Qu, X., Zhu, W. et al. (2016). *Proceedings of the National Academy of Sciences of the United States of America* 113 (8): 2206–2211.

27 Miri, A.K., Nieto, D., Iglesias, L. et al. (2018). *Advanced Materials* 30 (27): 1800242.

28 Zhao, Y., Yao, R., Ouyang, L. et al. (2014). *Biofabrication* 6 (3): 035001.

29 Chen, Y., Gao, D., Liu, H. et al. (2015). *Analytica Chimica Acta* 898: 85–92.

30 Huang, T.Q., Qu, X., Liu, J., and Chen, S. (2014). *Biomedical Microdevices* 16 (1): 127–132.

31 Jeon, J.S., Zervantonakis, I.K., Chung, S. et al. (2013). *PLoS One* 8 (2): e56910.

32 Meng, F., Meyer, C.M., Joung, D. et al. (2019). *Advanced Materials* 31 (10): 1806899.

14

Protocol of Typical 3D Bioprinting

1. Cell Culturing [1]
   (1) Prepare Dulbecco's modified Eagle medium (DMEM), supplemented with 10% fetal bovine serum (FBS) and 1% penicillin/streptomycin, used to culture human breast cancer cell (MDA-MB-231) lines and human umbilical vein endothelial cell (HUVEC) lines.
   (2) Prepare DMEM with L-glutamine (DMEM/F-12), supplemented with 10% FBS and 1% penicillin/streptomycin, used to culture bone marrow mesenchymal stem cell (BMSC) lines.
   (3) Set the culturing environment as 37 °C and 5% CO_2. Culture MDA-MB-231, HUVEC, and BMSC, and passage the cells in a 1 : 2 ratio when 90% confluence is reached.
2. Bioprinting of GelMA Microspheres [1]
   (1) Print the fixture as shown in Figure 14.1 with polylactic acid (PLA) on a fused deposition modeling (FDM) printer. Place two metal ring electrodes in the fixture.
   (2) Connect the two metal ring electrodes with the ground and positive poles, respectively. Place the metal plate connected with the high voltage below the ring electrode and place a petri dish with silicone oil on the metal plate as a droplet receiver (Figure 14.1).
   (3) Dissolve freeze-dried GelMA (5% w/v) and photoinitiator LAP (0.5% w/v) in Dulbecco's phosphate-buffered saline (DPBS) as the bioink (10 ml). Filter the bioink through a 0.22 μm filter for sterility and heat it in a 37 °C water bath for 15 minutes.
   (4) Detach MDA-MB-231 cells with 3 ml of 0.25% trypsin-0.02% EDTA solution for three minutes at 37 °C. Centrifuge cells in a 15 ml centrifugal tube at 100× g for five minutes to obtain a cell pellet.
   (5) Remove the supernatant. Mix the cell pellet with 1 ml of prepared bioink by slowly pipetting it to prevent the production of bubbles.
   (6) Place 1 ml of bioink (MDA-MB-231) into a 3 ml sterile syringe. Feed the bioink by the force of compressed air (~0.5 kPa). Put the syringe on the fixture. (*Note:* The printing environment should be strictly controlled at a temperature of 30 °C and humidity of 50%.)

Cell Assembly with 3D Bioprinting, First Edition. Yong He, Qing Gao, and Yifei Jin.

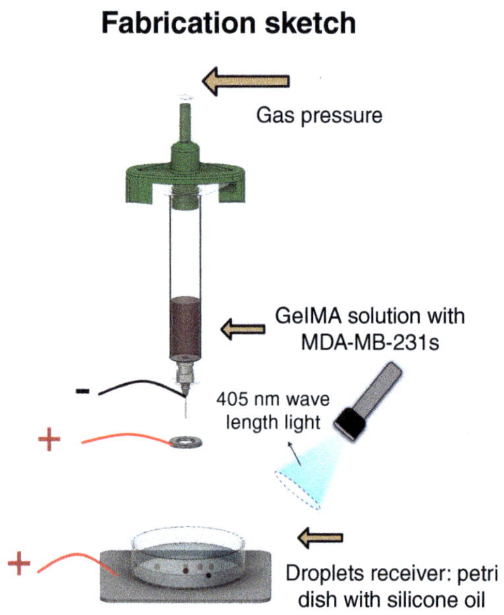

Figure 14.1 Bioprinting sketch of GelMA microspheres. Source: Xie et al. [1].

(7) Switch on the high voltage power and set the voltage at 0–4 kV. Simultaneously, turn on the 405 nm wavelength light to cross-link the GelMA droplets in 5 ml of silicone oil.

(8) Pour the most of the silicone oil away by decanting the petri dish. Transfer the remained silicone oil and the GelMA microspheres into a 15 ml centrifugal tube using a spoon.

(9) Add 5 ml of DPBS and shake the mixture uniformly. Centrifuge the tube at 100× g for five minutes and remove the supernatant fluid.

(10) Repeat Step 2 (9) for three times.

(11) Take out the GelMA microspheres with a spoon and culture them in DMEM in a petri dish at 37 °C and 5% CO_2 for three days.

3. Bioprinting of GelMA fibers [1]

(1) Prepare a coaxial nozzle. Fix an inner nozzle (25G, OD = 510 μm, ID = 250 μm) and outer nozzle (18G, OD = 1200 μm, ID = 900 μm) by soldering. Connect a glass tube (length = 50 mm, inside diameter = 1.2 mm) to the end of the coaxial nozzle.

(2) Dissolve sodium alginate (Na-Alg) powder that is sterilized under ultraviolet (UV) light for 30 minutes in deionized water at 2% (w/v).

(3) Prepare a sterile bioink solution following step 2 (3). Heat the GelMA bioink and Na-Alg solution in a 37 °C water bath for 15 minutes.

(4) Detach BMSCs cells with 3 ml of 0.25% trypsin-0.02% EDTA solution for three minutes at 37 °C. Centrifuge cells in a 15 ml centrifugal tube at 100× g for five minutes to obtain a cell pellet.

(5) Remove the supernatant fluid. Mix the cell pellet with 2 ml of prepared GelMA bioink by slowly pipetting it to prevent the production of bubbles.

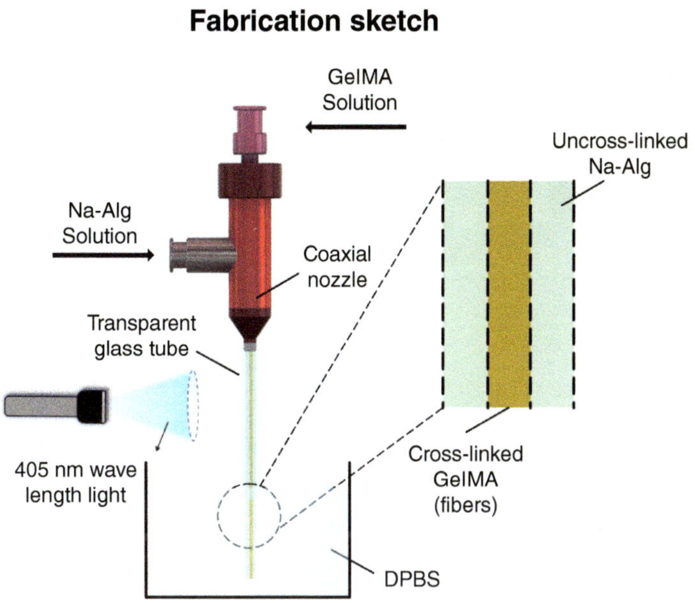

Figure 14.2 Bioprinting sketch of GelMA fibers. Source: Xie et al. [1].

(6) Place 2 ml of bioink (BMSCs) into a 10 ml syringe. Place 2 ml of Na-Alg solution into another syringe (10 ml). Feed them with two syringe pumps, respectively (here, bioink at 50 μm/min and Na-Alg solution at 350 μm/min). (*Note:* The printing environment should be strictly controlled at a temperature of 30 °C and humidity of 50%.)

(7) Turn on the 405 nm wavelength light to irradiate the transparent tube to cross-link the GelMA fibers. Use a petri dish with DPBS to receive the fibers (Figure 14.2).

(8) Take out the GelMA fibers with a spoon from DPBS and culture them for three days in the prepared DMEM/F-12 at 37 °C and 5% CO_2.

4. Bioprinting of complex 3D GelMA structures [1]

(1) Wipe the DLP bioprinter with 75% alcohol and expose it to UV irradiation for 30 minutes for sterility.

(2) Dissolve the freeze-dried GelMA (10% w/v) and LAP (0.5% w/v) in DPBS. Add magenta edible pigment into the solution (3% v/v) to improve the printing accuracy.

(3) Filter the solution through a 0.22 μm filter for sterility and heat it in a 37 °C water bath for 15 minutes.

(4) Build the 3D models with computer-aided design (CAD) software. Import the model documents to the upper software (EFL) of the applied DLP bioprinter.

(5) Add 10 ml of prepared bioink into the trough of the DLP bioprinter (Figure 14.3).

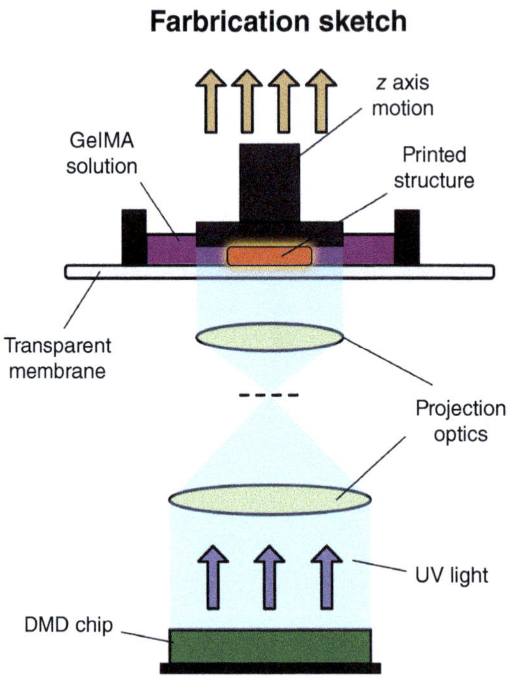

Figure 14.3 Bioprinting sketch of complex 3D GelMA structures. Source: Xie et al. [1].

(6) Set the printing parameters in the upper software as follows: light intensity = 12 mW/cm^2, irradiation duration = 30 seconds, and slice height = 100 μm. Start printing.
(7) Remove the printed structure from the bioprinter and immerse it in DPBS in a petri dish.
(8) Detach the MDA-MB-231s cells with 3 ml of 0.25% trypsin-0.02% EDTA solution for three minutes at 37 °C. Centrifuge cells at 100× *g* for five minutes in a 15 ml tube to obtain a cell pellet.
(9) Remove the supernatant fluid and mix the cell pellet with 2 ml of DMEM.
(10) Add 100 μl of cell suspension on the printed structures. Culture them for three days in the prepared DMEM at 37 °C and 5% CO_2.

5. Fabrication of GelMA-based microfluidic chips [1]
(1) Dissolve the freeze-dried GelMA 10% (w/v) and LAP (0.5% w/v) in DPBS. Filter the GelMA solution through a 0.22 μm filter for sterility.
(2) Sterilize the gelatin powder under UV light for 30 minutes and add it to the GelMA-LAP solution prepared in step 5 (1) to a final concentration of gelatin of 5% (w/v). Heat the mixture in a 37 °C water bath for 15 minutes.
(3) Design a group of molds with CAD software and manufacture them with photopolymer resin on a DLP printer.
(4) Fill the molds fully with the prepared bioink (Figure 14.4).
(5) Put the molds into a 4 °C refrigerator to cross-link the gelatin for 30 minutes.
(6) Remove the molds and demold with a blade the partially (physically) cross-linked hydrogel sheets from the molds.

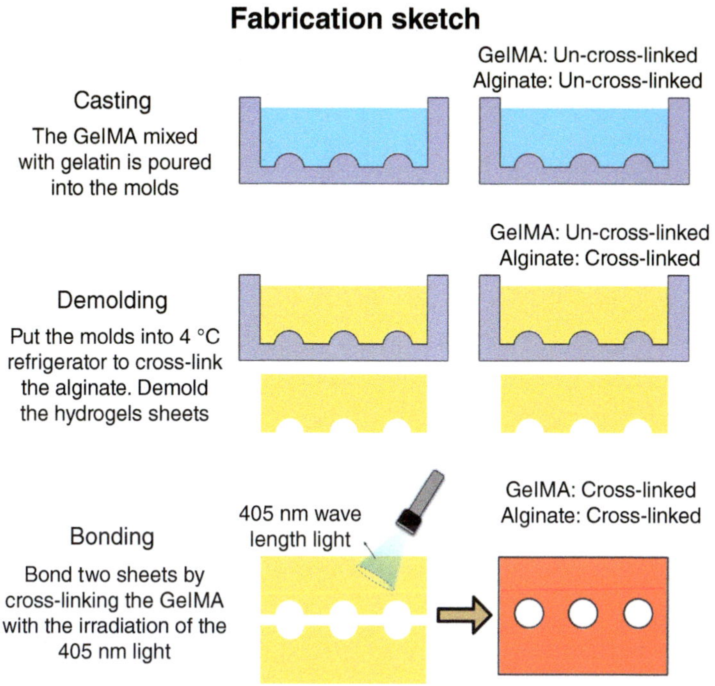

Figure 14.4 Fabrication sketch of GelMA-based microfluidic chips. Source: Xie et al. [1].

(7) Combine the two demolded hydrogel sheets and bond them with the help of GelMA by irradiating at 405 nm for one minute.
(8) Detach the HUVECs cells with 3 ml of 0.25% trypsin-0.02% EDTA solution for three minutes at 37 °C. Centrifuge cells in a 15 ml centrifugal tube to obtain a cell pellet at 100× g for five minutes.
(9) Remove the supernatant fluid and mix the cell pellet with 2 ml of DMEM.
(10) Fill the microchannel fully by injecting the cell suspension with a nozzle and syringe.
(11) Flip the chip upside down every 15 minutes during the next three hours to achieve a uniform and complete cell seeding. Culture the chips in the petri dish for three days in the prepared DMEM at 37 °C and 5% CO_2.

Reference

1 Xie, M., Yu, K., Sun, Y. et al. (2019). *JoVE* (154): e60545.

Index

Cell Assembly with 3D Bioprinting, First Edition. Yong He, Qing Gao, and Yifei Jin.

d

l

m